WAR STARS

WAR STARS

★★★★★★★★★★★★★★★★★★★★★★★★★★★★★★

The Superweapon and the American Imagination

H. Bruce Franklin

New York   Oxford
OXFORD UNIVERSITY PRESS
1988

Oxford University Press

Oxford   New York   Toronto
Delhi   Bombay   Calcutta   Madras   Karachi
Petaling Jaya   Singapore   Hong Kong   Tokyo
Nairobi   Dar es Salaam   Cape Town
Melbourne   Auckland

and associated companies in
Berlin   Ibadan

Library of Congress Cataloging-in-Publication Data
Franklin, H. Bruce (Howard Bruce), 1934–
War stars: the superweapon and the American imagination /
by H. Bruce Franklin.
p.   cm.   Bibliography: p.   Includes index.
ISBN 0-19-505295-1
1. Munitions—United States—History.   2. Weapons systems—
United States—History.   I. Title.
UF533.F73 1988
355.8'2'0973—dc19   87-34734   CIP

2 4 6 8 9 7 5 3 1

Printed in the United States of America
on acid-free paper

For Jane

Acknowledgments

Material assistance for this project was provided by a Research Grant from the William Joiner Center for the Study of War and Social Consequences at the University of Massachusetts at Boston and three Research Awards from the Rutgers University Graduate School in Newark. I am grateful to both institutions for their timely support.

Many people helped my research into this enormous subject and the shaping of the book. Sam Moskowitz and John Seelye made vital suggestions. Martha Bartter, Paul Brians, Stuart Teitler, A. Langley Searles, and Robert Madle supplied valuable material. Extraordinary assistance was rendered by John Newman, Archivist of the preeminent Imaginary Wars Collection at Colorado State University in Fort Collins. Among many other exceptionally helpful librarians and archivists, I am especially indebted to Mimi Bowling and Erik Olsen of the Edison National Historic Site, John Skillen of the Montclair Public Library, Charles Cummings of the Newark Public Library, Terry Geesken and Mary Corliss of the Museum of Modern Art's Film Study Center, and Wanda Gawienoski and Patricia Ruggles of the John Cotton Dana Library. Karen Franklin, Gretchen Franklin, Robert Franklin, and Carolyn Karcher each read the entire manuscript and made countless invaluable suggestions. Jane Franklin's participation was crucial at every stage of composition, and her contributions are essential to whatever success the book may have.

Contents

Introduction: Imagining Our Weapons   3

I   Beyond Manifest Destiny

   1. Robert Fulton and the Weapons of Progress   9
   2. Fantasies of War: 1880–1917   19
   3. Thomas Edison and the Industrialization of War   54

II   Victory Through Air Power

   4. Peace Is Our Profession   81
   5. Billy Mitchell and the Romance of the Bomber   91
   6. The Triumph of the Bombers   101
   7. The Final Catch   112

III   Chain Reactions

   8. Don't Worry, It's Only Science Fiction   131
   9. Atomic Decision   149
  10. The Rise of Nuclear Culture   155
  11. The Baruch Plan: American Science Fiction   162
  12. Nuclear Scenarios   166
  13. Early Warnings   170
  14. Triumphs of Nuclear Culture   180

IV   Final Solutions

  15. Arms Control?   191
  16. War in Space?   199
  17. The Age of the Automatons?   204
  18. Recall?   211

Notes   213   *Bibliography of Fiction Discussed*   231
Films Discussed   237   *Index*   239

Illustrations follow page 77

WAR STARS

. . . I call upon the scientific community in our country, those who gave us nuclear weapons, to turn their great talents now to the cause of mankind and world peace, to give us the means of rendering these nuclear weapons impotent and obsolete.

—President Ronald Reagan
March 23, 1983

INTRODUCTION

Imagining Our Weapons

Some readers of this book can remember when we human beings were unaware of any threat to the survival of our species. Global catastrophes were confined to cosmological speculation, apocalyptic religion, and science fiction. Though we knew that each of us would die as an individual, we lived secure in the belief that no danger menaced our race or even our civilization.

Especially in the United States of America, nothing challenged our security. No foreign power had the ability to invade or devastate our country. Our science was continually developing more potent means to protect our health, to extend our lifetime, and to improve the quality of our material life.

But then we ourselves devised mechanisms eventually capable of wiping out our own nation, global civilization, and possibly the human species. At the same time, we set up a worldwide confrontation in which these doomsday machines may very well be used. So now all of us, in this nation and in the world, live day by day under the shadow of extermination by our own weapons, with no escape in sight.

How did we get ourselves into this plight? When we look back, we see that as we took each step toward the tyranny of superweapons, somehow we always thought that we were making ourselves more secure and the world more free. We built and used the first atomic bombs for peace and democracy. Protected by the two largest oceans, we invented and deployed the first intercontinental bombers capable of nullifying that geographical security. Unsatisfied with manned bombers, we initiated a race for intercontinental missiles that could be launched in minutes and that we could then neither recall nor destroy. We devised thermonuclear bombs, a thousand times more powerful than atomic bombs. By inventing and building submarines capable of firing long-range thermonuclear warheads while submerged, we transformed our old security moat, the oceans, into an enveloping menace. We equipped our missiles with self-guided multiple warheads, so that even a single accidental launch could waste several targets, killing possibly millions of people and virtually guaranteeing wholesale retalia-

3

tion. Then we deployed new missiles that could reach the other side's heartland within six to eight minutes, thus goading it to launch its missiles at us as soon as its computers display indications of an attack. And now we are spending billions of dollars annually to develop an automated hair-trigger weapons system to be placed literally over the heads of all of us on the planet—again, of course, in the name of defense, security, and freedom.

Today these superweapons of our own creation not only imperil but dominate America. The main purpose of our national government is to achieve "defense," not against the forces that actually threaten our lives and living conditions—such as disease, polluted air and water, urban decay, ignorance, unemployment, industrial and highway accidents, domestic violence, teenage suicide, alcoholism and other drug addiction, toxic and radioactive wastes—but against certain foreign nations, particularly one whose only military engagements with the United States have been as an ally and a victim of U.S. invasion. So of course "defense" has become our nation's greatest budgetary expense. Ever more elaborate superweapons, and the military forces to use them, have become so immensely expensive that they must be financed with deficits that grow like cancer, mortgaging the foreseeable future of our nation. The glorification of war is a principal business of not just one but several multi-billion-dollar industries, including movies, television, advertising, and the manufacture of toys and games for both children and adults. To be against militarism, messianic anticommunism, and the reign of superweapons is to be perceived by some as un-American.

How can we possibly explain what seems to be a pell-mell rush toward self-annihilation? Inventing and constructing this planetary suicide machine might appear so irrational as to be called insane. Indeed, in the first months after the atomic bombing of Hiroshima and Nagasaki, many people, foreseeing the arms race toward what we now call Mutually Assured Destruction, labeled our behavior "mad." Lewis Mumford's eloquent 1946 jeremiad entitled "Gentlemen: You Are Mad!" called our leaders "madmen" arranging global suicide while solemnly convinced that they are rationally and responsibly working for security and peace. There can be only one reason, Mumford tells us, why we acquiesce in such madness: "We are madmen, too. We view the madness of our leaders as if it expressed a traditional wisdom and a common sense."[1]

But suppose that creating mechanisms for universal extermination does indeed express "traditional wisdom" and "common sense." To focus on the apparent madness of individuals might then be misleading, for the source of their irrationality—and the matrix of those devastating mechanisms—would be the culture itself.

To create the objects that menace our existence, some people first had to imagine them. Then to build these weapons, a much larger number of people had to imagine consequent scenarios—a resulting future—that seemed desirable. Thus our actual superweapons originated in their imagined history, which forms a crucial part of our culture.

I have tried to locate and describe this history of the imagination from which has emerged our nemesis. It is difficult to disentangle our own thinking from this history, for we are partly creatures of both our own imagination and the imagination of the past. And we are also creatures of the material environment that shaped our imagination.

Since culture itself both expresses and influences the material conditions of society, historical processes cannot be understood without comprehending the interplay between material and cultural forces. While emphasizing cultural aspects, I do not mean to imply that they have been the main source of the empire of superweapons, for the cultural rationalization of these weapons is itself a product of the technological and industrial potential for producing them.

By focusing on cultural phenomena primarily in America (and peripherally in Britain), I do not mean to suggest that America is the center of the world, or that similar ideas were not emerging in other industrialized societies, or that there were not vital transnational connections among these cultures. Yet it is American culture of the past century that has most clearly shaped the imagination of a human destiny dominated by superweapons. Indeed, the cult of the superweapon originated as a distinct phenomenon between 1880 and World War I, in the form of future wars imagined by American authors of fiction. These stories and novels offer panoramic vistas of the early formations of modern America's ideological terrain.

Fascinating as they may be as expressions of psychology and culture, American fantasies about superweapons are not primarily fantasies at all. For when they shape the thinking of inventors and leaders and common people, they become a material force. Ever since the dawn of the nation, Americans have been actually trying to build their imagined superweapons, with more and more success. No matter how bizarre they may once have seemed or how unforeseen their consequences, American innovations in weaponry have fundamentally transformed not only warfare and geopolitical relationships but also the human condition. As the creator of awesome weapons, America has surpassed all rivals, becoming the great pioneer nation of modern warfare, especially in the oceans and the skies.

In the midst of the revolution against Britain, it was an American who conceived and built the first submarine used as a weapon of war. The world's first steam warship was built in 1814 to defend American ports in the nation's second war against Great Britain. During the Civil War, ironclad steam warships made their debut, and for the first time a warship was sunk by a submarine. Direct descendants of these prototypical naval weapons are another American first: nuclear submarines, each capable of annihilating every major city on any continent.

From the observation balloons of the Civil War to Kitty Hawk to the thousand-plane raids of the Second World War to Mutually Assured Destruction to Star Wars, America has led the way in transmuting the skies above us into the deadliest medium. The first multicity aerial firestorm raids, first atomic bomb, first intercontinental bomber, first thermonuclear bomb, first "Multiple Independently targeted Reentry Vehicle" (MIRV), first "Maneuverable Reentry Vehicle" (MARV), first laser-guided bomb, first automated system for launching thermonuclear war—these are all certainly spectacular American achievements of technological imagination and production. Inextricably intertwined with the development of these American superweapons have been those fantasies and illusions essential to their conception.

But the rise and dominion of these weapons has not been unchallenged. The same interplay between material and cultural forces that created them also created their antithesis. Our success in analyzing and manipulating the material components of our existence, allowing us to turn the very forces that shape matter into catastrophic weapons, has demonstrated that our minds are capable (within limits) of comprehending and changing objective reality. This ability depends on our capacity to imagine a wide range of future possibilities. So the technological, industrial, economic, and political environment that has led us to produce and rationalize superweapons has also stimulated a dynamic opposition, making the cultural history of the superweapon in America anything but one-sided. Indeed, some of the greatest expressions of resistance to the empire of these weapons have been made in America. As this empire has grown, so have the efforts of the American imagination in the struggle to win our freedom from it.

American weapons and American culture cannot be understood in isolation from each other. Just as the weapons have emerged from the culture, so too have the weapons caused profound metamorphoses in the culture. Comprehending this process may show us how we got into our current predicament. It might even help us find our way out.

I

BEYOND MANIFEST DESTINY

We go not to conquer, but to free mankind.
          —BENJAMIN RUSH DAVENPORT, *Anglo-Saxons Onward!*, 1898

In our hands has been given by a miracle the most deadly engine ever conceived, and we should be delinquent in our duty if we failed to use it as a means of controlling and thereby ending wars for all time.
          —ROY NORTON, *The Vanishing Fleets*, 1907

1

Robert Fulton and the Weapons of Progress

The Liberty of the Seas will be the Happiness of the Earth.
—Epigraph, ROBERT FULTON,
Torpedo War, and Submarine Explosions

Throughout the eighteenth century, vast formations of the most deadly weapons of the age—wooden vessels propelled by sails—fought for control of Europe and the world. Although the awesome fleet of the British Empire had not yet won supremacy over the seas, in 1776 it seemed to have little to fear from the rebels in the agricultural and mercantile colonies thinly scattered along part of the Atlantic seaboard of North America. Possibly half the colonists were still loyal to the king, and the rest had no way to construct vessels capable of any serious threat to His Majesty's navy. The ragtag "army" of the rebel leader George Washington was being driven back toward the town of New York, many of whose twenty-five thousand citizens were Loyalists. Meanwhile, a mighty British force was being massed in New York harbor to drive a wedge between the rebels in the north and the south. Well over forty thousand professional soldiers and sailors waited on board the invasion fleet of 350 ships, which included such fearsome vessels as Admiral Lord Richard Howe's flagship, the sixty-four-gun HMS *Eagle*.

Fighting for national independence against the overpowering war machine of the colonial occupiers, the American rebels were relying on their daring and ingenuity. The remnants of General Washington's forces, defeated on Long Island by General Sir William Howe, audaciously slipped across the East River to Manhattan on the night of August 29, 1776. Then an entirely new type of weapon, built to "pulverize the British navy," entered the fray.[1]

On the night of September 6, the mighty *Eagle* was attacked by the world's first combat submarine, the tiny *American Turtle,* designed and built by David Bushnell of Connecticut. The crew of the walnut-shaped *Turtle*—Sergeant Ezra Lee—industriously cranked the propellers while maneuvering to implant a "torpedo," an underwater explosive charge, into the hull of the *Eagle*. Though the *Eagle's* copper sheathing frustrated the attack, much of the British fleet was thrown into confusion by the explosion, as alarmed officers slashed anchor cables so their ships could drift away from the unseen attacker toward open water.[2]

9

While ineffective in its time, the *Turtle* introduced into warfare the basic features of the military submarine: ballast tanks that could be emptied with a pump or filled; means of propulsion and steering; longitudinal stability; a prototypical conning tower; an illuminated instrument panel; underwater explosives. More than a century later, John P. Holland, the designer of the first practical combat submarines, traced his concepts directly back to what he called "Bushnell's remarkably complete vessel."[3]

Like other early American efforts, Bushnell's submarine was conceived as a desperate challenge to superior military and manufacturing might. This little invention was more than the first operational submarine: it soon became the apparent model for the first American attempt to create an ultimate weapon that would change history. For when Robert Fulton began to search for some technological advance in weaponry that would bring about the reign of reason, he turned to that ingenious American design, the submarine.[4]

Celebrated now mainly as the man who developed the first commercially viable steamboat, Robert Fulton also holds a unique status in the history of the superweapon. Naval historians recognize him as the great "pioneer of undersea warfare." The prototypes of ultimate war vessels that Fulton engineered and built were technological marvels far ahead of their time. Beyond that, Fulton has the dubious distinction of being the first person to articulate the modern ideology for rationalizing superweapons. What he did and how he thought provide astonishing insights into our current crisis.

Fulton's unending quest for fame and fortune drove him from occupation to occupation, project to project, and nation to nation. His many claims to new inventions—all disputed—look to some observers like mere vehicles for personal ambition. Nowhere does his motivation seem more suspect than in his obsession with designing spectacular new mechanisms to revolutionize, or to end, naval warfare: the submarine; floating and anchored "torpedoes" (actually mines); torpedo boats; an invulnerable steam warship. For almost two decades, Fulton tried to persuade one government after another, each intermittently at or almost at war with each other, to adopt one of his various versions of the ultimate naval weapon—in order to bring about peace, prosperity, and general human happiness.

Cynics find it easy to debunk Fulton's idealism as mere rationalization of his hunger for wealth and glory. Certainly, his motives were not untainted by greed and egoism. Yet his mixed motives actually identify him as a representative man of a new age, incarnating ideological contradictions inherent in the transition from mercantile to industrial capitalism: Robert Fulton sought fame and riches as recognition and reward for becoming a great human benefactor through the exertion of his individual genius, initiative, and industry.

Thus, understanding Fulton as an individual is far less important than comprehending him as a man emerging from his own time to embody and generate forces for the future. As the inventor of the first economically practical steamboat, Fulton helped unleash immense productive powers. As the inventor of prototypical naval weapons, he helped unleash potentially even more enormous destructive powers. And as a remarkably early, if not the earliest, theoretician to link the aspirations of industrial capitalism with weapons technology, he formulated the first coherent statement of the ideology that evolved into the American cult of the superweapon.

Although a militant supporter of republicanism, Fulton left the United States in 1787, four years after independence had been won and while the Constitution was still being written, to spend ten years in England. Then an aspiring artist, he sought the training, milieu, and audience that only the Old World could provide. Yet even while hobnobbing with the social and artistic elite of England, he continued to deepen his radical republican creed. Then began his fateful transition from a competent but undistinguished artist into one of the most inventive engineers of his epoch. Neither his republican ideals nor his artistic skills were ever divorced from his engineering imagination.

When he imagined the future that progress would bring, Fulton saw it in precise detail, realized through specific technological advancement. His first material images of this future were interwoven with his first grand engineering projection: an extensive network of small inland canals. His beautifully illustrated *A Treatise on the Improvement of Canal Navigation* (1796) promises that his proposed engineering designs would not only advance agriculture and commerce, but, by expanding transportation and communication, would also extend knowledge, obliterate "local prejudices," reduce labor while increasing the general standard of living, promote international harmony, make "happiness spread," and lead to a future society of all-around peace and prosperity.[5] How is all this possible?

Fulton's projections must be understood in their intellectual and material context. By the 1790s, the physical and cultural conditions of England and Europe were being transformed swiftly by an ever-accelerating process of industrialization. Before this radical transformation, rationalism had projected unfettered human reason as both the means and the goal of progress. Between such preindustrial thinking and the ideology of developed industrial capitalism, a transitional stage of thought, contemporaneous with Fulton, linked human reason to what we now call science and technology in an ever-advancing chain of progress. That is, human reason can make inventions and discoveries that will not only improve material conditions but also help free human reason for more inventions and discoveries, and so on. Of course, such thinking did not emerge ab ovo in the late eighteenth century; its essence can be found in the early seventeenth century (for example, in Francis Bacon), and by the early eighteenth century it was already the object of satiric derision (by Jonathan Swift, for instance). However, it was only in the midst of the Industrial Revolution that such thought began to shape the material conditions and prevailing ideology of a world system. Here Fulton played his role.

The ideology expressed by Fulton, in both his writings and his practical activities, projects an interchangeable series of causes and effects as illustrated in the following diagram:

unfettered human reason

free trade   mechanical improvements

education   the end of armies,
                                   navies, and war

sanctity of
private property   abolition of monarchy

industry   individual freedoms

general prosperity

There is no particular order in the arrangement of all these good things. Any can be the cause or effect of any or all of the others. Accordingly, small inland canals, a product of the unfettered human reason of a republican American, constitute a mechanical improvement that will facilitate free trade, promote industry, enlarge individual freedoms, raise general prosperity, undercut the basis for armies, navies, and warfare, and thus (though this is only implicit in *A Treatise on the Improvement of Canal Navigation*) lead to flourishing republicanism.

The year after this remarkable publication, Fulton moved to revolutionary France. In the fall of 1797, he addressed to the French Directorate a pamphlet entitled "Thoughts on Free Trade with Reasons Why Foreign Possessions And all Duties on Importation is Injurious to Nations." Here he argues that the revolutionary French government could by specific acts of economic policy, proposed in the pamphlet, achieve universal free trade and so attain "the eternal honour of Removing the obstacles between man and Happiness": "Thus equal rights being established we should in the Spirit of Liberty and Philanthropy rejoice to See Each nation peaceably enjoy the Fruits of their Virtuous Industry."[6]

The same year, Fulton composed a fifty-six-page treatise denouncing war and proposing means to bring about perpetual peace. Entitled "To the Friends of Mankind," and evidently also intended for the French government, it opens with a denunciation of war, including an elaborate argument exposing its irrationality. Especially vehement is Fulton's condemnation of the parasitic role of armies, navies, and "the manufacturers of cannon, muskets, Swords, Bayonets, Powder, ball and Soldiers apparel, the manufacturers of the numerous materials of Camp equipage, the Builders of Ships of War. . . ."

Then follows "The Republicans Creed," a passionate statement of his nine articles of personal faith, including these beliefs: people "receive their Ideas and prejudices" from the culture that surrounds them; "perpetual peace" can be established through education and "by investigating and explaining the errors which misguide the mind"; everyone has "a right to apply his talents to everything which does not Injure society and he has a right to Sell or exchange the produce of his Industry in any country without any Restriction whatever . . .";

unrestrained free trade forms "the true Rights of nations"; governments exist solely "to promote *education and industry*" and "to establish a Social intercourse with each other and give a free circulation to the whole produce of Virtuous industry." Merely by implementing this republican credo, Fulton concludes, governments may end war, eliminate poverty, and bring about a global triumph of "harmony" and "abundance":

> Thus persevere and virtue will prevail and peace commence her reign, then war shall cease to desolate the world nor burning cities mark its dreadful track by blackened clouds which lift its crimes to heaven; Then education bending the young mind to science, to agriculture, and to arts will beautify all nature, harvest will spread her yellow fields where Ignorance and war gave naught but famine.[7]

There was nothing especially novel, in 1797, about the radical republicanism of "To the Friends of Mankind." But precisely at this time, Fulton conceived a startling new way to bring about its vision of peace and prosperity: he would invent a weapon to end war. In a letter to his friend Edmund Cartwright, he half-jokingly explained that he was hard at work contriving "a curious machine for mending the system of politics."[8] He proposed to build a submarine vessel of war designed to destroy the main obstacle to free trade and revolutionary republicanism: the British navy.

With this one conception, Fulton radically altered the configuration of those interlinked republican causes and effects. Others were already hard at work contriving mechanical improvements that would accelerate human progress by revolutionizing industry, as Fulton intended with his small inland canals and as he would later achieve with his engineering advances in steamboat technology. But what was unprecedented was his new conception of specific mechanical advances in weapons technology as a means—perhaps the primary means—to accelerate human progress.

On December 13, 1797, Fulton submitted a historic proposal to the French government, outlining the terms under which his "Nautulus Company" would construct and operate a fleet of submarines that could "annihilate the British navy" and thus establish complete liberty of the seas. For each British warship destroyed by his "Mechanical Nautulus" the company would receive a handsome sum, while captured British vessels and cargoes would become "the Property of the Company." Fulton urged the revolutionary government to accept his offer at once, so that "the terror" of his invention would facilitate the planned French invasion of Britain.[9]

Fulton made some extraordinary claims in his campaign to persuade the French Directorate of his invention's revolutionary potential. For example, in a letter written to Director Paul Barras on October 27, 1798, he predicted that the submarine would not only devastate the British navy but would also precipitate the collapse of the British monarchy:

> . . . [when] warships [are] destroyed by means so new, so secret, and so incalculable, the confidence of the sailors is destroyed, and the fleet rendered worthless in the age

of the Jeremiahs of fright. In this state of things, the English republicans will rise to facilitate the descent of the French, to change their government themselves, without shedding much blood, and without any cost to France.[10]

Writing to the marine minister on October 5, 1799, Fulton appended a long essay entitled "Observations on the Moral Effects of the *Nautilus* Should It Be Employed with Success." He predicted the unchecked spread of peace and republicanism until even Ireland and England would enter a "millennium," eager to "suppress old hatreds and treat each other like Sisters."[11]

Less than a year later, Fulton was commanding the three-man crew of his *Nautilus,* a remarkably engineered forerunner of the modern war submarine. After the *Nautilus* passed its first tests in the Seine in the summer of 1800, Fulton took it into the English Channel, where it proved capable of lurking at a depth of twenty-five feet, operating fully submerged for over six hours while oxygen was supplied by both compressed air and a snorkel-like device, and actually hunting British ships. The British, alarmed by this invention, posted extra lookouts and surrounded some of their warships with small flotillas of rowboats, so Fulton's hunt bagged no quarry.[12]

Although Fulton's submarine had no significant military effect, not managing to attack even a single British vessel, it demonstrated impressive potential. Nevertheless, just as Napoleon was expressing his interest in seeing this marvelous innovation in warfare, Fulton dismantled the *Nautilus* and permanently abandoned further experiments with the submarine. For the next ten years, he was to ascribe to "torpedoes"—that is, anchored, floating, or maneuverable mines—the same powers of global liberation that he heretofore had claimed for the submarine. Then, in his final years, while not abandoning torpedoes, he built a steam warship—the *Demologos* (the voice of the people)—to liberate the world from the powers of kings, navies, and other relics of the past.

A few months after dismantling the *Nautilus,* Fulton reaffirmed his political vision and the role of his new weapons in bringing it about. Writing from Paris to William Reynolds in England on February 4, 1802, he asserted that "for the civilization of Europe and to secure the peace of America one principle must be understood, established and adhered to": "useful industry," to be achieved through free trade. However, certain forces were blocking this peace and prosperity, and Fulton was about to become the great benefactor of the human race by removing them:

> But priests, privileged order, and the Military naval System are the three great obstacles to such a simplification and Improvement of society—hence I have spent near two years in experiments to find a means of terminating the system of military marines; And I am convinced as well as many of my friends that the principles which I have established will go directly to their total destruction, and with them terminate all wars by sea and consequently give to commerce an uninterrupted circulation.

Fulton passed over the fact that he was developing these weapons for France to destroy the British fleet so that Napoleon could invade the homeland of the man to whom he was writing. After all, this was in the long-run interest of all maritime nations, including Britain: "But while I am about this work it will be well to prove to each Maritime nation that I am doing them a service and not an injury. . . ."[13]

Perhaps in that spirit, Fulton in 1804 switched his base of operations from France to its enemy, Great Britain. Using a secret agent, a clandestine rendezvous, and the suggestion of handsome rewards, the British government enticed Fulton into its service, where he was to operate under the code name "Robert Francis." On May 22, apparently no longer seeing his weapons as the vanguard for a French invasion of Britain, "Francis" wrote to his new employers to propose a plan "which will be prompt in execution and if Successful will forever Remove from the mind of Man the possibility of France making a descent on England. I propose a submarine expedition to destroy the fleets of Boulogne and Brest as they now lie. . . ."[14]

The British government made its new naval weapons expert a rich man, as he put his proposals into action. Fulton personally planned and led two of his submarine expeditions—actually attacks with small boats attempting to implant underwater explosives—that resulted in slight but not entirely insignificant losses to the French fleet. Then in 1805 Nelson's victory at Trafalgar visited upon the French fleet the devastation Fulton had been hoping his torpedoes would inflict; the inventor was no longer of any use to the British.

Soon after he finally returned to America in late 1806, Fulton switched the target of his weaponry back to Britain. In the midst of the semiwar of 1807 between Britain and the United States, he proposed, to an enthusiastic President Jefferson, novel methods of deploying his torpedoes to neutralize the British fleet. The fighting ended before he could implement his plan. But in 1810, as war with Britain loomed once again, he began new torpedo experiments.[15]

In this context, Fulton published his handsome 1810 volume, *Torpedo War, and Submarine Explosions,* once again combining his talents as artist, engineer, and republican ideologue. After detailing and illustrating his various past and projected modes of torpedo attack, he presented his "thoughts on the probable effect of this invention." His argument is familiar to those who have heard the rationalization supporting the American "Strategic Defense Initiative," or "Star Wars" plan, for the late twentieth and twenty-first centuries.

The world, according to Fulton, was being menaced by modern offensive weapons that "spread oppression" across the globe and "carry destruction to every harbour of the earth."[16] Since it was "science" that equipped these ships of war with tremendous explosive power and increased "their present enormous size and number," he theorized, "then may not science, in her progress, point out a means by which the application of the violent explosive force of gunpowder shall destroy ships of war, and give to the seas the liberty which shall secure perpetual peace between nations. . . . ?" His answer would need only minor stylistic revisions to be incorporated into a speech advocating Star Wars:

My conviction is, that the means are here developed, and require only to be organized and practiced, to produce that liberty so dear to every rational and reflecting man; and there is a grandeur in persevering to success in so immense an enterprise—so well calculated to excite the most vigorous exertions of the highest order of intellect, that I hope to interest the patriotic feelings of every friend to America, to justice, and to humanity, in so good a cause. (33)

In the concluding section, "On the imaginary inhumanity of Torpedo war," Fulton goes on to explain that since his weapons are purely defensive in character, they will prove of "immense advantage" to all nations and peoples, providing the "specific remedy" for the "political disease" of "military marines," which are merely "remains of ancient warlike habits." Since his torpedoes "will be an effectual cure for so great an evil," Fulton reaches this startling conclusion: "To introduce them into practice, and prove their utility, I am of opinion, that blowing up English ships of war, or French, or American, were there no other, and the men on shore, would be humane experiments of the first importance to the United States and to mankind" (45).

Possessed by messianic faith in the beneficent power of his weapons, Fulton could imagine exploding ships full of men as "humane experiments" to liberate the United States and mankind. (His intoxicating enthusiasm is betrayed by his clumsily added afterthought—blowing up American ships only "were there no other" and excusing American crews from the experiments.) This disjunction between imagination and reality is characteristic of the history of the super-weapon in American culture. Even those responsible for the atomic bombing of Hiroshima and Nagasaki convinced themselves that these were humane acts, taken to save lives and establish the reign of peace. As for Fulton, a few months after publishing this passage, he tried to implement its vision of a world saved by torpedoes from militarism, kings, and war. How? By once again offering to sell torpedoes to Napoleon, who now ruled openly as a military dictator and crowned emperor.[17]

This is not to say that Fulton's weapons could not have defensive applications. When war against the British Empire came again in 1812, the United States had about twenty warships, while the British had over a thousand. As British forces burned and pillaged defenseless settlements along the coast, one of the few obstacles they encountered was the handful of Fulton's torpedoes that had been produced. Though they sank no ships, their explosions managed to check some of the invaders' more brazen forays.

Meanwhile, Fulton, who took time off from defending his steamship patents and would-be monopoly on steam navigation to defend the nation's seaboard, was commissioned to build his last great naval weapon, the world's first steam warship. Originally called the *Demologos*, and renamed (not by Fulton) *Fulton the First*, the vessel was launched in 1814. Although *Fulton the First* was not yet quite ready for combat when Fulton died in January 1815 or when the war ended six months later, it embodied a number of remarkable innovations in naval weaponry.[18] Its steam engine was protected by a twin hull, which itself was fortified with heavy logs and decks specially designed to make hostile boarding exceedingly difficult. Although apparently not armed with Fulton's underwater cannon, which he called "Columbiads," it had an impressive range of batteries, mechanisms to make its shot red-hot, and a special force pump designed to throw a solid stream of cold water for a distance of about two hundred feet (giving rise to the rumor among the British that the ship was prepared to boil enemy crews in clouds of live steam). Designed, like all Fulton's naval weapons, purely for defense and to inaugurate the age of universal peace, the *Demologos* proved to be the forerunner of modern offensive warships.

Fulton's thinking dramatically charts the formative psychology and ideology of the superweapon. Though his military projects may have seemed fantastic to his contemporaries, and were perhaps decades ahead of the industrial capacity of his time, his conceptions and engineering were truly dazzling. If anything, Fulton understated rather than exaggerated the destructive potential of submarines, underwater explosives, and steam-powered warships. Typical of the many projectors of new weapons who were to follow in his footsteps for the next two centuries, Fulton's flaw was not that his inventions would not work, but that they would work too potently, and with unanticipated consequences.

To modern eyes, the fallacies seem all too obvious. Why could Fulton not see that the submarines could be used against commercial shipping, perhaps even more easily than against warships, thus jeopardizing rather than strengthening his beloved free trade? With all his engineering foresight, why could he not see that each of his weapons would necessarily evoke counterweapons, thus increasing rather than eliminating the roles of the standing navies and weapons manufacturers that blocked the road to peace, prosperity, and liberty? Could he not imagine the actual forms of naval warfare that were to arise as nations developed and embellished his inventions?

The ironies began to come home to Fulton's own American republic during the Civil War, when ferocious internecine naval conflict inaugurated major use of all of Fulton's inventions. It must have been difficult to see the peaceful potential of these weapons as steam-powered warships fought deadly duels, and Confederate submarines, torpedo boats, and naval mines exacted a terrific toll on the blockading Union fleet. With American ships actually blowing up before any attack was detected, sometimes losing their entire crews as a result, even Fulton might have lost faith in these "humane experiments."[19]

When the United States decided to enter World War I, a powerful motivation was the German slaughter of American civilians with two of Fulton's purely defensive superweapons: the submarine and torpedo. In 1954 appeared a new version of Fulton's *Nautilus:* the world's first nuclear-powered submarine, the USS *Nautilus.* By the 1980s, each nuclear-powered Trident-class submarine was armed with thermonuclear missiles packing an explosive force equivalent to over six times the total tonnage of all U.S. bombs dropped in World War II. How did Fulton's dream turn into this nightmare?

The interchangeability of superweapons in Fulton's imagination is a clue to their real significance. Fulton consistently maintained his vision of a prosperous, peaceful, free world liberated from war, kings, standing navies, and the fetters of archaic beliefs and customs. Only the means to attain this world kept shifting, from small canals to steamboats, from submarines to torpedoes to steam warships. Psychologically, any particular invention could serve as an infantile wish fulfillment fantasy. Ideologically, any invention might promise to materialize the bourgeois vision of a grand design of unending progress flowing from the dynamic of industrial capitalism.

This belief that particular mechanical improvements in warfare will bring about the bourgeois millennium is merely an extension of the faith in teleological progress inherent in capitalism. The steam engine, railroad, telegraph, telephone, automobile, airplane, television, computer—each would bring about a new world of wonder, of freedom, of prosperity. So why not the submarine, torpedo, steam-driven warship, balloon, machine gun, warplane, atomic bomb, or Star Wars?

Fulton's wonderful weapons were imagined means to achieve a political vision he articulated before expressing any special interest in weapons. They were, simultaneously, imagined means to achieve the personal fame and fortune that were also part of the bourgeois global vision. So the contradictions posed by the ideology associated with these weapons are not primarily personal, psychological, or unique to Fulton as an individual. Rather, they are central to industrial capitalism, especially as it came to dominate the theory and practice of warfare.

What was new about Robert Fulton was not the invention of novel weapons. That had been going on nonstop from the stone ax to the swift warships that were extending British rule around the globe. What Fulton inaugurated was the fusion of the ideology of emerging industrial capitalism, marked by its faith in mechanistic, teleological progress, with the "improvement" of weapons as the means to achieve the goals of this ideology—for the individual, the nation, and the human species.

2

Fantasies of War: 1880–1917

Now there would be no more exhibitions of the powers of the instantaneous motor-bomb. Hereafter, if battles must be fought, they would be battles of annihilation.
—FRANK STOCKTON, *The Great War Syndicate*, 1889

The city of doom became a place of a thousand fires. A great mushroom of smoke grew and billowed above it. Beneath its somber pall the Japanese planes still flew.
—J. U. GIESY, *All for His Country*, 1914

The Imagined Wars of America

Modern technological warfare—the war of the factories and machines and organization inevitably generated by the Industrial Revolution—burst upon the world between 1861 and 1871 in the U.S. Civil War and the Franco-Prussian War. Gigantic armies were now transported by railroad, coordinated by telegraph, and equipped with an ever-developing arsenal of mass-produced weapons designed by scientists and engineers. From the repeating rifle, primitive machine gun, observation balloon, and steam-powered, ironclad warship of the Civil War would evolve the juggernaut of twentieth-century weaponry, with its power to devastate the planet.

The literary expression of this fateful period in the history of warfare was a new genre—fiction that imagines future wars.[1] This genre began to take shape in 1871, as an aftermath to the stunning German technical and managerial innovations in the Franco-Prussian War. Most of the early novels and stories were warnings of military unpreparedness, propaganda aimed at the newly emerging mass reading audience, which, like the new forms of warfare, was also produced by industrial capitalism. In the typical British and European fiction, the readers' defenseless homeland was invaded by a likely enemy, often armed with some deadly new weapon, emphasizing the alleged failure of the home nation to keep up in the technological arms race. When American authors began to produce future-war fictions in the 1880s, they added characteristically American ingredients to these formulas—with ominous implications for the culture and history of the twentieth century.

Between 1880 and America's 1917 entry into World War I, novels and stories imagining future wars became an influential part of American popular culture. Projecting the causes, forms, and consequences of war fought years or centuries hence, the literature expressed and helped to shape the apocalyptic ideology prominent in America's wars from 1898 through the waning years of the twentieth century. In this popular fiction, the emerging faith in American technological genius wedded the older faith in America's messianic destiny, engendering a cult of made-in-America superweapons and ecstatic visions of America defeating evil empires, waging wars to end all wars, and making the world eternally safe for democracy.

Plunging into this literature is a disturbing kind of time travel into the world that created our own. Compared with our culture, these imagined wars are like dreams stripped bare of the veils and distortions concealing their true content. To experience this period through its projections of future wars, dominated by imperialist illusions and fantasies of peace through technology, is to reexperience the formation of our own ideology and consciousness.

In reading through the literature, one may sometimes smile sadly at its naiveté, marveling at how the horrendous history of actual twentieth-century warfare has made some of these prognostications seem almost charmingly quaint and archaic. But any smugness that may come from looking backward in time is wiped away by startling resemblances between some of the fiction's most outlandish visions and the bizarre ''scenarios'' that we, a century later, have come to accept as a guide to national policy. It is somewhat dizzying to find in musty pre–World War I American novels the core of the ideology of the nuclear arms race.

One disturbing effect of this fiction is to make post–World War II readers discover how casually we take for granted the doctrines of total war, including the annihilation of civilian populations. Almost all the works written before America's entry into World War I regard the sanctity of the lives of noncombatants—so long as they are white—as a touchstone revealing the moral character of warring nations. And yet some of this literature advocates and enthusiastically describes total genocide, down to the extermination of the very last person, of black and Asian peoples.

Surveying the future-war novels of these four decades is like staring into a nightmarish kaleidoscope. The identity of America's imagined enemies—Britain, Spain, blacks, the Yellow Peril, the Russian czardom, Germany—shifts and somersaults in revealing patterns. But looming above all, no matter who the imagined enemy, appear avatars of the superweapon in terribly modern and familiar shapes. Before we know it, we are face to face with atomic weapons in 1908, only twenty-eight years after the emergence of American future-war fiction, and thirty-seven years before the bombing of Hiroshima and Nagasaki.

Since the years between 1880 and 1917 were precisely those in which the United States finished its consolidation as a transcontinental nation and embarked on a new Manifest Destiny as a global power, it is no surprise to find in this fiction a dynamic transformation from the imagination of a republic to that of an empire. But contrary to what one might expect, although the earliest future-war novels customarily refer to the United States as "the Republic," this republic is seen not as an expanding but as a collapsing state, not as a newly emerging empire but as the feeble prey of old empires. Bearing such titles as *Last Days of the Republic* (1880) and *The Fall of the Great Republic* (1885), the prophecies of doom seem at first curiously irrelevant to the actual emergence of America as an industrial and military colossus. But close inspection may reveal that what seem to be the attitudes of a defenseless ugly duckling are in reality characteristic traits of the American imperial eagle, a bird that habitually views its own behavior as "defense" against its prey.

The earliest American fiction to imagine an alien invasion actually satirized this self-image so central to American culture and American history. Back in 1809, just as the republic was being established, Washington Irving published this imagined invasion in his popular *A History of New York by Diedrich Knickerbocker.*[2] Exposing the ethnocentric core of the argument that Europeans "discovered" "America," a land inhabited by ferocious "savages" called "Indians," Irving presents a parallel case, an invasion by "the Men from the Moon." The "Lunatics" approach us with the same cultural imperialism that Europeans and their colonial descendants bring to nonwhite peoples, and they come armed with a technology vastly superior to our own, including directed-energy beams (precisely the superweapon being sought by the U.S. government in the final years of the twentieth century). Irving thus revealed America itself as a nation that originated as an act of conquest by alien invaders.

Unlike Irving's satire, the American future-war fiction that emerged as a body of literature in the 1880s turned America's colonial history inside out, establishing what was to become a conventional pattern: the invasion of defenseless America by aliens from across the seas. With unintended and revealing irony, this literature often perceived the victims of domestic oppression—Chinese "coolies," blacks, Indians, European immigrants—as these foreigners' confederates treacherously lurking inside the nation.

By the end of the nineteenth century, American fiction had projected invasions or attacks by China, Spain, Britain, Japan, and Russia, as well as genocidal uprisings by various nonwhite peoples inside the United States. In the first fifteen years of the twentieth century, still more invaders came—from Germany, France, Italy, Mexico, and Africa. Since then, American culture has continued to imagine conquering hordes of assorted foreigners, as in the 1984 film *Red Dawn,* which dramatizes the heroic struggles of a band of teenage American guerrillas against the colossal occupation armies of the Soviet Union, Cuba, and Nicaragua, or the 1985 film *Invasion: USA,* in which a lone superhero saves the nation from conquest by a ragtag mob of several hundred sinister, Russian-led, dark-complexioned invaders.

Of course, throughout this century of unceasing imagined invasions, no forces from these or any other nations have actually set foot in the United States (though Japan did attack American possessions in the Pacific). Rather, U.S. forces have attacked, invaded, "liberated," or occupied Cuba, the Philippines, Puerto Rico, Panama, Nicaragua, Costa Rica, Honduras, Venezuela, El Salvador, the Dominican Republic, Haiti, Guatemala, the Soviet Union, Germany, France, the

Netherlands, Belgium, Denmark, Italy, Japan, Korea, Lebanon, Vietnam, Cambodia, Laos, and Grenada, to name a few.

One of the earliest fantasies of alien attack was "The End of New York," a short story by former naval officer Park Benjamin, by then a prominent editor and writer on naval affairs. Though fairly typical of tales that had already become commonplace in Britain and Europe, "The End of New York" created a minor sensation when it appeared in the October 31, 1881, issue of *Fiction Magazine*. It soon launched an influential subgenre of American future-war fiction: stories intended to scare the public into a vast military preparedness campaign, aimed especially at making the United States a formidable naval power. (From 1884 on, the story was regularly reprinted in volume 5 of Scribners' popular *Stories by American Authors*.) New York meets its sad fate in a bombardment by a Spanish fleet of five ironclads, impregnable to the archaic U.S. shore batteries and wooden ships. Amid the ruins of landmark buildings and faced with the "utter destruction" of New York, the United States prepares to surrender, but at the last moment is saved by three modern warships sent by Chile. This rescue by "an insignificant Republic of South America" underscores the humiliating American impotence. Fictions such as "The End of New York" played an important part in the military preparation for U.S. imperialism, unleashed in the actual conflict with Spain seventeen years later. By then, of course, the relative military strength of the combatants was the reverse of that projected in Benjamin's 1881 tale.

Throughout the 1880s, American future-war stories incited the public to yearn for a large peacetime war fleet. Although the fiction usually presented a standing navy as a means of defense against foreign invasion, it actually was a precondition to America's ascent to global power, which of course would also require overseas colonies. Indeed, the navy and the possessions were inseparable, for a military fleet would be necessary to seize and hold colonies, which in turn would provide bases indispensable to maintaining such a fleet. Amid this fiction, in 1890, appeared the great theoretical treatise on the role of the navy in establishing global power, Captain Alfred Thayer Mahan's *The Influence of Sea Power upon History, 1660–1783*, which "was to have as profound an effect on the world as had Darwin's *Origin of Species*."[3]

American aspirations to global empire did not emerge abruptly in the 1890s. But the struggle between imperialist and anti-imperialist ideology did wax more and more intense until the outcome was decided in 1898, when the United States grabbed key pieces of the disintegrating Spanish empire in both hemispheres, including Cuba, Puerto Rico, Guam, and the Philippines. Animated by the emerging imperial ambitions, the future-war novels of the 1880s and early 1890s clashed with the two main forces obstructing their fulfillment: the ideology upon which the United States had been founded, and Britain, the dominant rival military power that any rising empire would have to confront. American republicanism, anti-imperialism, and antimilitarism all had deep roots in a nation generated by a revolution against the British Empire, standing armies, the taxes necessary to finance global power, and the autocratic authority needed to wield such power and command such armies. And any attempt to assert American influence onto or across the seas risked immediate confrontation with the nation's oldest enemy, then the mightiest empire the world had ever seen, with its colossal navy and its bases almost encircling the United States.

War with Britain, Our Oldest Enemy

How strange it would have been for Americans a century ago to hear Great Britain referred to—in a phrase commonplace among today's politicians—as "America's oldest ally." Looking with late-twentieth-century eyes at Great Britain, we generally see only our ally in World War I, in the 1917–21 intervention in the Soviet Union, in World War II, and in the ensuing global crusade against Communist revolution and national liberation movements. But Great Britain looked very different to many Americans in the late nineteenth century.

Not only was the very foundation of American national identity built on a revolution against the British Empire, but that empire had remained for a century unreconciled to the revolution. In the War of 1812, caused partly by British impressment of American seamen and partly by American attempts to extend the benefits of its revolution to Canada, Britain punished its rebellious child by burning Washington, the capital city. During the Civil War, Britain conspired directly with the Southern secessionists and even constructed their fleet. In the 1880s and 1890s, conflicts between the world's dominant empire and the latest aspirant to imperial power frequently led to the brink of war. Many American views of Britain during this period resemble twentieth-century Soviet views of the United States. Great Britain was portrayed in both future-war fiction and leading newspapers and magazines as a hostile power that sent its fleet and army to suppress the revolution, attacked again within decades, armed and financed a secessionist movement, established a string of bases around the country, and penetrated U.S. coastal defenses with warships.

Indeed, for eleven years after Park Benjamin's 1881 story of war with Spain, virtually every American novel that imagined a future war with foreign powers presented Britain as the primary, if not sole, enemy. These novels imagining imminent war with Britain merely extrapolated the warnings and forebodings rising throughout the society into the early 1890s.

For example, the *New York Times,* hardly an instigator of Anglophobia, ran such ominous analyses as "A Stranger at Our Gate," a two-thousand-word article by William Drysdale warning on July 27, 1890, that "Great Britain is no longer a distant power across the seas, but a powerful nation with intrenchments thrown across our front yard, ready to interfere with our ingress and egress whenever opportunity may offer." Drysdale traces the line from the fortifications along the Canadian border to the island fortress at Bermuda down through the British colonies strung across the Caribbean to British Honduras:

> This offensive line—for I can hardly regard it as a defensive line—is maintained at almost fabulous expense, and there is a reason for its existence. . . . A map of

America with all the British possessions printed in black would be as dark as a thunder cloud. . . . Is England a friendly nation? I read every day of the war ships she is sending up to Behring Sea, and of the fleet she is concentrating at Halifax, and again ask the question. I see those guns she is pointing at us all along our shore, and leave the question for you to answer. . . . The British army officer can sit down and tell by the hour why the Confederate Army should have whipped us, and explain how it happened that they didn't, without mentioning that their not doing it was no fault of the British government.

Throughout this period, armed hostilities often seemed imminent. Especially dangerous were the confrontations in the Bering Sea, where Canadian ships manned by British subjects were seized by the United States in waters it claimed, while British warships steamed to the area and American warships, hastily rigged with new armor and long-range cannons, raced to meet them.

On January 5, 1891, for instance, the banner headline of the *New York Times* blared "IS IT WAR OR BLUSTER," while below appeared a box juxtaposing the number of guns, the number of men, and the speed of each ship in the opposing war fleets. Two days later, the lead story discussed Canadian preparations for war against the United States and compared the possible military scenarios with the War of 1812—a favorite exercise of the future-war novels. On January 18, an article entitled "England in Our Waters" warned, like much of the future-war fiction, against our lapses in naval power, and traced the history of our conflicts with Great Britain back to "the deadly apathy" that allowed the "riveting of the chains of British oppression" in the eighteenth century.

Some of the fictions imagining wars with Britain, such as *The Stricken Nation*, an 1890 pamphlet by "Stochastic," were little more than preparedness tracts warning of British subversion and imminent naval attack. But one of these tracts, Samuel Barton's 1888 *The Battle of the Swash and the Capture of Canada*, is worth close attention.[4]

Ostensibly written in 1930, *The Battle of the Swash* begins by sketching the background to a war that broke out in 1889 between Britain and the United States. By 1860, Barton points out, the Northern "mercantile and manufacturing States . . . had become formidable rivals to England" and "the United States occupied a front rank among the maritime powers of the world" (11), with 67 percent of the vessels entering U.S. ports flying the American flag. But thanks to the British, "our commerce was swept from the ocean during our civil war" (27), so that by 1887 a mere 15 percent of the vessels entering U.S. ports were American. For shortly after the start of the Civil War, "a number of so-called 'Confederate cruisers,' which had been built and fitted out in English ports with English money, were scouring the ocean, capturing and destroying American merchant vessels wherever they could find them" (16). Barton reminds his readers in 1930 that this is not mere allegation; after the Civil War, the arbitration conference in Geneva had forced Britain to pay fifteen million dollars for the damages it had inflicted. So then America, which had failed to arm itself, had to face the colossal "British empire, upon whose possessions it is the Englishman's boast that 'the sun never sets,' that 'her drum beat encircles the world,' that 'her ships fill every sea'" (54).

In the fictive war, Britain decides to abandon Canada but attack the virtually defenseless United States with its vast "hostile fleet—representing the most formidable naval power in the world, and presumably containing all the best and most improved offensive weapons known to modern science" (73). After destroying the puny U.S. fleet, the British armada gives forty-eight hours for noncombatants to evacuate "the cities of New York and Brooklyn" (101), which are now abandoned "to the idle and criminal classes," who unleash "an epidemic or carnival of crime" (107). The British ignore U.S. "protests against the inhumanity and barbarity of bombarding a defenseless city" (109). To the fictive narrator and his imagined audience of 1930, such savage bombardment seems almost unimaginable. Yet to a reader today, all that may seem improbable are the advance warning and moral outrage:

> To us of the present generation, who have never experienced any of the horrors of war, it seems almost incredible that civilized and Christian men, could thus coldly arrange the details of the destruction of life and property on such a vast scale, and calmly count the seconds on their watches as they ticked away the few remaining moments which separated the two great cities from destruction. (115)

Then follow scenes of devastation that were soon to become conventional in the future-war novels up until World War I, scenes that recent history has made altogether too familiar to us, though only in fiction have they taken place in the United States. "The whole of the lower portion of New York" is reduced to "a confused heap of ruins," with rubble making the streets almost impassable (117). The Brooklyn Bridge, which is destroyed over and over again in these novels, is shelled until "the massive granite towers gave way, and the whole magnificent structure fell into the river beneath where for many months it remained an absolute barrier to navigation" (116). Two more British fleets are dispatched to bombard the defenseless cities on the South Atlantic and Gulf coasts, and the British seem victorious.

But during the naval battle of "the Swash," a waterway leading into New York harbor, Barton conjures up a world-shaking event. "Two insignificant looking little boats" that "had been built by private subscription" head toward the British armada. Their only armament is "a gigantic tube or cartridge, containing two tons of dynamite" carried in a long steel ram and detonated by an electric timer (87–88). After aiming the boat, each crew jumps overboard, and two great English warships are "blown to atoms" (91). Though the British recover and go on to devastate the U.S. cities, "this American invention of self-destroying torpedo boats" (121) permanently nullifies British naval power. Great Britain cedes Canada and its West Indian possessions to the United States, and soon loses control in Europe. Russia takes over India, Australia declares independence, and "liberal expenditures" soon make the U.S. navy "superior to that of any other nation." England is reduced to an impotent island kingdom, and we Americans "now occupy unchallenged . . . the foremost position among the nations of the earth" (126).

Barton was here dramatizing—albeit in an almost ludicrous form—one of the earliest versions of a characteristically American conception with truly earthshaking implications. Many of these early American future-war novels at first seem to be just preparedness propaganda tracts in the European tradition. But unlike the typical British and Continental fiction, in which the leading industrial nations raced neck and neck to apply existing technology to new weapons of war, American writers often conjured up some leap in weapons technology that would suddenly transform the entire history of the world. A new weapon, evoked by America's technological wizardry, would either bestow global hegemony on the United States (sometimes with Britain as a junior partner) or immediately put an end to all armies and navies.

The quick technological fix fantasized by this fiction has turned out to be what is now called the fallacy of the last move, the will-o'-the-wisp that the United States has pursued in plunging the planet into the colossal arms race of our age. Faster and faster we chase this mechanical rabbit, always believing that American technological ingenuity is capable of creating an ultimate weapon that can grant perpetual world peace through either universal disarmament or American global hegemony. This fallacy characterizes the cult of the superweapon, which would soon receive its first full-blown expression in this fiction, as it moved in a few years from self-destroying torpedo boats to atomic and beam weapons.

In 1889, the year after *The Battle of the Swash*, Frank Stockton's *The Great War Syndicate* appeared, featuring not only bombs as powerful as the atomic weapons dropped on Japan but a military-industrial complex to create and use them—in the interests of peace. When war breaks out between Britain and the United States over fishery disputes, twenty-three "great capitalists" form themselves into a "Syndicate, with the object of taking entire charge of the war" (12) in order to achieve both victory for the nation and enormous profit for themselves. The Great War Syndicate finances and deploys two superweapons, the "Crab," a steel-clad submersible vessel with gigantic pincers that can disable the mightiest warships, and the "Motor Bomb," which has the explosive force of a thermonuclear warhead. Of course the United States doesn't actually use such a dreadful bomb: the mere demonstration of its power brings victory without a single person being killed in combat.

Britain, though defeated, is permitted to join as a junior partner in what becomes the "Anglo-American Syndicate of War" (180). Since any future battles would be "battles of annihilation" (191), the world submits to the peaceful and enlightened rule of the Anglo-American syndicate. In the final words of the novel: ". . . all the nations of the world began to teach English in their schools, and the Spirit of Civilization raised her head with a confident smile."

It is easy to laugh at the naiveté of *The Great War Syndicate*. Didn't it occur to Stockton that if one war was immensely profitable for the "great capitalists," they might have an irresistible urge for a few more? Why wouldn't the War Syndicate use its invincible weapons to extract superprofits from other people at its mercy? Why would a private armaments industry, once it had become an equal of national governments, relinquish any of its power? But the last laugh might be on us if it has turned out that, for the ensuing century, the United States has been acting on the assumptions that seem so blatantly naive in Stockton's novel.

Novels foreseeing war with Britain were by no means unanimous in projecting fear, much less hate, of its empire and its values. Indeed, some antipopulist authors used the future-war form mainly as a vehicle to attack socialism, feminism, Irish republicanism, and non-Anglo-Saxon immigrants. A few even looked to invasion from Great Britain as the best chance for rescue from "the socialists, the communists, the anarchists" ("Coverdale" 64), who impose mob rule by such lowly types as Italian-Americans, German-Americans, and Irish-Americans, referred to as "the hyphenated Americans" (Barnes 36).

Samuel Rockwell Reed's 1882 *The War of 1886, Between the United States and Great Britain* predicts that England will be goaded into attack by the "blustering, filibustering, Canada-invading, Irish Republican, whip-the-whole-world-in-arms, jingo party" (18) of the ignorant masses. Ironically, the successful invasion overexpands Britain's "imperial foreign policy," thus bloating its "military establishment," leading to increased taxes and national debt, industrial decline, and the degeneration that comes from "the admission of the ignorant majority into her governing class" (24). Meanwhile, the American defeat leads to "the mighty paradox that our nation's ruin became the source of her rise to that imperial position . . . for which nature had given her all the elements" (25).

The nation has a less glorious destiny in *The Fall of the Great Republic* (*1886–1888*). Actually published in 1885, this novel bears a title page stating that it was published "By Permission of the Bureau of Press Censorship" in 1895 by "Sir Henry Standish Coverdale," "Intendant for the Board of European Administration in the Province of New York." The book paints a hideous picture of the "poisonous ferment" of "socialism and feminism" (14), the machinations of the "utterly illogical" Irish, "coarse of feature and coarser of mind" (35), and "the immigrants from the most dangerous classes in Europe" (12), whose communist revolution defiles the American republic, originally established solely for those with the "sterling virtues . . . which characterize the Anglo-Saxon race" (32). Led by Britain, the European powers agree that "the revolution must be crushed at any cost, and so completely that there should be no danger of its reviving," lest all the established powers face revolution (160). So the "allies" intervene, as indeed they were to do in actuality against the Russian Revolution, though in the novel they successfully crush the American socialist government, occupy the country, and establish a beneficent conservative state.

Similar antipopulist novels that envisioned a demagogic popular government provoking war with Britain continued until James Barnes' *The Unpardonable War* was published in 1904—the year before the Russo-Japanese War was to change conceptions of modern naval warfare and initiate shifts in the conflicts between the opposing empires that would culminate in World War I.

Many of the themes of *The Unpardonable War* had by this time become conventional. War with England is provoked by the triumph of "the People's Party," a coalition of left-wing forces. To Barnes this means the uneducated, the incompetent, and the vicious, everyone from the antimilitarists who had prevented adequate military preparation to the immigrant "hyphenated brethren" (94), such as the Anglophobic Irish-Americans and the socialist-leaning

German-Americans. (Prior to the pre–World War I propaganda campaign against the Germans as the arch-militarists of the planet, authors both left and right identified the British as the most potent aristocratic force against incipient German socialism.) Although in conception and ideology *The Unpardonable War* is at first virtually indistinguishable from the novels of Reed, "Coverdale," and Stockton, its vision of modern war is far less benign, despite the happy ending in which Great Britain and the United States become peaceful "dictators of the world" (351), eradicating populism and enlightening the backward planet.

In a terrific naval battle, during which two lines of opposing submarines are annihilated in a colossal explosion, the U.S. navy is all but destroyed by the vastly superior British armada. But fumes from a burning British ammunition ship mingle with those from a new American explosive to form a poisonous fog that asphyxiates most of the survivors, turning the victorious British ships into a ghost fleet. When at first it seems that the Americans had intentionally used poison gas, the outraged response measures the distance between 1904 and the gas-filled European battlefields of the First World War: ". . . the awful aftermath was ascribed at first to Yankee invention and devilish cunning. The widespread, unwritten, but well-understood, rules so-called, of civilized warfare, had been broken. The loosening of deadly vapours was as bad as poisoning wells. . . ." (226).

A lone genius in Britain invents a new smokeless powder to forestall such accidents, but a lone genius in America, "the Wizard of Staten Island" (230), overcomes this with "the Force," a field of "intense electrical vibrations" (239) at precise frequencies that automatically detonates all the British cartridges and shells under its influence. As British scouts are cut in two by the cartridges exploding in their own bandoliers, small units of Americans, at grave personal risk, chivalrously advance with old muskets and black powder to warn the British of their peril. With science in charge, the U.S. campaign is now run "on a business basis" (328), and the officers in charge of the opposing sides negotiate in a friendly offhand manner, for it's no longer a "military affair; it's commercially scientific" (327). Since "the Force" has thus deglamorized war, the British and American commanding generals agree that this is "the end of our profession" (338).

After all, such a horrendous weapon certainly could not be used as an instrument of mass destruction, for people's minds would be "revolted." And so science has made war impossible (338, 340). When the flags of the United States and Britain are hoisted together amid the mingled singing of "God Save the King" and "America," the Anglo-American alliance, thanks to the American superweapon, can assure "the rule of peace, the sway of justice, and the dawn of common-sense" (356).

Progressive America Versus Reactionary Russia: Populist Imperialism

To antipopulists such as Reed, "Coverdale," and Barnes, the rise of the vulgar masses in America would lead inevitably to war with Britain, the great conservative power of the world. When the populists imagined themselves coming to power, they too sometimes saw war with Britain as a corollary. And like their conservative opponents, they too foresaw beyond war a mission of global enlightenment for the reunited English-speaking peoples.

Indeed, Ignatius Donnelly, a founder and leader of the People's Party, published in 1892 a future-war novel projecting scenarios remarkably similar to those of the antipopulists, though approaching the burning domestic issues of American society from the opposite side. *The Golden Bottle*, written in 1890 while Donnelly was campaigning as the People's Party candidate for governor of Minnesota, fantasizes the ultimate triumph of the populist vision, with the American flag becoming "the banner of mankind" (222).

Opening with realistic scenes of the poverty crushing the small American farmer, *The Golden Bottle; Or, The Story of Ephraim Benezet of Kansas* imagines Ephraim, an impoverished young Kansan, endowed with a miraculous process for converting iron into gold. With this power, he speedily realizes a central tenet of populist ideology, demonetizing silver and gold and establishing a fiat paper currency, thus freeing the farmers and workers from the power of finance capital and unleashing tremendous forces of production and progress.

Elected president, Ephraim discovers that "the Plutocracy" is using its control of the newspapers to prevent his inauguration. But taking the advice of his wife, Sophie, Ephraim uses his almost limitless personal wealth to buy up all the newspapers, which then support his program and power. Once in office, Ephraim Benezet is able to lead America into becoming a virtual paradise of people's power, with boundless productivity and freedom.

When President Benezet calls for the liberation of the European workers, Britain—backed by most of the Continental powers—declares war; and so, "The battle of the ages, between liberty and despotism, was at hand" (207). Just as in the antipopulist *The Unpardonable War,* the United States takes Canada. But then the war is speedily decided by revolution. As the British masses rise up, chants of "America! America! America!" are heard all over the British Isles, and "everywhere among the troops, all over England and Scotland, the flag of the United States was improvised" (222). When the Americans march, in segregated "black and white" units, against the remaining kings, the revolutionary "contagion" sweeps across the Continent: The German army revolts; the masses rise up in Hungary, Austria, Poland, Italy, and Spain; and soon President Benezet decrees the establishment of "a new nation, to be called 'The United Republics of Europe'" (237, 246). The fleeing "enemies of mankind, emperors, kings, princes and aristocrats," are forced to regroup in Russia—"that land so utterly given over to ignorance, superstition, fanaticism, and despotism that it was to be the last abiding-place of the devil of injustice on earth" (247).

For the new American social order to triumph, the Russian czardom must be vanquished: "Either we must wipe out that colossal wrong or we must fall before it" (247). Thus comes "Armageddon," the global showdown between socialist America and reactionary Russia. It is Sophie, the all-American woman from Kansas, who leads "the great Army of Liberation" to victory over Russian despotism:

Yes! A Yankee woman had won Armageddon! A Western girl had achieved the Millennium!

The thousand years of peace and happiness and love had begun, amid the corpses of that bloody battle-field; the last battle-field of the ages. (263)

Although Ephraim awakes to discover that this glorious vision had been a dream, he learns that the dream is an allegory pointing the way toward political and economic liberation. Fiat currency is the true "golden bottle" that could lead to prosperity for the workers of the United States and Europe, the overthrow of the reigning monarchs and capitalists, and the establishment of a grand socialist Pax Americana.

This marriage of populism and imperialism was hardly unique to Ignatius Donnelly. Indeed, it was crucial to the triumph of imperialist ideology in the America of the 1890s—and the century to follow. In 1898, the very year that the United States definitively committed itself to global imperialism, appeared S. W. Odell's *The Last War; Or, The Triumph of the English Tongue,* a novel projecting this ideology seven centuries into the future when "Armageddon" takes the same form as in *The Golden Bottle*—a showdown between the progressive forces led by America and the reactionary forces lurking in Russia.

By the mid-twenty-sixth century, United America, consisting of 185 states in North and South America, is a prosperous egalitarian society, happily restored to racial purity by the emigration of the blacks. It is an integral part of "the Allied Anglo-American Nations," which includes the entire English-speaking world— that is, Britain, western Europe, India, Australia, Africa (thanks to British colonization), and just about everything but Russia, eastern Europe, Japan, China, and some contiguous countries controlled by the Russian Czar (who has now also become the Pope, and thus stands for everything evil and reactionary). Since the English-speaking peoples "are approaching human perfection," "the gulf between evil and right has widened to such an extent that only war remains—war to the end—war which will result in the annihilation of the forces now guarding evil" (74). There can be no other solution for the forces of Anglo-Saxon virtue: "For the onward march of enlightened civilization has met inertia; it can neither go around nor through—it must crush and destroy and pass over" (10).

Thus comes "Armageddon," in which the progressive English-speaking armies fight the reactionary Russian-Asian forces with awesome new explosives and fleets of aluminum-armored dirigibles. The American air armada manages to turn the tide of battle by dropping a form of inextinguishable fire, much like napalm (152). After suffering nine million casualties, the Czar's hordes of Moslems, Buddhists, Catholics, and other infidels surrender; a chapter entitled "Pacification of the Conquered Territory" documents the thirty-five-year occupation that succeeds in imposing the English language and "the customs of civilization" on the "benighted minds" of "the ignorant and savage inhabitants" of Russia, eastern Europe, and Asia (160). Finally, in the year 2600, the United States of the World is formed, and *The Last War* concludes with these words: "The dream of the ages had been realized and peace assured to the human race forever."

The seventeenth-century American Puritans had been fond of imagining their little colonies as a beacon on a hill, illuminating a path for the dark Old World of Europe. In these late-nineteenth-century American fantasies of future war waged by forces deemed progressive, we see this vision becoming truly messianic as the old beacon light begins to assume the proportions of a global conflagration. Just as fierce class struggle seemed about to tear America apart, the ideology of "civilizing" imperialism became a force capable of uniting key sectors of the working class with the capitalists in wars of conquest waged, with nothing but the highest of motives, against the dark forces and races of the planet.

Anglo-Saxons, Onward!—to Armageddon

When the United States actually made its decisive venture into global imperialism in 1898, it moved not with a provocation or challenge to the British Empire but in tacit collusion with British global aims and with the blessing of the British government. By preempting anticolonial revolutionary movements against the doomed Spanish empire in Cuba, the Philippines, and Puerto Rico, the United States managed to establish its own imperial bases while strengthening rather than weakening the existing world hegemony of European peoples. In so doing, it also laid the foundation of an Anglo-American alliance that was to become a major force in the twentieth century.

Along with *The Last War; Or, The Triumph of the English Tongue,* two other future-war novels published at that critical moment in 1898 dramatize the imaginative core of the triumphant American ideology. One seems to yearn romantically for medieval forms of warfare, while the other worships in the cult of the superweapon. Each, however, imagines the victorious Armageddon of the English-speaking alliance less as the triumph of progressive and populist forces over czarist feudalism than as the victory of the superior Anglo-Saxon race over the inferior races led by the Russians.

Looking backward from an imagined future to the 1898 of its publication, Benjamin Rush Davenport's *Anglo-Saxons, Onward! A Romance of the Future* hails the Spanish-American War for opening "a new vista . . . of grandeur and glory" in which "Anglo-Saxons, Onward! becomes the watchword of the Americans" (37–38), as the United States conquers overseas colonies and begins to fulfill its divinely intended destiny:

> By the capture of the Spanish islands the United States found itself compelled to enter a larger policy. . . . [I]t would be manifestly contrary to all that was righteous, to resist the obvious destiny for which the All-wise brought the nation into being to accomplish.
>
> Hence, dating from 1898, the expansion and extension of America's influence, power, wealth and territory had been phenomenal. . . . (40)

Since the Spanish-American War, "the mighty brother nations" of "England and America have been bound together by a sentimental semi-alliance" (8, 13), which the other nations seek to disrupt: "All the nations of earth clearly comprehended that a permanent and firmly held alliance between America and Great Britain . . . gave to the Anglo-Saxons the domination of the whole of this terrestrial sphere and enabled them at will to change the map of the world" (114). When Russia declares war on Britain, America declares war on Russia. The righteous Anglo-Saxons need no newfangled weapons, as their cavalry charges the Russian artillery, cutting down the gunners with swords. Russia surrenders, and the Anglo-American alliance achieves its global destiny. Of course this hymn to the glories of imperialism has as its refrain: "We go not to conquer, but to free mankind" (66).

Stanley Waterloo's *Armageddon* also looks back from the imagined future to the year of its publication; it foresees the Spanish-American War leading, during a few deceptively peaceful decades, to the formation of new alliances that fight a world war to end all wars in the early twentieth century. The "Anglo-Saxon alliance" (aided, rather surprisingly, by Japan) must engage in the Armageddon of the title against a vast host of the inferior races, especially Slavs and Latins, led by the Russians, who have somehow also cajoled the Germans into betraying their "Teutonic blood."

On their side the Anglo-Saxons have their transcendent martial spirits and discipline, as well as the canal the Americans have built through Nicaragua. As Richard Gid Powers has pointed out:

> *Armageddon* is permeated with a popular version of Captain Alfred Thayer Mahan's theory that sea power is the determining force in world history. . . . Nearly a fifth of *Armageddon* is devoted to the construction of the Nicaragua Canal. . . . It was because popular journalists had so indoctrinated the public with Mahan's doctrine of naval determinism that Waterloo could rely on his readers to see the point of what a modern reader might judge an unaccountable digression into the mechanics of canal building.[5]

The conflict ends, predictably enough, with world rule by the Anglo-Saxons, who magnanimously allow their German brothers to join in the global hegemony of the master race, while the United States prudently prohibits immigration by any of the inferior races, those "ignorant, helpless millions, hopelessly pauperized, alien in race, language, and affiliations" (243).

But fighting spirit, superior blood, and awesome sea power would not by themselves guarantee victory in this ultimate war: the deciding factor is a superweapon invented and built by an American lone genius. Subtitled *A Tale of Love, War, and Invention*, the novel reserves its deepest eroticism not for the rather flimsy romantic subplot, but for its passionate love affair with weapons of war:

> . . . we came in close view with our glasses of the iron monsters of the British navy. As the grim line of battle-ships gave forth their din of salute to our flag-ship my heart jumped into my throat, and tears found an unaccustomed place in my eyes. It was beautiful, but with the beauty of terror, that assembly of naked metal fighting machines lying there on the strongly heaving yet unbroken sea of blue water. (204–5)

This awsome beauty of naked metal is about to be overmastered by the "Wild Goose," a torpedo-shaped aluminum "air-ship," a "dirigible" superweapon powered by some mysterious "driving force" understood only by the American technological wizard who invented it. When the two vast armadas meet for their Armageddon on the high seas, the Wild Goose decides the contest, single-handedly wiping out the Russian flagship, aptly named the "Czar," by dropping high explosives on it from the sky. The omnipotent sea power exalted by Captain Mahan must yield to the supremacy of air power.

At the end, with "the Anglo-Saxon . . . now dominating the world" (240), this aerial superweapon has brought the promise of eternal peace, for "Civilization has reached a point where war is suicide." The mere threat of "aerial warfare," by placing kings, parliaments, and congresses in as much danger as the private soldier, makes war impossible. Predictably naive, the sweetheart of the genius inventor responds to this argument with a typically feminine question: " 'Why do we make these killing machines then, if they are not be used?' " The inventor patiently answers with the dominant ideology of his age—and ours:

> "To have a world at peace there must be massed in the controlling nations such power of destruction as may not even be questioned. So we shall build our appliances of destruction, calling to our aid every discovery and achievement of science. When . . . war means death to all, or the vast majority of all who engage in it, there will be peace." (259)

But while such genocidal weapons may make war seem impossible among the "civilized" nations that possess them, such restraints may not apply when these powers face the menace posed by nonwhite peoples. And even Russia might not be as great a threat to civilization as the Yellow Peril.

The Yellow and Black Perils

Chinese, Africans, and Other Imperialists

The first American novel imagining a future war with a foreign nation seems to be Pierton Dooner's *Last Days of the Republic,* copyrighted in 1879 and published in 1880. At the fountainhead of what within two decades would become a torrent of Yellow Peril literature,[6] *Last Days of the Republic* grimly warns of "the impending catastrophe, which, at this period, menaces not only our civilization, but indeed, our very existence as one among the nations of the earth." Dooner points across the Pacific to the source of this imminent national doom: "To rule the World, is a dogma, a creed, a holy tradition of China" (4, 22).

The immediate threat is embodied by the "coolies." Dooner's fantasy transmutes that industrial army of Chinese laborers—in reality brutally sacrificed to forge the railroad networks that would soon help transform the United States into a dominant power in Asia—into a Trojan horse of Chinese imperial ambitions. The capitalists of the western states are so greedy for cheap labor that they allow Chinese workers to win their strike for "civil rights," resulting in Chinese control of the state governments of California, Nevada, and Oregon. The landowners of the South, seeking replacements for black chattel slaves, and the industrialists of New England, eager for the cheapest wage slaves, encourage swarms of the ever-fecund Chinese to overrun these sections of the nation. Resistance by the white workers merely paves the way for Chinese imperial intervention and control of the railroads. Too late, the whites take to arms; China invades, and the novel ends with bloodcurdling pictures of Asian hordes rampaging and pillaging their way into Washington:

> The Republic had fought its last battle; and the Imperial Dragon of China already floated from the dome of the Capitol.
> The very name of the United States of America was thus blotted from the record of nations and peoples. . . .

The "annihilation" of America—which once "cultivated the arts of peace and gave to the world its brightest gems of literature, art and scientific discovery"— devastates "the prestige" of "the one race of man" that "had borne upon its crest the emblem of sovereign power since the dawn of history" (256–57). We are left with the mournful conclusion: "Thus passed away the glory of the Union of States, at the dawn of the Twentieth Century."

Remarkably similar to Dooner's novel is "The Battle of the Wabash," a short story published, also in 1880, under the pen name "Lorelle" in the *Californian*. (The *Californian,* which ran only from 1880 through 1882, was published by the same San Francisco group that published the *Overland Monthly* from 1868 to 1875 and 1883 to 1935; both journals were influential voices in the anti-Chinese campaigns of the nineteenth century.) This tale pushes the ultimate Chinese conquest of America to the year 2081, when, after two centuries of unrestricted immigration and intermarriage, Chinese-Americans outnumber whites by three to one. The whites belatedly rise up in armed struggle, but are no match for the Chinese hordes, aided by five million "imperial soldiers" from China and sustained by little boxes of condensed food. The Chinese crush all opposition in an apocalyptic battle along the Wabash River, exterminating "the grand American army" and instituting an unprecedented "reign of horror." The nightmarish conclusion has a clear moral:

> From the Gulf of Mexico to the St. Croix, from Boston to San Francisco, the flames of unbridled passions rioted upon defenseless people; each midnight sky photo- graphed in its angry reflections the conflagrations of a thousand cities. The blow was decisive; the Republic fell on the Wabash, broken into a million fragments; her people passed into Asiatic slavery. The fruit-time of folly had come. By vice-regal order, rapine, murder, arson, and all the devils of human passion were to be unrestrained for a hundred days. Revolting at this sickening corollary of a people's folly, I turned away, murmuring to myself, When will the world learn that milksop philanthropy is not statesmanship?

Still more alarmist was Robert Woltor's 1882 novel, *A Short and Truthful History of the Taking of California and Oregon by the Chinese in the Year A.D.*

1889, which predicted an imminent Chinese invasion, led by Prince Tsa, who is compared explicitly to Satan. In a plot that was to become a mainstay of Yellow Peril future-war fiction, Chinese immigrants rise as a fifth column to greet the invaders, who seize the West coast and menace the rest of the United States.

The fictional visions of Chinese conquest expressed and helped to inflame the anti-Chinese hysteria that led to the Chinese Exclusion Act of 1882, which banned virtually all Chinese immigration. With this success, anti-Chinese future-war fiction seems to have dried up during the ensuing decade, when future wars with Britain dominated the genre. But when it reappeared in the 1890s, it had lost none of its virulence.

Indeed, the reemergent Yellow Peril literature now combined occasionally with another subgenre of future-war fiction that had been emerging in America: Black Peril literature. A characteristic example of this latter mode is King Wallace's 1892 novel, *The Next War: A Prediction*. Writing explicitly to warn white America of a treacherous invasion from within by the "vast horde" of "eight million enumerated Africans" in the South, together with "fully twelve million Mulattoes, Quadroons and Octoroons," Wallace reminds his readers that the glorious future of America could be easily defiled, for just "one drop of black blood flowing in whitest veins ten generations hence will bear the stamp of shame and universal hatred" ("Author's Preface").

In *The Next War*, "The four divisions of the colored race" in America (Africans, Mulattoes, Quadroons, and Octaroons) are secretly planning "for the extermination of the white race" (15) on December 31, 1900. Their conspiracy centers on a campaign to outbreed and overrun the whites—"*A war in the dark!*" But there is a happy ending. The whites discover the plan, strike first, and drive all those with Negro blood into extinction in the swamplands of the South: "No extermination could be more complete" (216). Cleansed of its darker population, America becomes a more wholesome society, reaping unanticipated benefits: "Housework is not the toil for slaves, and may be made again, as it was once in the good old days, the noblest occupation of women" (216).

Yellow Peril and Black Peril visions merge in a most revealing story, William Ward Crane's "The Year 1899," published in the June 1893 issue of the *Overland Monthly*. (Crane later authored a little survey and favorable assessment of the future-war genre, "Fanciful Predictions of War," in the November 1898 issue of *Lippincott's Magazine*.) In the story, war among the European nations precipitates a widespread socialist revolution, which in turn leaves them vulnerable to rebellions by nonwhite and other colonized peoples. In the British West Indies, the blacks massacre the whites and "set up governments of their own." The Chinese rise up, exterminate all whites in China, seize the foreign ships in their ports, and organize a grand alliance of Asians, Africans, and nonwhite Americans into a global war of annihilation "against the white race everywhere."

"Two immense armies" of "Indo-Chinese, Hindoos, Afghans, Beloochees, Persians, Turks, Arabs," Chinese, and Africans sweep across Europe: "The two great human floods rolled on unchecked, living on the country, butchering all who could not escape, and leaving desolation behind them." In America,

Chinese infiltrators organize the Indian tribes and millions of blacks, who are soon issuing "demoniac yells and shrieks" as they dance around fires "like their savage ancestors," while chanting such blood-chilling insurrectionary songs as:

> One, two, tree! All de same!
> White, black, red! All de same!

Meanwhile, "Negro risings" place Cuba and Puerto Rico "into the hands of the insurgents, who in the usual way massacred all the whites they could find."

North America is invaded from two directions. "Swarms of half naked negroes" embark in an armada of small boats from Cuba to seize Florida, while a Chinese fleet transports a colossal army, resembling "the migrations of African locusts," across the Bering Strait. But the West Indian invasion fleet is devastated by American warships; the immense horde of blacks and Chinese who emerge from the swamps of the South, "though fully three times as large as the white force opposed to it," is handily defeated by the disciplined white soldiers; the Indian uprisings are easily dispersed; and the Chinese invasion on the West Coast is forced to retreat when the grand pan-Asian global alliance disintegrates into a religious war among its various fanatical sects.

Then Japan allies with Britain and the United States and invades Korea. At last report, "American, British, and Japanese ships have bombarded all the principal Chinese ports, and devastated nearly the whole coast of China proper." The remaining Asiatic forces in Europe succumb to pestilence; those who survive the plague are exterminated by the returning Europeans, who "kill the Asiatics wherever they find them, treating them everywhere like noxious reptiles."

Published during the very period in which the European powers were ruthlessly conquering what remained of the uncolonized nonwhite world, "The Year 1899" blatantly converts the victims into the invaders. In its perverse way, however, the tale is also remarkably prescient. From the vantage point of 1893, in the calm eye of the storm almost precisely midway between the last and the next great European wars, it foresees the intra-European conflicts and revolutionary upheavals that would unleash the national independence movements of the world's nonwhite peoples.

Also revealing is the role that "The Year 1899" allots to Japan. Although there is a tendency to lump anti-Chinese and anti-Japanese visions together in the category "Yellow Peril," there were actually two quite distinct modes of anti-Asian literature. Before the twentieth century, Yellow Peril literature was almost entirely anti-Chinese, and William Ward Crane's perception of Japan as an ally of the western powers was fairly commonplace. It was not until after the shocking defeat of Russia by Japan in 1905 that Japan became the nemesis in Yellow Peril literature.

Fantasies of China as a world-conquering empire continued to assume new forms into the late 1890s. For example, Oto Mundo's bizarre 1898 novel *The Recovered Continent; A Tale of the Chinese Invasion* imagined the Chinese hordes led by Toto Topheavy, a mentally retarded white man who has been surgically transformed into a supergenius. After marching into Southeast Asia and Russia, Toto leads hundreds of millions of Chinese "raging savages" to overwhelm Europe and the United States.

The snarling racism of the Yellow Peril literature expresses cultural furies that have shaped some of the ugliest features of American history: the savage exploitation of "coolie" labor; the World War II incarceration of Japanese-Americans in concentration camps and expropriation of their property; the incineration of Japanese cities with incendiary and atomic bombs; the genocidal assault on the Vietnamese rural population, using everything from extermination raids on villages to indiscriminate poisoning of the countryside with chemical weapons. In the interests of impartiality, though, one should also note that British future-war fiction was every bit as virulent in its fond fantasies of genocide for Asiatics, and British colonial policies in Asia seem to have borrowed many pages from these books.

The leading British purveyor of Yellow Peril hysteria was M. P. Shiel. His 1898 novel *The Yellow Danger* (reissued in 1899 as *The Yellow Peril*), immensely popular on both sides of the Atlantic, elaborated in gory detail the fantasy of Chinese conquest already familiar in American literature. Shiel contributed an original method of perceiving a victim of European and American imperialism as a wicked aggressor: the Chinese are offering pieces of their own country to the colonizing nations merely as a sinister trap designed to "plunge the world into war by working upon the rapacity and selfish greed of the nations" (25). When the European nations fall for this unscrupulous plan and begin to war against each other, the Chinese hordes sweep across the European continent. However, a British mastermind devises a twofold strategy for massacring the invaders. Twenty million Chinese are towed out to sea and drowned; then, perhaps following the hint about pestilence dropped in Crane's "The Year 1899," the British employ germ warfare, using the "injected Chinamen" (380) to spread the plague among their compatriots until over 150 million are killed, temporarily turning Europe into "a rotting charnel house."

Perhaps the most ominous work among all this anti-Chinese future-war fiction is Jack London's short story "The Unparalleled Invasion," written in 1906 and published by *McClure's Magazine* in 1910. Though appearing after anti-Japanese works had displaced the anti-Chinese, and thus in a sense a kind of throwback to an earlier mode, "The Unparalleled Invasion" is most definitely a work of the twentieth century. Indeed, by allowing full play for the racist aspects of American populism, by extending racism into genocide, and then merging genocide into the cult of the superweapon, London's story achieves archetypal significance for our own times.

London's conception of "The Unparalleled Invasion" emerged from his experience as an on-the-scene reporter in the Russo-Japanese War, where he was startled to discover an Asian people efficiently using the most up-to-date Western management methods and weapons of mass slaughter. His speculations about what might happen if Japan were to "awaken" the 400 million people of China were published in two very similar essays, "The Yellow Peril" (*San Francisco Examiner*, September 25, 1904) and "If Japan Awakens China" (*Sunset Magazine*, December 1909). Long passages from "The Yellow Peril" are incorporated verbatim into "The Unparalleled Invasion," in which London envisions Japan taking over "the management" of China, rousing it from slumber to become "the colossus of nations," thoroughly industrialized, modernized, and prepared to conquer the world.

The Japanese are expelled from China and are crushed militarily when they atempt to reassert control over their Frankenstein's monster. But "contrary to expectation, China did not prove warlike," so "after a time of disquiet, the idea was accepted that China was to be feared not in war, but in commerce." The nations of the world fail to apprehend that "the real danger" from China "lay in the fecundity of her loins."

The Chinese take over French Indochina through a massive wave of immigration and colonization, with a "monster army" of militia and their families simply "brushing away all opposition." Soon this tidal wave sweeps through southwestern Asia, pushes relentlessly into Russia, and threatens to overwhelm the entire western world. Against this force, "war was futile," for "there was no combating China's amazing birth rate": "Never was there so strange and effective a method of world conquest." When European war fleets attack, they are simply ignored, and European invasion armies are swallowed up when they go ashore. By 1975, the world seems doomed to be overrun by the teeming hordes of Chinese.

But now a savior arises—who else but a brilliant American scientist with a secret weapon? The nations of the world join together to send tens of thousands of warships to the coast of China and many millions of troops to seal its land boundaries. China, confident in the overwhelming strength of numbers, merely "smiled" at the massed fleets and armies. But then, on May 1, 1976, according to this tale published a few years before World War I, comes the first strategic attack from the air.

Fleets of "tiny airships" shower China with what appear to be "strange, harmless missiles, tubes of fragile glass that shattered into thousands of fragments." These missiles have been loaded with "a score of plagues," "every virulent form of infectious death":

> Had there been one plague, China might have coped with it. But from a score of plagues no creature was immune. The man who escaped smallpox went down before scarlet fever. The man who was immune to yellow fever was carried away by cholera; and if he were immune to that, too, the Black Death, which was the bubonic plague, swept him away. For it was these bacteria, and germs, and microbes, and bacilli, cultured in the laboratories of the West, that had come down upon China in the rain of glass.

The bacteriological warfare imagined in Shiel's *The Yellow Danger* seems almost mild by comparison, for this is, as London so aptly says, "ultra-modern war, twentieth-century war, the war of the scientist and the laboratory."

The Chinese Yellow Peril is expunged forever: "All survivors were put to death wherever found." "And so perished China," leaving in its wake a virtual utopia for the victorious forces of progress:

And then began the great task, the sanitation of China. Five years and hundreds of millions of treasure were consumed, and then the world moved in—not in zones, as was the idea of Baron Albrecht, but heterogeneously, according to the democratic American program. It was a vast and happy intermingling of nationalities that settled down in China in 1982 and the years that followed—a tremendous and successful experiment in cross-fertilization. We know today the splendid mechanical, intellectual, and art output that followed.

"The Unparalleled Invasion," Jack London's exaltation of the superweapon and Asian genocide, appeared after most of America's imagined wars with China had already been fought. From nightmares about China's devastation of the United States, they had metamorphosed into utopian fantasies about the extermination of the Chinese.

The Japanese Attack

In the wake of Japan's 1905 defeat of Russia, a tidal wave of novels and stories envisioning war with Japan swept across America. It is of course no coincidence that this flood of anti-Japanese fantasies arose shortly after the United States, which had become established as one emerging power in the Pacific, faced a victorious Japan as its most likely rival for domination in the Pacific Basin. As early as 1907, this fiction suggested a prophetic final solution. America would use two superweapons in combination to defeat Japan: air power and radioactivity.

The earliest work I know of that portrays Japan as America's enemy is J. H. Palmer's *The Invasion of New York; Or, How Hawaii Was Annexed*, published in 1897, just as the United States was consciously preparing to transform itself into a global imperialist power. This was the year President McKinley submitted to the U.S. Senate his treaty for the annexation of Hawaii, while the nation prepared psychologically and militarily to wage war against the disintegrating Spanish empire.

Palmer's novel foresees the United States annexing Hawaii in 1898, with the battleship USS *Maine* providing protection on the scene. Suddenly, Japan launches a sneak attack while a Spanish war fleet heads for the U.S. Atlantic coast. Unlike Park Benjamin's 1881 "The End of New York," *The Invasion of New York* projects the destruction of the Spanish fleet by torpedoes launched from shore defenses in New York harbor. The American Pacific fleet defeats the Japanese armada that has destroyed San Francisco and goes on to devastate the coastal cities of Japan, forcing it to surrender. So this page of American history is "one of pride and of glory," proving the wisdom of the axiom, "In time of peace prepare for war" (248).

Most of the fiction published after Japan defeated Russia was less sanguine. Marsden Manson's 1907 *The Yellow Peril in Action* imagines a combined Japanese-Chinese force victoriously invading the United States in 1912. Ernest H. Fitzpatrick's 1909 novel, *The Coming Conflict of Nations; Or the Japanese American War*, does predict the eventual defeat of Japan by an alliance of

America and Britain, but not until after millions of Japanese troops, aided by a fifth column of Japanese workers in California, have conducted apocalyptic battles against the American army in Idaho and Montana. (Here one gets a close look at the cultural roots of the World War II internment camps for Japanese-Americans.) By far the best known of the works warning of the Japanese peril was General Homer Lea's immensely influential *The Valor of Ignorance,* an invasion fantasy published in 1909 and reprinted in 1942, months after the attack on Pearl Harbor, with a long introduction by Clare Boothe (Luce) hailing it as brilliant prophecy.

Lea, an apostle of Anglo-Saxon racial purity, American nationalism, and the military spirit, was a central figure in the military preparedness campaign leading up to America's entry into World War I. Though advocating ruthless combat by the Anglo-Saxon nations against other peoples, Lea did not disparage the abilities of the races and nations he saw as threats. Indeed *The Valor of Ignorance* warns that the "Yellow Peril" is not China but Japan, a nation that incarnates for Lea many of his ideals of racial purity and militant nationalism. Later Lea, with the model of Japan still in his mind, was to spend years as a key military adviser to Sun Yat-sen in the campaign to forge China into a unified modern nation.

America, according to Lea, is being menaced by the rise of "feminism," "commercialism," and "the heterogeneous masses that now riot and revel" (25). Since "a nation can be kept intact only so long as the ruling element remains" racially "homogeneous" (124), America is threatened by contamination from its own colonies of the Philippines, Puerto Rico, and Cuba, the concentration of Negroes, especially in a number of southern states where "the negro outnumbers the white inhabitants," and the increasing percentage of "foreign-born"—particularly Germans, Russians, and Italians—in the northern cities (127). Since it is absurd to expect "patriotism" from a man who deserts his own country, "a naturalized citizen is an anomaly" (127). Revering "military activity" as a prime purpose of national existence, Lea disdains "feminism" and "commercialism" as corrosives of the manliness and purity necessary to a healthy state. Devotion to industrialism and commercialism threatens to make America "a glutton among nations, vulgar, swinish, arrogant": "It is this commercialism that, having seized hold of the American people, oveshadows and tends to destroy not only the aspiration and worldwide career open to the nation, but the Republic" (26–27). Industrialism is mere "national alimentation," while "military activity" is a "primordial element in the formative process and ultimate consummation of the nation's existence." Thus Lea arrives at a protofascist vision that places militarism at the heart of both national purpose and true peace:

> Commercialism is only a protoplasmic gormandization and retching that vanishes utterly when the element that sustains it is no more. Military or national development, on the other hand, is not only responsible for the formation of all nations on earth, but for their consequent evolution and the peace of mankind. (27)

Most of *The Valor of Ignorance* consists of systematic description of the inevitable war Lea predicts between Japan and the United States for control of the Pacific. Japan, uncorrupted by America's weaknesses, seizes Hawaii and the Philippines as a prelude to the invasion of the U.S. West coast. Lea draws up detailed plans, complete with maps, charts, and timetables, for the successful Japanese seizure of California and the other Pacific states, and seeks to prove with mathematical certainty that it would be impossible for the United States either to repel or to expel the invaders. Though praised as uncanny prophecy, *The Valor of Ignorance* merely systematized already familiar materials. It also might be viewed as self-fulfilling prophecy: made required reading in the Japanese military academies, the book may have helped convince Japan's military leaders that if they did attack, their victory would be certain. This message was passed to the populace by a Japanese reprint issued in 1942, the same year the American reprint appeared.[7] And so this ultrapatriotic work that helped militarize America also, in an ironic twist, offered—to the very nation that it envisioned as America's most formidable enemy—a detailed how-to-do-it plan for the conquest of America.

During these years prior to America's entry into World War I, fictions of future war against the Yellow Peril took forms far more fantastic than Homer Lea's textbook treatise on how Japan could invade the United States. J. Hamilton Sedberry's 1908 *Under the Flag of the Cross* imagines an apocalyptic showdown with the "Yellow Peril" in the twenty-first century: "One-half of the world was arrayed against the other, in this mighty struggle between the yellow and the white races." "Thousands of air ships" from each side hurl upon the armies a rain of terrifying weapons. Fortunately for the "Caucasians," some of their "aerial monsters of war" are armed with "the wonderful electro bombs invented by Thomas Blake," and "these inhuman machines of destruction," "exploding with an awe-inspiring sound of thunder, killing and wrecking everything within hundreds of feet," devastate the ranks of the "Mongolians," helping to save the world for "Christianity" (448–49).

Frederick Robinson's 1914 *The War of the Worlds; A Tale of the Year 2,000 A.D.* foresees an even more outlandish danger. A Russian prince, furious because an American woman spurns his amorous advances, organizes a grand alliance of Japan, the Chinese Empire, Africa, India, Persia, Chile, Argentina, and the Martians to bring the United States to its knees. A colossal aerial armada, led by a gigantic Chinese aerial flagship named "The Yellow Peril" (69), sends New York "up in flame and smoke" (90), and the arrival of "the war flyers of the Martians," "marvels to earthly eyes" (96), seems to portend victory for the forces of darkness. But America is saved by its "automobile sky guns," "capable of firing heavy projectiles straight upward" (97).

Less bizarre were two widely read novels about a future war with Japan: Roy Norton's 1907 *The Vanishing Fleets* and John Ulrich Giesy's 1914 *All for His Country*. Each connects American culture before World War I with American history in World War II.

The Vanishing Fleets was serialized by leading newspapers across the nation in their Sunday insert, *Associated Sunday Magazines,* in 1907 and published as a book in 1908. Its introduction extols and acknowledges the assistance of Hudson Maxim, the inventor and manufacturer of innovative high explosives whose writings and 1915 invasion movie were shortly to play a crucial role in the campaign to prepare America for war. But this is not just another variation on the hackneyed theme of defenseless America. The novel reveals part of the cultural matrix of the Superfortresses and atomic bombs of World War II, offers an eerie forecast of events that were to open and close America's war with Japan in the 1940s, and may have had an infuence on decisions that led to these events.

As *The Vanishing Fleets* opens, trouble of a "racial character" (3) has been brewing between Japan and the United States for some years when Japan, allied with China and Great Britain, sends its fleet on a sneak attack. The United States surrenders the Philippines and Hawaii without a fight and mysteriously withdraws into total isolation, cutting off communications with the rest of the world, ordering its Pacific fleet to sail to neutral European ports, and sealing the Canadian border with a solid line of troops.

Why? Because of a secret weapon invented by "Old Bill" Roberts, a typical turn-of-the-century scientific wizard, and his brilliant, modern daughter Norma. In love with Norma, the Secretary of the British Embassy fears that "no human affection could drag her away from those crucibles and retorts with which she wrought through the days and nights in silent companionship with her queer old sire" (10). Here we have a twist on the familiar theme from early nineteenth-century science fiction—such as *Frankenstein,* Hawthorne's "The Birthmark" and "Rappaccini's Daughter," Melville's "The Bell Tower," and Fitz-James O'Brien's "The Diamond Lens"—where the only female figures in contact with the lone genius are passive and submissive objects to be manipulated or adored by the scientist. Norma is every bit the equal of her father, that old inventor who "seemed possessed of Titanic power" (181).

History is about to be rearranged by these two geniuses and their friend in the White House. As for the President himself, "all his hope was founded on a war for peace" (183).

One of the remarkable features of this pre-World-War-I novel is its glorification of unchallenged presidential rule, backed up by official secrecy designed to develop new weapons to be placed at the President's disposal. The President must, of course, deeply "dip into the nation's treasury" (187) to finance his hush-hush research and development project, and, just as with the Manhattan Project a few decades later, national security depends on keeping these expenditures and their aim secret from the American people. The cult of the superweapon requires the rites of official secrecy.

The cult also draws upon the myth of the Red Menace, even in this novel published ten years before the Russian Revolution. The main instruments of the potent Japanese "secret service" in America turn out to be none other than U.S. "communists" or "Reds" (55, 58, 59). To save the nation, American counter-intelligence must smash the espionage network woven of the Yellow and Red Perils. The Japanese master spy discovers too late—just before he is killed—the ostensible "devilish ingenuity of the Americans, who had led the whole world to believe them defenseless when they were in reality only lulling other nations on to their doom" (69).

When the huge Japanese invasion fleet prepares to attack the West coast, it mysteriously disappears on its way to Hawaii. The world infers that the United States must have some secret weapon of mass annihilation: "It seemed impossible that a civilized nation should have chosen deliberately to exterminate its enemies wholesale; and yet there was no other conclusion tenable." Horrified at this apparent massive slaughter of the crews of an entire war fleet, the world believes "it was time the United States ceased to exist as a nation, when peopled by inhuman monsters." Japan asserts "that had she possessed such monsters of destruction as were evidently owned by the United States, she would have scorned to use them without notifying the whole world of her power" (88–89). For modern readers, the ironies are dizzying.

When Great Britain tries to intimidate the United States by sending a gigantic naval armada toward Canada, this fleet also vanishes. Taking advantage of the opportunity, Germany threatens to attack Britain, but the Kaiser mysteriously disappears. Soon an American admiral tells his old friend the British king that the United States indeed has the ultimate weapon.

A flashback allows readers to observe the great experiment, when two "monster" machines become "Frankensteins under control!" The father-and-daughter genius team has discovered "the most powerful force the world has ever known." Having "wrested from Nature one of her greatest powers," the old inventor prepares "to harness it for all time, a slave to peace, progress, and the welfare of his fellows." The two scientists have forged a new alloy made "intensely radioactive" by electric current; this gives them control over a radioactive force, "corpuscles of radioactive matter," that can be driven in any direction by electricity. Possessing this, Norma and her father design and supervise the building of gigantic antigravity "radioplanes," thus bestowing upon the United States "the greatest engine of war that science has ever known" (210–23).

Now appears the most characteristically American element of the fantasy. The President, "his Americanism exceeded only by his humanitarianism," realizes that his solemn duty to humanity lies in using this weapon in war—in order to end war. He explains that this is why secrecy is imperative: " 'If our secret becomes known, there will be no war, and war is a necessity for our purpose' " (237). Since the President has vowed " 'to give my life to peace' " and his overriding purpose is to " 'put-an-end-to-war!' " (314), he explains that " 'our duty' " is to use this " 'most deadly machine ever conceived . . . as a means for controlling and thereby ending wars for all time' " (237).

The radioplanes, with their simple power source and their tiny crews of six, embody the triumph of reason in human affairs: "Science was bringing an end to brute force, and the last battle against barbarism was at hand" (243). An elite corps of American fliers thus prepares for "the last great battle in history" (244).

The flashback narrative of the battle reveals the actual fate of that "invincible" (248) Japanese invasion fleet that had disappeared so mysteriously. The American aerial armada, capable of flying almost six hundred miles per hour, is led by its flagship "the Norma," flown by none other than Norma herself. To an old admiral, she seems "the incarnation of the Goddess of War" (256).

Norma smashes the superstructure of the Japanese flagship, then picks up the twenty-thousand-ton battleship and flies away with it (260–61). The other radioplanes emulate this exploit, and soon the Japanese fleet is deposited on Lake Washington, near Seattle. A similar fate befalls the British armada, which is carried to Chesapeake Bay. So the Americans, contrary to the world's first opinion, turn out to have been most peaceful and humane warriors.

The President and the King of Britain next plan the fate of the world. Britain allies with "the most powerful nation history had ever known," and "the Anglo-Saxon race" is reunited "after its separation of nearly a century and a half" (320, 323). The Kaiser, having been "taught many lessons in democracy" while stranded in the Canadian woods, accepts peace (329). Japan surrenders, abandoning "its hope of competing with the other great Powers" (338).

The U.S. government announces that "in the interest of perpetual peace the secret of the radioplane would be maintained inviolable." The President issues a declaration: since the United States has "the power to conquer the world," it outlaws war and makes all nations agree to maintain existing boundaries; "The United States, having faith in the Anglo-Saxon race as representing one of the most peaceful and conservative, has formed an offensive and defensive alliance with Great Britain" (339–41). Thus comes universal disarmament and the end of war—thanks to the radioplanes, now known as "the peacemakers" (349). (In the closing days of World War II, the Boeing Company advertised a new name for its B-29 Superfortresses that incinerated the cities of Japan: "Peacemakers.") Norma and the British diplomat marry in a world where American-Anglo hegemony and perpetual peace are guaranteed by one thing—the eternal American monopoly on the superweapon, produced in secret by American ingenuity and merging air power with control over radioactivity.

America and Japan fight a truly modern war, with eerie foreshadowing of 1945, in John Ulrich Giesy's *All for His Country,* serialized by Frank A. Munsey's *Cavalier Weekly* in 1914 and published as a book in 1915. We cross the line into our own epoch when the air attack on a city becomes "an unrestrained slaughter of unarmed beings": "The city of doom became a place of a thousand fires. A great mushroom of smoke grew and billowed above it" (167). Flying through "the vast billowing cloud of the burning city," an airman from the nation being bombed thinks that the attackers are "fools" because "the whole civilized world will turn on them" for this outrage against humanity (174). The mushroom cloud rises from Japanese bombs; the devastated city is New York.

All for His Country offers the usual scenario of "local Japs" (88) rising up to seize California, a Japanese sneak attack and invasion, the "insulting" proposal that the Japanese should be allowed "full rights of intermarriage," and desperate guerrilla warfare by the heroic American defenders of their race and nation. The invaders' weaponry, which includes deadly surface-to-surface and surface-to-air missiles, is unusually modern. But of course it is soon overmatched by a superweapon invented by the good individual scientist-capitalist, Professor Stillman, who has been victimized by ruthless monopolists selling second-rate warplanes to the U.S. government.

The inventor's son, Meade, an innocent "wild man" of the West, perfects the Stillman aero-destroyer, a gigantic vertical-takeoff airship powered by radioactive antigravity screens, virtually invulnerable to gunfire, and armed with vast numbers of metal-seeking "magnetic bombs." Both a brilliant inventor and a master engineer, Meade saves the country by building the first three "monster air-ships" (256) in three months.

Stillman's first ship devastates the army of Japan on the East coast. The second "great destroyer" annihilates the entire Japanese fleet in Chesapeake Bay. Under the threat of the third, the Japanese on the West coast surrender. Although the happy ending includes Japan yielding Hawaii and the Philippines, the triumph of the good capitalist over the evil monopolist, and a successful romance for young Meade, *All for His Country* has a few surprising insights into the triumph of the superweapon. Amid a fine description of the carnage wrought by one of the airships, Meade experiences "a sick revulsion" at the massacre and orders it stopped, exclaiming: " 'My God, we're worse than they were!' " (285).

The Germans Are Coming

One of the most spectacularly successful propaganda campaigns in history took place in the United States between the outbreak of World War I in 1914 and America's entry into the war in 1917.[8] This campaign accomplished its three major goals: massive popular support for joining the war on the side of the grand alliance of world empires (Britain, France, czarist Russia, Japan, Portugal, Italy, the Netherlands, and Belgium); crippling the surging movements of populism, socialism, feminism, antimilitarism, and anti-imperialism; and transforming the United States into an industrialized military state with enormous armed forces. Achieving these goals would require mobilizing "the total power of the state, and to be ready for that meant regimenting the whole population," so the American people had to be won "to acceptance of the conscript army and the rudiments of the garrison state." As the preparedness drums beat, militarization became a dominant theme in American life: by mid-1916, "monster 'preparedness parades' were to fill the streets of American cities from dawn to dusk."[9] Crucial to this campaign were those nightmares of invasion and daydreams of glory popularized by American future-war fiction.

Before 1914, virtually no American fiction projected Germany as an invader or likely enemy. (Possibly the only exception is *Bietigheim;* published in 1886 under the pseudonym "John W. Minor," the novel predicted a war to end all wars in 1890–1891, with an alliance of the United States, Britain, France, Italy, and Spain defeating the Continental empires of Germany, Austria, and Russia, leading to an epoch of global democracy, prosperity, and peace.) But in the wave of future-war novels published between 1914 and 1917, all of America's earlier imagined enemies virtually disappeared beneath torrents of anti-German images. Readers were exposed to so many graphic descriptions of the German invasion of defenseless America that they might have found it hard to believe that these terrifying visions were merely products of the imagination.

No segment of the reading public was left out. Thinly fictionalized tracts, filled with facts and figures and told in the style of men conversing over after-dinner cigars, were aimed at substantial citizens; romances shot for the hearts of women; series of boys' books inflamed the passions of the youth who would soon be called to wage the war to end all war.

Scientific American ran an influential series of proarmament pieces entitled "Our Undefended Treasure Land"; in 1915 its editor, J. Bernard Walker, wrote his own anti-German novel, *America Fallen! The Sequel to the European War.* Walker imagined Germany, defeated in the European war and forced to pay massive indemnities, ruthlessly invading the United States to collect the money from the defenseless Americans. New York is bombarded and seized, Boston and Washington are captured, and the new capital at Pittsburgh is overrun, as America is forced to pay the dreadful price for harboring antimilitarist illusions.

Although Donal Hamilton Haines' books for boys call the invaders "the Blues," the illustrations and background clearly identify them as Germans. In both *The Last Invasion* (1914), told from the point of view of a boy swept up in the land battles, and *Clearing the Seas* (1915), narrating the experiences of a young sailor, the invaders' surprise attack gives them possession of Maine, Costa Rica, and part of Texas. But after colossal air and sea battles, an aroused American populace defeats the aggressor, thereby ending all war. The war "has proved that trained fighting men cannot conquer a great nation" (*Last Invasion* 336), for "nowadays, a whole people must be beaten" (*Clearing the Seas* 256). So, somewhat to the regret of the young fighting men, whose lives are boundlessly romantic, "this war has taught the whole world that fighting doesn't pay" (*Clearing the Seas* 281).

By 1916, the German invasion had become so thoroughly popularized that it could be taken for granted in juvenile fiction. That year the four-volume set called The Conquest of the United States Series by H. Irving Hancock, an astonishingly prolific author of books for boys, was published by the Henry Altemus Company, whose other boys' books included The Submarine Boys Series, The Battleship Boys Series, and The Boys of the Army Series, all by Hancock. The first three volumes of The Conquest of the United States—*The Invasion of the United States; Or, Uncle Sam's Boys at the Capture of Boston, In the Battle for New York; Or, Uncle Sam's Boys in the Desperate Struggle for the Metropolis,* and *At the Defense of Pittsburgh; Or, The Struggle to Save America's "Fighting Steel" Supply*—describe the harrowing, desperate struggle of our youthful heroes against the apparently overwhelming might of the treacherous German invaders. In the final volume—*Making the Stand for Old Glory; Or, Uncle Sam's Boys in the Last Frantic Drive*—American heroism, inventiveness, and miraculous production of airplanes and submarines, together with timely aid from Brazil, defeat the Germans. Young readers of the series learn that even though America's intentions are entirely noble, the nation must prepare for war to maintain peace.

Adults were getting a similar lesson from Cleveland Moffett's 1916 *The Conquest of America; A Romance of Disaster and Victory.* Moffett's smorgasbord offers all the standard dishes: bloodcurdling images of the terroristic German hordes; vitriolic attacks on socialism, pacifism, and feminism; rhapsodic glorification of American capitalism, inventive genius, potential armed might, and victory through air power.

Striking unexpectedly, the Germans destroy the U.S. Caribbean fleet, isolate the Pacific fleet by blowing up the Panama Canal, wipe out the East coast defenses with aerial and naval bombardment, and land an invasion army on Long Island, where they launch the same campaign waged by fictional invaders since the 1880s to seize Brooklyn and Manhattan, providing the customary scenes of devastation of familiar landmarks. As usual, the Brooklyn Bridge is destroyed: "Seen through the darkness at the moment of its ruin the vast steel structure of the Brooklyn Bridge, with its dim arches and filaments, was like a thing of exquisite lace. In shreds it fell, a tangled, twisted, tragically wrecked piece of magnificence" (66). German immigrants provide a sinister army of spies to betray the country and aid in its destruction. A gigantic German airship carries out a "mission of frightfulness," bombing the helpless city of Baltimore (192). Any town or village offering armed resistance is swept by a rain of "fire bombs dropped from the sky" (110).

The socialists and feminists continue to campaign for international proletarian solidarity and universal love, but fortunately there are Americans with more honor, courage, and resources, including the leading capitalists and a couple of technological geniuses. Just as in Frank Stockton's *The Great War Syndicate* (1889), these are the forces of true American greatness:

> "Have you ever heard of the Committee of Twenty-one? No? Very few have. It's a body of rich and patriotic Americans, big business men, who made up their minds, back in July, that the government wasn't up to the job of saving this nation. So they decided to save it themselves by business methods, efficiency methods. There's a lot of nonsense talked about German efficiency. We'll show them a few things about American efficiency. What made the United States the greatest and richest country in the world? Was it German efficiency? What gave the Standard Oil Company its world supremacy? Was it German efficiency? It was the American brains of John D. Rockefeller, wasn't it?" (184)

So the nation is saved by the genius inventor Lemuel Widding, who gets the necessary capitalist resources, as well as some help from Thomas Edison. Widding's aerial-launched torpedoes devastate the invader's fleet, forcing Germany to surrender. This is a triumph not only of American business acumen and technological mastery, but also of American air power: "For the first time in history an insignificant air force had conquered a great fleet" (305).

The happy ending of *The Conquest of America* is a vision not of universal peace but of America as a mighty military giant, prepared for a future of war. With a large standing army, an awesome naval armada, and "an aerial fleet second to none in the world" (309), the United States is ready to fight and win the wars that the author sees "for centuries to come, as an inevitable part of human existence" (26).

Ultimate Peacemakers

American technological wizards and their superweapons were commonplace in these pre–World War I novels of wars between the United States and its rapidly shifting cast of enemies. Before seeing how these conceptions were beginning to influence America's creation and apotheosis of actual weapons, it is worth looking at four novels that carry such images to their limits. In these visions, the United States wages no war; instead, the lone American genius achieves mythic stature by using his superweapons to save the world all by himself.

Simon Newcomb embodied the myth of the self-made American wizard. With little early formal education, he became one of the nineteenth century's leading astronomers, establishing still-current methods of calculating celestial motion, holding forth as a distinguished professor at Johns Hopkins University, winning numerous international prizes, and retiring from his official position as U.S. naval astronomer with the rank of rear admiral in 1897. Newcomb came to regard himself as a kind of all-around genius, turning out tracts on how to resolve the greatest financial, economic, and political problems of modern society. In 1900 he published a novel, *His Wisdom, the Defender,* dishing out his panaceas in a fantasy of how a mastermind much like himself could save the world from itself and lead it into a "Golden Age."

In 1941 the hero, Harvard Professor Archibald Campbell, invents not one but several superweapons, all based on two of his discoveries: the antigravity substance "Etherine," a new form of matter whose vibrations "react on the ether of space in such a way as to fly through it as a bird flies through air" (21), and the power source "Therm," "an agent somewhat akin to electricity" (87). Using Etherine, Therm, and a new superhard alloy of aluminum (his own invention), the wondrous professor constructs a host of airships and spaceships of assorted shapes, sizes, and functions. To take control of human destiny, he sets up a multimillion-dollar private corporation, builds two gigantic bases in the United States and Europe, and establishes an elite officer corps called "The Angelic Order of Seraphim," recruited from football players and fraternity members of the leading American colleges.

Unlike the usual patriotic geniuses who materialize in the nick of time to save America from foreign enemies, Professor Campbell is such a complete "individualist" (177) that he has "the point of view of an impartial looker-on belonging to no one country, and not even bound to any one stage of civilization" (199). He places himself above all nations to save them all. Realizing that "I am the possessor of a power which, if made public, would result in a disaster to the human race" (59), Campbell assumes "the responsibility of a god" (163), demanding of course that "my power must be absolute" (111). His assumption of power is justified by his intelligence and good intentions, since he has decided "to put an end to war" (59) and to establish a new system of international law (192).

For this godlike individual, with his superweapons and private elite army, it is almost child's play to disarm and disband the armies of Germany, Austria, and Russia, sink the combined fleets of Britain, Italy, Germany, and Austria, and, by

1946, force all the nations, including his own, to accept him as the beneficent dictator of the world, with the titles of "His Wisdom" and "The Defender of the Peace." His glorious story is narrated from "the Golden Age" of peace and prosperity brought by His Wisdom to the human species.

Published three years before the Wright brothers' 1903 flight at Kitty Hawk, *His Wisdom, the Defender* expressed a vision of air power characteristic of the contemporaneous imagination of both novelists and inventors. Professor Campbell is able to drive "the armies of Europe into caves and dens to hide from his power" (306) because "no defence of person or property against an army flying the air where it chose, and pouncing down at any moment, was possible" (296). Within a dozen years after Kitty Hawk, military air power would be a reality, but a far more ambiguous one than Simon Newcomb had imagined, as airplanes from several nations fought each other and dropped small bombs, with no influence on the stalemated battlefields of World War I.

Even in the midst of this peripheral and ineffective combat, the religion of air power was gaining converts. In 1915, John Stewart Barney published *L.P.M.: The End of the Great War*. At first glance it seems little more than a rewrite of Simon Newcomb's 1900 novel—another lone genius uses his privately constructed airship to end all war and make himself the beneficent ruler of the world. Yet the differences are telling, and ominous.

The name of Barney's individualistic mastermind, John Fulton Edestone, evokes the two most legendary American inventors. This genius also embodies another great American hero: the man of awesome wealth. Since he not only has "the mind of a superman" but also is "the richest young man in the world" (11), Edestone has no difficulty constructing the ultimate weapon: the "Little Peace Maker" (L.P.M.), a steam-powered, forty-thousand-ton armored antigravity airship.

Edestone, like "His Wisdom, the Defender," is so dedicated to achieving his "dream of universal peace" (348) and a "Utopian" (412) world order that he, too, renounces his allegiance to the United States to become a beneficent "outlaw,"openly declaring, "I now become a law unto myself" (343, 345). But he has some additional goals, including the defeat of socialism and feminism, the isolation of the Jews, and the complete segregation of the darker races, for, as he proclaims, "I dedicate my life to my people, the Anglo-Saxons" (102). Though the immediate target of his war machine is the German soldiers who have become "mere machines" in "the hands of a Master . . . sick with a mania which took the form of militarism, imperialism, and pan-Germanism" (358), modern readers may find as many foreshadowings of fascism in this superman as in his adversaries.

After his Little Peace Maker annihilates the German fleet and conducts a week-long "rain of destruction" (405) on German ports, Edestone brings the now-tottering heads of the world's leading states aboard his colossal vessel, which also serves as a splendid private yacht. Here he disposes the new world order. National governments are replaced by an organization modeled on the world's most rational institution, the giant American corporation, run by a "Board of Directors" with himself as "Chairman of the Board." He scorns majority rule, turns socialism into a joke, and will admit on board the L.P.M. no representative of the women's rights movement "who had not borne and raised twelve children" (410–13).

He establishes "a very limited ruling class" named "the Aristocracy of Intelligence" to run the world under his directorship. Blaming the outbreak of war on competition among groups of capitalists, Edestone resolves all contradictions by creating on a global scale the ultimate in monopoly and state capitalism, the perfect corporate state of fascism: " . . . in the new form of government competition would be eliminated, the interest of the whole being controlled by one head with power to police, and great profits to all would accrue by the elimination of waste of time and money and by the efficiency of a single administration" (413, 416).

In this remarkable linkage between the cult of the superweapon and the ideology that later became known as fascism, Barney dramatizes the quest for global domination through some ultimate weapon. But by 1915, there was already clear evidence that no airship would be likely to provide such a final solution. World War I was intimating that many nations could develop formidable air power, and that no single aircraft, no matter how much genius and wealth its creator possessed, could be the ultimate weapon leading to the new world order. Only the *true* ultimate weapon—such as a radioactive beam that could disintegrate everything in its path—could achieve world peace.

As early as 1895, in the midst of the excitement aroused by the discovery of radioactivity, British novelist Robert Cromie's *The Crack of Doom* had imagined a mad scientist unleashing the power of the atom. After warning that it is "not wise to wreck incautiously even the atoms of a molecule," since "one grain of matter contains sufficient energy . . . to raise a hundred thousand tons nearly two miles," this genius is barely stopped from destroying the planet with his atomic experiments.[10]

A molecular disintegrator beam is invented by Thomas Edison to defeat the Martians in Garrett P. Serviss's 1898 novel *Edison's Conquest of Mars*. A similar device is contrived as a personal weapon in Jack London's 1899 short story "A Thousand Deaths"; to liquidate his scientist-father and two attendant "blackies," the narrator uses a force field that reduces all organic molecules to inorganic atoms.[11] And, as seen earlier, Roy Norton's 1907 novel *The Vanishing Fleets* used radioactivity as a motive power for the stupendous "radioplanes" that establish American world rule and peace. But the dubious distinction of inventing the first nuclear weapon of warfare seems to belong to Hollis Godfrey's 1908 novel, *The Man Who Ended War*.

The man who ended war is an American scientific mastermind who creates the ultimate weapon, a focused beam of "radio-active waves" that instantaneously disintegrates the atoms of all metals into subatomic particles. He issues an ultimatum demanding immediate universal disarmament. When the nations of the world fail to comply, he begins sinking their battleships, at one point annihilating eighty-two warships from two opposing armadas by firing his radioactive beam weapon from a submarine. All the countries worth mentioning—Britain, the United States, Japan, France, Russia, and Germany—disarm.

"The man who stopped all war" then destroys his machine, his secret, and himself—so that no one will obtain this weapon and the world, therefore, will be permanently at peace. Of course no other individual and no nation will be capable of duplicating this feat of the scientific superman.

A similar plot but a far more modern outlook appeared a few years later in *The Man Who Rocked the Earth* by Arthur Cheney Train and Robert Williams Wood. Serialized in the *Saturday Evening Post* in 1914-15 and published as a book in 1915, the novel opens with a prediction of a hideously stalemated World War I, with millions slaughtered and catastrophic devastation wrought by a multitude of novel weapons and counterweapons created by "the inventive genius of mankind, stimulated by the exigencies of war" (4). Then appears the American scientific wizard who calls himself "PAX."

PAX has created the ultimate weapon, a radioactive beam that can annihilate mountain ranges or armies. He fires this atomic weapon from his Flying Ring, an airship powered by "atomic energy" (100) from uranium forced into rapid disintegration.

This is not the fantasy, so typical of these novelistic superweapons, of sanitized, humane bombs and forces that do not actually kill many people. Some bystanders unlucky enough to witness PAX's destruction of the Atlas Mountains, intended as a gentle message to the warring nations, die from radiation sickness: in a few days they "began to suffer excruciating torment from internal burns, the skin upon their heads and bodies began to peel off, and they died in agony within the week" (68). When a German general, in defiance of PAX's cease-fire ultimatum, prepares to annihilate Paris with the "Relay Gun," considered "the most atrocious engine of death ever conceived by the mind of men" (117), his entire artillery division is wiped out by an atomic attack from the Flying Ring, leaving nothing but "the smoking crater of a dying volcano" in which the giant Relay Gun is reduced to "a distorted puddle of steel and iron" (171). One survivor describes a scene interchangeable with that later depicted by survivors of Hiroshima:

> And then the whole sky seemed full of fire. . . . the earth discharged itself into the air with a roar like that of ten thousand shells exploding all together. The ground shook, groaned, grumbled, grated, and showers of boards, earth, branches, rocks, vegetables, tiles, and all sorts of unrecognizable and grotesque objects fell from the sky all about him. It was like a gigantic and never-ending mine, or series of mines, in continuous explosion, a volcano pouring itself upward out of the bowels of an incandescent earth. . . . Great clouds of dust descended and choked him. A withering heat enveloped him. . . .
>   . . . he could not find the village. There was no village there; and soon he came to what seemed to be the edge of a gigantic crater, where the earth had been uprooted and tossed aside as if by some huge convulsion of nature. Here and there masses of inflammable material smoked and flickered with red flames. (134–35)

In the glare of these atomic weapons, PAX, unlike so many of his counterparts in other novels, seems utterly credible to modern ears when he declares that "either war or the human race must pass away forever" (142).

Of course it is war that becomes extinct. Threatened by the atomic arsenal of the lone peacemaker, who calls himself "The Dictator of Human Destiny" (228), the nations destroy their weapons, abolish armies, form a world government outlawing armaments, and set up an International Police to guarantee perpetual peace. Freed from "the fear and shadow of war," "the nations grew rich beyond the imagination of men" (225). As for the United States, the name of "the old War Department" is changed to "the Department for the Alleviation of Poverty and Human Suffering" (228). PAX, destroyed by his own weapons, does not live to see this happy ending, much less the actual history of the U.S. War Department, renamed the Department of Defense shortly after it acquired the first actual nuclear weapons and just as it initiated the most stupendous arms race of all.

These four novels are not as blatantly nationalistic as those that fantasize about America attaining world hegemony through superweapons concocted by native technological genius, such as the military-industrial syndicate's superbomb in Frank Stockton's *The Great War Syndicate,* the aluminum dirigible in Stanley Waterloo's *Armageddon,* or the miracles of radioactivity in Norton's *The Vanishing Fleets* and Giesy's *All for His Country.* Yet all four virtually caricature identifying features of emerging American ideology. Indulging without restraint in the myth of the superweapon, they become intoxicated by the myth of the omnipotent individual that also addicted America between the Civil War and the First World War.

At least in *The Man Who Ended War* and *The Man Who Rocked the Earth,* these pipe dreams take the relatively innocuous form of scientific geniuses who sacrifice themselves to save the world. *His Wisdom, the Defender* and *L.P.M.* take American ideological ambitions much further, to visions of a global dictatorship by an American hero embodying the nation's cultural ideals. Such fantasies would persist well into the nuclear age. For example, Robert Heinlein's 1941 story, "Solution Unsatisfactory"—a tale familiar to some of the physicists who devised the first atomic bombs (see pages 141–46)—argues that when nuclear weapons come into existence, the world will be doomed unless it is ruled by a beneficent American dictator.

These four novels about individuals who determine the history of the planet with their ultimate peacemakers sharply focus the American ideology running throughout these fictions of future war. The more extreme the fantasy, the more obviously it projects the archetypal interlinked fallacies: technological advance comes not from the dialectic between productive forces and consciousness, but from the unfettered imagination of American technological genius; a single brilliant invention could change the whole course of history; even the greatest human problems can thus be solved by a technological miracle, particularly if the miracle takes the form of something that just might turn out to be the ultimate weapon.

These novels and stories were a main dish in the cultural diet of what is now called middle America. Before their publication in bound volumes, many were serialized in newspapers and popular magazines. For example, *The Vanishing Fleets* was first serialized in 1907 by the *Associated Sunday Magazines,* which was inserted weekly in major newspapers from coast to coast. "The Last Conflict" appeared in *American Magazine* in 1914, the *Saturday Evening Post* serialized *The Man Who Rocked the Earth* in 1914–15, and *McClure's* printed Jack London's "The Unparalleled Invasion" in 1910 and serialized *The Conquest of America* in 1915. Who read these magazines? A fairly typical reader was a young Missouri farmer and businessman named Harry S. Truman, who throughout this period avidly devoured the fiction in *American Magazine,* the *Saturday Evening Post,* and *McClure's.* As he wrote in a 1913 letter to Bess Wallace, the young woman he was courting: "I suppose I'll have to renew my subscription to *McClure's* now so I won't miss a number."[12]

3

Thomas Edison and the Industrialization of War

The war of the future . . . will be a war in which machines, not soldiers, fight. . . . the new soldier will not be a soldier, but a machinist; he will not shed his blood, but will perspire in the factory of death at the battle line.
—Thomas Alva Edison, 1915

We are confronted with a new and terrible engine of warfare in the submarine. . . . with your own wonderful brain to aid us, the United States will be able . . . to meet this new danger with new devices that will assure peace to our country.
—Letter to Edison from Secretary of the Navy Josephus Daniels, 1915

My solution for war . . . is preparation. . . . this preparation or preparedness may one day involve the discovery of some terrific force, some engine of war the employment of which would mean annihilation for the opposing forces. The way to make war impossible is for the nations to go on experimenting, and to keep up to date with their inventions, so that war will be unthinkable, and therefore impossible.
—Thomas Alva Edison, 1922

"A Very Pretty and Destructive Toy"

Robert Fulton and Thomas Edison bridge the cultural history of American weaponry from the end of the eighteenth century, when the republic of thirteen agricultural and mercantile states was forming, to the early decades of the twentieth century, when America was becoming the most powerful industrial nation on earth. Each advanced the myth of the lone American inventine genius in his respective epoch. Fulton and his steamboat helped launch a century of faith in the individual inventor. Edison—the self-educated, homespun tinkerer—became the mythic embodiment of this figure in American culture. With his endless inventions, such as the incandescent electric lamp, phonograph, and motion-picture projector, Edison altered the patterns of daily life and came to be popularly perceived as a "wizard" single-handedly responsible for much of the industrial transformation of American society.[1] Yet paradoxically, Edison also became a force in transmuting that relatively naive age when individualism supposedly reigned supreme into an epoch dominated by the gigantic, complex institutions of industrialized militarism.

Edison shifted steps and patterns in his long dance with the rising American weapons industry. He billed himself sometimes as a man of peace who would never invent killing devices, at other times as the master inventor of marvelous weapons that could destroy whole armies with the push of a button. Despite his often wildly contradictory pronouncements on the nature of modern warfare, he was chosen by the government to bring order to the nation's technological war-making capacities. In a period of dizzying transition, Edison was a living expression of the ambiguities and contradictions in America's military potential.

While Fulton's weapons were eventually to prove even more terrible than he had imagined, they had little or no practical significance until decades after his death, when rapid advances in industrial technology began to realize their awesome potential. Edison, on the other hand, actually invented no new weapons. Yet in his lifetime he was viewed by the nation as a military genius whose deadly inventions could win wars, change history, and even save the planet.[2]

Submarines, torpedoes, torpedo boats, a steam warship—these were the weapons that Fulton was designing, in the dawn of industrialism, to launch the rule of reason. In 1886, almost a century after Fulton first conceived of these ultimate weapons of progress, the *New York Times* announced the formation of the Sims-Edison Electric Torpedo Company "to manufacture, sell, and use torpedoes, torpedo boats, submarine vessels, and warships."[3] Was Edison about to bring forth, into an age of global industrialized warfare, titanic progeny of Fulton's conceptions?

The curious tale of the Sims-Edison Electric Torpedo dramatizes both Edison's insignificance as a maker of actual weapons and his profound cultural importance as a godfather of the weapons industry. Modern biographies accept this minor refinement of one of Fulton's devices as Edison's sole legitimate invention of a functional weapon, amid assorted bizarre claims by himself and others that he had originated various grandiose tools of mass carnage. But in fact Edison did not invent even this ingenious contraption. His association with it is more revealing than if he had, for it was the great inventor's *cultural* prominence that helped sell novel weaponry.

After the public announcement of the formation of the Sims-Edison Electric Torpedo Company, Edison's credentials were bound to keep its operations newsworthy. Expectations of a revolution in naval technology coming from Edison's weapons factory were teased along by the *New York Times*' progress report a few months later:

> At the Edison machine shops, No. 104 Goerck-street, the third of the electrical torpedoes in process of manufacture for the Government was finished yesterday. The torpedo is 45 feet long and 30 inches in diameter, and is made of polished copper. Its motive power is a small dynamo, which is inclosed in an iron case in the centre

of the torpedo, and which operates a propeller. The torpedo will run 10 minutes at the rate of 10 1/2 miles an hour, and is controlled by two wires connected with the shore. These are rolled up in an apartment in the torpedo and uncoil as it goes along. The contrivance is shaped like a cigar, pointed at both ends, and is suspended from a float. The float moves on the surface of the water and the torpedo is 4 feet below. The charge is 350 pounds of dynamite. The torpedo will cost $5,000 and will demolish a million-dollar man-of-war. The Government will have eight of them.[4]

In 1890 *Scientific American* featured the ''Sims-Edison Torpedo'' as the cover story for its July issue; along with many other journals, it enthused about recent successful tests. In a sensational 1892 *New York World* interview reprinted in *Scientific American* and newspapers across the country, Edison boasted of a fabulous electric superweapon he was creating (which will be discussed shortly), and cited this torpedo as a harbinger of his dreadful powers:

It is true I have invented an electric torpedo, the Sims-Edison torpedo, which we have sold out to the Armstrong Gun Company. It is a very fine thing. It is put on a wire, as of course you understand, and moved by electricity. It can be run out two miles ahead of a man-of-war's bow and kept at that distance ready to blow up anything in reach. It is a very pretty and destructive toy.[5]

In fact this ''toy'' was invented by W. Scott Sims, at least as early as 1879, seven years before Edison's association with it. The early history of the torpedo is outlined in *Sketch of the Sims-Edison Electric Torpedo, Historical, Descriptive, & Illustrative of Its Efficiency for Harbor and Coast Defense, and Its Applicability to Naval Warfare. . . .* This remarkable twenty-seven-page bound document, published by the Sims-Edison Electric Torpedo Company in 1886, supplies a kind of missing link between Robert Fulton's handsome early-nineteenth-century volumes, illustrated by his fine drawings and presenting a sales pitch for his naval weapons in the form of utopian philosophic speculation, and the lavish glossy publications of the late twentieth century promoting the wares of the giant aerospace companies.[6]

The brochure, obviously intended for military customers, opens with projections about the form of future naval wars, similar to some of the contemporaneous future-war novels. Since steam-powered, steel-armored warships with long-range rifled cannon of ''enormous calibre'' render all coastal forts obsolete, ''the torpedo is destined to enter largely into all future naval wars'' and become ''one of the most potent weapons, both offensive and defensive'' (7). What will be needed is a long-range torpedo, faster than warships, combining ''the qualities of being *movable,* easily *portable, invisible* to an enemy, *indestructible* by shot or shell, *certain* of operation,'' and ''*intelligently controlled*'' (7–8). Needless to say, no torpedo then in existence—including the surface torpedo (spar-torpedo boat), the self-propelled torpedo, and the projectile torpedo (such as the Whitehead)—could claim all these marvelous capabilities.

''The Sims-Edison Electric (Fish) Torpedo'' (referred to interchangeably as a ''fish'' or ''fish torpedo'' and an ''Electrical Torpedo Boat''), on the other hand, is presented as invulnerable for coastal defense and almighty as an offensive weapon. Its "submarine boat" has "*inexhaustible*" electric power and is supported by an "*indestructible float*" (9–11). The company's product can be operated from shore batteries or can travel "*with its own power,* about 100 feet ahead of or off from the side of a steam war-vessel, attached to the vessel by electric snap-cables." This torpedo "is the only one that is driven by a power *not* within itself," and can be controlled flawlessly from a "key-board" on ship or shore (11).

In narrating the history of the device, the pamphlet does not even mention Edison's name: "As early as 1879, the Sims Torpedo has engaged the serious attention and study of the most distinguished military men in the United States" (13). Included are detailed records and charts of tests of the Sims torpedo conducted by the U.S. Army's Battalion of Engineers in 1880, 1882, and 1885, showing the same specifications and performance data as the 1886 version. After noting that Congress has appropriated $187,500 for eight torpedoes, making a total of ten already under contract, the company boasts of its connection with Edison, whose Edison Machine Works of New York has a manufacturing subcontract. The pitch to other governments rings clear: "The Sims-Edison Torpedo Company of New York . . . has secured patents . . . from every European and American Government, and is now prepared to negotiate for the sale of its torpedo or for certain of its patent rights" (26).

A search through the Edison archives shows no record of Edison actually working on the design or any modifications of the Sims-Edison Torpedo during the dozen years of its existence. What does show up, however, is the inventor's great value as a public-relations asset. Although Edison is listed on the company's letterhead as "Consulting Engineer," the extant correspondence between him and the other officers relates only to his holding of thirty shares of stock, his serving as a trustee, and, most of all, his presence at public functions. Whenever the company arranged an exhibition of the torpedo for potential foreign buyers, Edison received a telegram stressing the importance of his presence.

One such display took place in July 1890 for the benefit of a number of South American government officials and the U.S. press. Everett Frazar, then president of the company, wrote personally to Edison to make sure he would attend. There is no evidence that the exhibition induced any foreign governments to buy the torpedo, but it certainly sold it to the press. Besides the cover story of *Scientific American,* glowing accounts appeared in the *New York Tribune,* which marveled that the "torpedo boat" had reached twenty miles an hour, *Public Opinion,* and the *Boston Journal,* which cheered that "this fragile, death-dealing little machine worked perfectly." A long, analytic article in the *New York Herald* did dash some cold water on the Sims-Edison Torpedo, comparing its limited range and speed unfavorably with current models of the highly accurate Whitehead torpedo, capable of at least thirty knots. Even this skeptical piece, however, indicated that the Sims-Edison version was still in favor with the War Department.[7]

Amid the public fanfare, no one outside the Sims-Edison Company seemed aware that the company's files contained an official document from the War Department, dated February 16, 1889, stating unambiguously that "it is not advisable, nor to the best interests of the service, to purchase more torpedoes of this design."[8] No wonder, as Edison so casually mentioned in the 1892 *New York World* interview, that the company had quietly "sold out to the Armstrong Gun Company" his "very pretty and destructive toy."

So the one actual weapon attributed to Edison was not his invention, did not match the capabilities of similar devices, was impractical for combat anyhow thanks to its giveaway flotation escort, and had been abandoned by the government. Edison's only other direct contributions to actual weapons came decades later, and consisted of such mundane business as modifying devices used for detecting underwater weapons, and devising a mixture of zinc dust and Vaseline to preserve submarine guns from rust. But such was not the myth, which was to become ever more important in American culture during the next three decades.

Electric War

For the last two hundred years, scientific discoveries and technological innovations have sent wave after wave of excitement sweeping through populaces and governments. One after the other, new means of manipulating the material environment have promised godlike powers of creation and destruction, with unlimited potential for prosperity or havoc.

Electricity—quasi-magical and usually invisible, used as thunderbolts by ancient gods and rapidly transforming daily life in the late nineteenth century— was bound to generate visions of dynamic weapons and lightning wars. The Sims-Edison Electric Torpedo was merely one minor artifact in a burgeoning cult of electrified warfare, promising global victory bestowed by electric submarines or electric airships or electric beams. Thomas Alva Edison was for many the apotheosis of this new religion.

When it suited his purposes, Edison was not reluctant to capitalize on this worship of himself as a divine artificer who could become a terrible god of electric war if America were imperiled. But in playing this role, he was more concerned with perils to his far-flung commercial enterprises. Neither the public nor various government bodies were convinced that electricity was safe, so any emphasis on its death-dealing capacities might interfere with sales, and Edison was nothing if not a master salesman. However, the main threat to sales of Edison's inventions and direct-current system was not public caution but a formidable rival—the alternating current system largely devised by Nikola Tesla and successfully marketed by George Westinghouse. Hence, Edison spent much energy trying to turn public apprehension about electricity into terror of alternating current.

Indeed, the only electric war in which he was actually involved was the so-called "war of the currents" that broke out in 1886. Edison had staked all his investments in electric-power distribution on a direct-current system that was restricted by the available technology to low voltage incapable of transmission beyond a few miles. Tesla's polyphase alternating-current system, on the other hand, allowed for stepping up the voltage to very high levels for long-distance transmission and then stepping it down for safe use. Equipped with Tesla's inventions, Westinghouse was rapidly displacing Edison's DC with the AC system that was to become standard in the United States. Edison's response was a frenetic campaign to prove that alternating current is intrinsically lethal.

In addition to articles, pamphlets, and rumors, Edison and his agents used public demonstrations in which cats, large dogs, and even horses were electrocuted with AC. Simultaneously, a powerful movement had developed in New York State against capital punishment, a movement that focused on the excruciating agony of hanging. Seizing this opportunity, Edison's lieutenants in 1889 bought three Westinghouse generators, resold them to three New York State prisons, and engineered the first use of the "electric chair." Staged to show that AC was so deadly that it would kill instantaneously, the execution of William Kembler on August 6, 1890, instead turned into a gruesome scene in which the condemned man writhed under repeated shocks.

Two years later Edison conjured up even more dreadful powers for alternating current. Now he imagined AC as a mass killer generating an invincible electric battlefield. Apparently, it was this vision of an entirely imaginary superweapon, elaborated in a *New York World* interview reprinted by hundreds of daily newspapers and *Scientific American,* that enshrined Edison as a potential demigod of war in the popular imagination.

Ostensibly responding to a question about the threat of war with Chile, Edison nonchalantly transformed one of the notions he was propagating in the war of the currents into an outlandish scheme for using AC as a defensive wonder weapon:

> "It is simple as A, B, C. . . . With twenty-five men in a fort I can make that fort absolutely impregnable. . . . This is not guess-work, but a matter of absolutely scientific certainty. . . . Some years ago, when the wires loaded with heavy electric charges began to go up everywhere, I predicted that there would be danger of the firemen receiving deadly shock by the electricity running down the streams of water which might cross the wires. The insurance people laughed at the idea. But I tried it on a cat, and the cat and I found my theory to be true. That is to say I did, and the cat found it out if there is another world for cats. . . ."

In terms aimed at the popular audience, Edison explained how his strategic defense initiative would make America impregnable by rendering all invading armies impotent and obsolete:

> "In each fort I would put an alternating machine of 20,000 volts capacity. One wire would be grounded. A man would govern a stream of water of about four hundred pounds' pressure to the square inch, with which the 20,000 volts alternating current would be connected. The man would simply move this stream of water back and forth with his hand, playing on the enemy as they advanced and mowing them down with absolute precision. Every man touched by the water would complete the circuit, get the force of the alternating current, and never know what had happened to him. The men trying to take a fort by assault, though they might come by tens of thousands against a handful, would be cut to the ground beyond any hope of escape.

Foreign soldiers undertaking to whip America could walk around any such fort as mine, but they never could go through it. It would not be necessary to deal out absolute death unless the operator felt like it. He could modify the current gently so as simply to stun everybody, then walk outside his fort, pick up stunned Generals and others worth keeping for ransom or exchange, make prisoners also of the others if convenient, or if not convenient turn on the full force of the current, play the hose once more, and send them to the happy hunting grounds for good."[9]

Without venturing an opinion on the feasibility of Edison's scheme, the *World* reporter drolly fantasizes even more bizarre countermeasures:

> The picture raised by Mr. Edison is certainly a most beautiful and attractive one. . . . Such a fort and such a warfare as Mr. Edison has planned would make old-fashioned generals . . . turn in their graves. We should have infantry moving on forts at a quickstep, dressed all in rubber, with chilled glass soles to their shoes and non-conducter handles to their swords and guns. Generals would look much funnier than a picture from *Punch,* charging at the head of their armies riding on horses shod with rubber arctics, the Generals themselves carrying large rubber umbrellas with gutta-percha handles over their heads.

After newspapers across the country reprinted the interview, a few skeptics challenged Edison more bluntly. For example, the *St. Louis Globe-Democrat* reported:

> W. E. Bailey, of the National Electric Manufacturing Company, is staying at the Laclede. Mr. Bailey follows the practical side of the electrical business. . . . "I was amused," he said, "to read of the great wizard's scheme to kill off the enemy in a war with Chili. Just think if what he proposed was taken seriously by the country! He said he would repel any attack by the use of alternating current. Let me tell you that Edison has never yet been able to produce an alternating current, and his suggestion shows that he does not understand its practical application. He was to attach the current to a hose, and so throw water on the enemy, one end of the current being ground. This is impossible, because you must have an open circuit on the electrical machine, and the moment you ground one end the fuse on the switch-board is burned out, rendering the production of a current an impossibility. To my mind the scheme mentioned was about as easy of accomplishment as would be a journey across the Atlantic on a pair of snow shoes."[10]

Edison's avowed faith in electric war, however, did not depend on the practicality of either the electric hose or the electric torpedo. Like many of the novels explored in the previous chapter, Edison seemed possessed with "martial ardor," according to the *World* reporter, as he contemplated an imagined invasion of the United States: " 'That is what I want to see, and I think that electricity will play such a part in war when that time comes it shall make gunpowder and dynamite go sit in humble obscurity with the obsolete flint and call him brother.' "

Edison continued to huckster in the newspapers about his fanciful electric weapons. For example, in 1895 he catalogued a whole menagerie of killer unicorns and jackalopes, including the electric hose, as "'only a few of the many inventions I have made for purposes of war.'" First came a new long-range version of his electric hose:

> "There is no need of a battle between two large armies in the field. A mere handful of men could sweep one of them away. The power to do this is electricity. Water can be made more deadly than bullets. I have invented a machine by which it can be hurled to a great distance, and water charged with 5,000 volts and then dashed on an army would sweep it away like chaff. The only question is how far the water can be thrown."

"I have also invented cables to be drawn around a besieged city, which would deal death to anyone who tried to cross them," Edison went on, moving next to his "electric chains, to be fired into an advancing army": "These are of different lengths, and are attached by one end to the wires of a dynamo, while the other is placed in cannon. When these are discharged the air will be filled with chains like great snakes, which would mean death to an advancing host." Then came his "aerial infernal machine," loosing a rain of bombs electrically timed for simultaneous detonation, so that "the force of the explosion would sink or disable every ship within a moderate distance" (thus more effective than the first atomic bomb tested at Bikini in 1946). If a nation preferred to do the same job from under water rather than from the air, it could use his "endless cable" hooked up to multiple torpedoes, capable of making any "harbour a mine of death."[11]

Electricity was indeed changing warfare, but not in such melodramatic ways. Thanks in part to the technology actually pioneered by Edison, armies and navies were rapidly increasing their communication, range, speed of action, and coordination. Advanced forms of Edison's incandescent light bulb were helping to extend large-scale combat throughout the night. Capacious electric storage batteries were converting the submarine—designed by Fulton in the eighteenth century as a purely defensive weapon of peace and progress—into an offensive nightmare for the twentieth century. Contrary to Edison's vision of the obsolescence of large armies, improvements in the telegraph and telephone were making it possible to deploy and organize land and naval armadas on an unprecedented scale. Electricity was to be not the magic weapon of a wizard but part of the inexorable industrialization of warfare. At the end of this process would lie the potential for a fifteen-minute global war of annihilation initiated by a pushed button or even an automated electronic impulse.

"When We Want to Blow up Our Civilization"

As Edison promoted an electric torpedo and fantasized an electric battlefield, the conditions for modern war were ripening. Armed with ever more potent weapons, driven by the forces of their own economies, and increasingly capable of global activities, the great colonial empires were carving up all that remained of the preindustrial world. From the dizzying perspective of Mark Twain's *A Connecticut Yankee in King Arthur's Court,* the deepening historical crisis of the age seemed grimly comic and finally catastrophic.

Twain published the novel in 1889, at the peak of the war of these currents. Unlike Edison, he saw at once the advantages of AC, noting in his journal on November 1, 1888 his belief that the alternating current "electrical machine lately patented by a Mr. Tesla & sold to the Westinghouse Company . . . will revolutionize the whole electrical business of the world"; "It is the most valuable patent since the telephone."[12] This adds a twist to the contest of rival wizards in *A Connecticut Yankee,* and makes Twain's choice of direct current for an ultimate war machine look like a sly poke at Edison's campaign against the dangers of alternating current. Perhaps Edison in turn was thinking of Twain's deadly electrified fortifications when he conjured up his 1892 electric fort or his 1895 concealed cables that "would deal death to anyone who tried to cross them." But these topical connections are far less meaningful than Twain's insights into the relations between industrial capitalism and modern war.

A Connecticut Yankee is about time travel—in many senses. Anglo-European thought arranges historical time in a spatial hierarchy. Each age is categorized by its prevailing technology, and "ahead" in time means more technologically, and therefore more socially, advanced. Thus the Bronze Age is "ahead" of the Stone Age, but "behind" the Iron Age. The age of industrial capitalism is more advanced than the age of feudalism. So when an industrial society encounters a feudal or prefeudal society in Asia or Africa, that is a form of time travel. (Twain himself used this concept of spatial movement in time when writing in his journal about the anti-imperialist message of *A Connecticut Yankee.*[13]) Another form of time travel was the eruption of industrial capitalism within a pre-industrial society—such as the transition taking place in nineteenth-century America. What might happen, Twain asks in *A Connecticut Yankee,* if a man embodying the latest technology and ideology of late-nineteenth-century New England were to appear suddenly in a feudal, slave society? And what if his special area of expertise were the industrial development of modern weapons?

In Twain's novel, Hank Morgan, an erstwhile foreman of a modern Connecticut gun factory, has command of the technology of his times. Dropped into sixth-century England, this knowledge makes him a lone genius, a "magician" single-handedly capable of carrying out an industrial revolution that transforms a feudal society into the material semblance of a modern industrial society. But without a thoroughgoing cultural revolution, human nature remains feudal, and the Church is able to mobilize the aristocracy and the commoners to overthrow the polity established by Hank, "the Boss." Left with only fifty-three boys he personally has trained and indoctrinated to be technocratic lads of his own historical age, Hank prepares to use his "deadly scientific war material" to eradicate the dark forces of feudal superstition, ignorance, and oppression. The great inventive genius Hank Morgan here stands as a true heir of Robert Fulton, with his visions of scientifically designed weapons capable of destroying the vestiges of feudalism and inaugurating the reign of reason and progress.

The culmination of *A Connecticut Yankee* in the "Battle of the Sand Belt" has long been cited as an eerie forcast of the trench warfare of World War I. But it is an even more uncanny projection of the ultimate war, in which the victors end up as victims of the universal death they have sown. Beyond that, it probes to the very core the ideology of warfare that was emerging in the late nineteenth century, to cast its spreading shadow over the planet for at least the next hundred years.

As brilliantly conceived by Twain, Hank and his youthful protégés are thorough pragmatists who hook "secret wires" to dynamite deposits under all their "vast factories, mills, workshops, magazines, etc." and connect them to a single command button. This is "a military necessity" so that nothing can stop them "when we want to blow up our civilization."[14] When Hank finally does initiate this instantaneous push-button war, his rationalization is appallingly ominous for late-twentieth-century readers: "In that explosion all our noble civilization-factories went up in the air and disappeared from the earth. It was a pity, but it was necessary" (476).

The second great use of electricity is in a fence designed in Hank's absence by his well-trained boys. Consisting of twelve large circles of wire powered by a direct-current generator, this death machine is designed to annihilate an army even bigger than the one Edison proposed to wipe out with his alternating-current electric hose. As Hank and his faithful young lieutenant Clarence discuss technical improvements in their electric fence, they become vehicles for Twain's savage satire on the crass materialism and pragmatism of modern industrialized war:

"The wires have no ground-connection outside of the cave. They go out from the positive brush of the dynamo; there is a ground-connection through the negative brush; the other ends of the wire return to the cave, and each is grounded independently."

"No-no, that won't do!"

"Why?"

"It's too expensive—uses up force for nothing. You don't want any ground-connection except the one through the negative brush. The other end of every wire must be brought back into the cave and fastened independently, and *without* any ground-connection. Now, then, observe the economy of it. A cavalry charge hurls itself against the fence; you are using no power, you are spending no money, for there is only one ground-connection till those horses come against the wire; the moment they touch it they form a connection with the negative brush *through the ground,* and drop dead. Don't you see—you are using no energy until it is needed; your lightning is there, and ready, like the load in a gun; but it isn't costing you a cent till you touch it off. Oh, yes, the single ground-connection—"

"Of course! I don't know how I overlooked that. It's not only cheaper, but it's more effectual then the other way, for if wires break or get tangled, no harm is done."

(467)

Beyond the electrified fence, these modern lads have prepared what Clarence rhapsodically calls "the prettiest garden that was ever planted," a belt forty feet wide entirely covered by concealed glass-cylinder dynamite torpedoes. When the first wave of many thousands of knights charges into this belt, the resulting explosion has disturbingly modern reverberations: "As to destruction of life, it was amazing. Moreover, it was beyond estimate. Of course we could not *count* the dead, because they did not exist as individuals, but merely as homogeneous protoplasm, with alloys of iron and buttons" (478).

But the most sensational part of the victory comes when Hank and his boys trap the rest of the feudal army inside the circles of their electric fence. Hank electrocutes the first batch: ". . . I shot the current through all the fences and struck the whole host dead in their tracks! *There* was a groan you could *hear!* It voiced the death-pang of eleven thousand men" (486). Then a flood is released on the survivors as the boys man machine guns that "vomit death" into their ranks: "Within ten short minutes after we had opened fire, armed resistance was totally annihilated, the campaign was ended, we fifty-four were masters of England! Twenty-five thousand men lay dead around us" (486). The conquerors themselves are conquered by "the poisonous air bred by those dead thousands." All that remains of this first experiment in industrialized warfare is a scene of total desolation, devoid of human life and marked by gigantic craters.

Hank's apocalyptic weapons resolve the paradoxes of time travel by destroying everything that modern technology has anachronistically introduced into the dark ages. Since this benighted past is partly a metaphor for Twain's present and the looming future, this resolution is fraught with grim ironies. The science and technology that mark progress, that distinguish forward from backward in time, become the means to annihilate all that humanity has created. Thus they display their potential to transform the future into the inchoate oblivion of the primeval prehuman past.

Edison Saves the Planet

Toward the beginning of the nineteenth century, Washington Irving's remarkable fantasy of an invasion of earth by "Men from the Moon" armed with directed-energy beam weapons (see page 21) satirized the ethnocentric arrogance of the emergent American republic, as it marched west from the Atlantic seaboard states, trampling the agricultural and hunting societies of the continent under its boots. At the close of the nineteenth century, H. G. Wells' *The War of the Worlds,* the classic nightmare of invasion by technologically superior aliens, assaulted the gospel of progress, imperialist pride, and faith in the manifest destiny of the British nation and the human species.

Wells' terrifying novel has become the fountainhead for a revealing stream of twentieth-century industrial capitalist culture. Out of the imagination of that culture come alien menaces in many shapes—giant slugs and ants, humanoids and robots, rays and microbes, people-eating plants and body-snatching pods, or just "creatures" and "things." New swarms of these invaders from space arrive ceaselessly in novels, pulp stories, comic books, movies, television shows, toys, and video games.

Both Irving and Wells were trying to make us confront our own hubris by imagining more technologically advanced civilizations that might also have expansionist aims and as much regard for humans as the white nations have shown for peoples of color. In the opening of *The War of the Worlds,* Wells suggests that his Victorian contemporaries, if they thought at all of extraterrestrial beings, considered them ''inferior to themselves and ready to welcome a missionary enterprise.'' He then confronts his readers with his now classic image of the conquerors from space: ''minds that are to our minds as ours are to those of the beasts that perish, intellects vast and cool and unsympathetic'' (123).

First serialized in 1897 in *Pearson's Weekly* and *Cosmopolitan, The War of the Worlds* appeared in the midst of the most aggressive expansion of the British Empire, part of the division and redivision of the nonwhite world by the colonial powers that was to culminate in the First World War. In the century leading up to the novel, Great Britain alone had seized Tasmania, Trinidad, British Guiana (Guyana), Saint Lucia, Malta, Dominica, Gambia, Sikkim, Singapore, North Borneo (Sabah), Malacca, Penang, the Gold Coast (Ghana), Assam, Mauritius, Sierra Leone, many Pacific and Atlantic isles including Ascension Island and the Falkland (Malvinas) Islands, Aden, New Zealand, Hong Kong, Natal, Sind, the Punjab, Burma, Bahrain, Nagpur, Nigeria, Baluchistan (southeastern Afghanistan), Basutoland (Lesotho), Fiji, Cyprus, British Honduras (Belize), Somaliland, the Seychelles, Bechuanaland (Botswana), Egypt, Southern Zambezia (Zimbabwe), Zululand (part of South Africa), Sarawak, Kenya, Zanzibar, Northern Zambezia (Zambia), Nyasaland (Malawi), and Uganda. Wells (who later was to analyze the relations among this imperialism, industrial capitalism, and World War I in Chapter 38 of *The Outline of History*) tries to jolt the readers of *The War of the Worlds* into confronting their own imperialist outlook. Before judging the Martian invaders too harshly, he cautions, ''we must remember what ruthless and utter destruction'' Europeans have wrought upon nonwhite peoples: ''The Tasmanians, in spite of their human likeness, were entirely swept out of existence in a war of extermination waged by European immigrants in the space of fifty years. Are we such apostles of mercy as to complain if the Martians warred in the same spirit?'' (125–26).

This leads to the second great theme of *The War of the Worlds.* Wells foresees the rapid advance of weapons technology, yoked to the ideology of the genocidal warfare waged by Europeans on nonwhites, potentially leading to a similar ''war of extermination'' waged by Europeans against other Europeans. Although ''the rout of civilization, the massacre of mankind'' (224) in this novel is the work of extraterrestrial invaders, Wells here is building a bridge between that post-1870 European future-war fiction, with its military preparedness propaganda, and his own predictions of catastrophic global conflict, such as *The War in the Air, The World Set Free,* and *Things To Come.*[15] When he compares the invaders' form of war with human wars of the past, he also hints of a possible future: ''Never before in the history of warfare had destruction been so indiscriminate and so universal'' (173). Viewed in this light, the Martians, with their armored war machines, poison gas, flying machines, and heat beams, are invaders not so much from the neighboring planet as from the approaching century.

What does all this have to do with the role of Edison in America's ever more torrid love affair with superweapons? Just a month after the final installment of *The War of the Worlds* appeared in the December 1897 issue of *Cosmopolitan* and the January 11, 1898 issue of the *New York Evening Journal,* up popped the first installment of a most revealing sequel: *Edison's Conquest of Mars.* Written by Garrett P. Serviss, astronomer, journalist, and popularizer of science, this marvelous expression of the American imagination ran in the *New York Evening Journal* in January and February 1898, that decisive year for American imperialism. *Edison's Conquest of Mars* uses the mythical Edison—that living embodiment of American technological genius, all-conquering optimism, and boundless destiny—to convert the jeremiad of *The War of the Worlds* into an effervescent advertisement for imperial aspirations, superweapons, and warfare of extermination.[16]

The American novel opens just after Wells' concludes. Even more dreadful than the terrible physical devastation left by "the merciless invaders from space" is "the profound mental and moral depression" produced by the encounter with our technological superiors. This "universal despair" becomes "tenfold blacker" when strange lights on the surface of Mars signal the imminence of a new invasion. But then the savior arises:

> Suddenly from Mr. Edison's laboratory at Orange flashed the startling intelligence that he had not only discovered the manner in which the invaders had been able to produce the mighty energies which they employed with such terrible effect, but that, going further, he had found a way to overcome them. (5)

This is "a proud day for America":

> Even while the Martians had been upon the earth, carrying everything before them, demonstrating to the confusion of the most optimistic that there was no possibility of standing against them, a feeling—a confidence had manifested itself in France, to a minor extent in England, and particularly in Russia, that the Americans might discover means to meet and master the invaders.
> Now, it seemed this hope and expectation was to be realized. (5)

But this is only the beginning for the miraculous powers of Mr. Edison, the incarnation of America's genius.

Edison now discovers an electric force that can overcome gravity, and uses it to design and build a spaceship. Faster and more maneuverable than the Martian space vehicles, Edison's ship in one quick jump leapfrogs the eons of Martian science and technology. Not content with this single-handed feat, Edison also invents (again all by himself) a weapon more potent than the deadly heat beam used by the Martians—a long-range disintegrator beam capable of reducing any substance into its constituent atoms.

In *The War of the Worlds,* Wells puts his readers in the position of those being colonized by the industrial capitalist nations; the most modern British firearms are mere "bows and arrows" compared with the beam weapons of the Martians

(177). Echoing this passage, *Edison's Conquest of Mars* reminds its readers that the Martians, having had many more epochs than humans in which to develop their science and technology, had used against us weapons "as much stronger than gunpowder as the latter was superior to the bows and arrows that preceded it." The moral of this American story, however, is precisely the opposite of that presented in Wells' tale: "But the genius of one man had suddenly put us on the level of our enemies in regard to fighting capacity" (102).

Upon learning of the American genius's inventions, the heads of all the nations and peoples on Earth convene—in Washington of course—to arrange financing for total war against Mars. Edison is put in charge of an interplanetary war fleet of one hundred of his electric antigravity spaceships, each manned by a crew of twenty male citizens of various nations. Since these two thousand men include all the world's great scientists, Earth will be entirely defenseless should they fail (104). Needless to say, they don't. "The Wizard" proves to be as brilliant at leading an invasion of an alien planet as he is in tinkering around in his New Jersey laboratory.

The message of *Edison's Conquest of Mars* must have been clear to those who read it as a *New York Evening Journal* serial in early 1898. Sensationalist news about a possible attack on the United States by Spain or about American naval preparations to carry the war to the enemy preceded each day's new installment. The departure of Edison's war fleet came in the January 15 issue as an installment entitled "To Conquer Another World." On the front page of this issue, under a giant "EXTRA," ran a banner headline: "MAINE MAY SAIL FOR HAVANA AT ANY MOMENT."

At first, the sheer numbers of Martian spaceships defending their home world stalemate the superweapons of the humans. Unfazed, Edison speedily devises a means to annihilate their whole civilization by flooding the planet. Leading a raiding party to the fortified building that controls the main floodgates, Edison is not even stymied by the alien technology. "Don't touch anything," he shouts, "until we have found the right lever":

> But to find that seemed to most of us now utterly beyond the power of man.
> It was at this critical moment that the wonderful depth and reach of Mr. Edison's mechanical genius displayed itself. He stepped back, ran his eyes quickly over the immense mass of wheels, handles, bolts, bars and levers, paused for an instant, as if making up his mind, then said decidedly, "There it is. . . ." (157)

Compared with this mythic Edison, the technological supermen who single-handedly achieve global peace in other American future-war novels are neighborhood handymen. *Edison's Conquest of Mars* extrapolates into gigantic fantasy the boundless optimism and self-aggrandizement characteristic of the bumptious upstart American empire in 1898. In place of Wells' somber reflections on the limits of ethnocentrism and anthropocentrism appears a pep rally for American nationalism. Wells' warnings to industrial society are transformed into mindless glorification of the cults of progress, empire, individualism, and technology. The sense of a world plunging into an arms race that may lead to wars of extinction yields to cheerleading for superweapons and hurrahs for the planetary annihilation of intelligent life.

Of course this genocidal war takes place against an alien species on another planet, and *Edison's Conquest of Mars* does envision a league of united nations fighting against these aliens rather than each other. But the rapturous descriptions of victorious combat, with atomic disintegrators blasting enemies out of existence and an American conqueror unleashing total war, appeared within the context of a nation whipping itself into a war fever, convincing itself that its manifest destiny was not merely continental but global. Within two months of the final installment of *Edison's Conquest of Mars,* the United States would begin its conflict with Spain for imperial possessions in the Caribbean and Pacific.

It must be admitted, though, that Serviss's wild fantasy about the development of new weapons has a kind of ironic accuracy compared with the human military impotence imagined in *The War of the Worlds.* Of course, Wells was not suggesting that the weapons of his time were harmless, and even before World War I he would project a cataclysmic war waged with ''atomic bombs.'' Yet the arms race of the twentieth century has been so spectacularly successful that if those Martians imagined by Wells in 1897 were to arrive today with their line-of-sight heat beams, clumsy three-legged armored fighting machines, ponderous automata, and primitive aircraft, we could dispose of them as easily as they were able to brush off the cavalry and artillery of Queen Victoria's empire. It is their weapons that would be mere bows and arrows against our thermonuclear bombs, intercontinental missiles, and automated guidance systems consciously designed to wreak the kind of planetary extermination cheerfully fantasized in *Edison's Conquest of Mars.* But in reality, unlike the fantasy, the species that the weapons have been designed for use against is not alien but human.

The Military-Industrial Organizer

Like those of fiction writers, war scenarios of governments and general staffs in the late nineteenth and early twentieth centuries envisioned a decisive victory going speedily to whoever possessed the most advanced machinery of war— dreadnoughts, tanks, submarines, machine guns, siege guns, airplanes, or some unexpected secret scientific weapon. Actual modern war turned out to be very different from the fantasies of novelists, generals, and rulers.

The eruption of the ''Great War'' in 1914 spread indecisive, grinding conflict, with unprecedented casuality rates, over hitherto unparalleled expanses of land and sea. The depth of the front ''line'' of land battles now had to be measured not in yards but in miles, thanks to the tank, the airplane, and the latest advances in artillery. Attacking forces of enormous scale suffered paralyzing losses against entrenched defenders armed with heavy machine guns and supported by their own long-range death machines. The colossal British surface fleet was able to pen up its counterpart in German coastal waters, only to have the Atlantic Ocean turned into a bloody playground for German U-boats (thus refuting Admiral

Mahan and the proponents of the dreadnought). Every spectacular new weapon seemed to be neutralized by some equally effective counterweapon. Modern industry on both sides displayed its ever-increasing capacity to deploy, organize, arm, and slaughter immense numbers of people.

Combatants and witnesses of the carnage filled journals with testimony about the machinelike, inhuman, inexorable features of modern industrial war. This early commentary was typical: "Modern battle is the cold, calculating work of science, largely shorn of the human element. Men mechanically load and unload artillery, firing in cold blood without enthusiasm, even without knowledge of results. . . ."[17]

The American nation—or at least its press—turned to Edison, as the oracle of technology, for interpretation. The Wizard, never one to disappoint his audience, routinely cranked out Delphic pronouncements about the defense of America, military theory, and the causes and future of war. The apparently mute, almighty, inscrutable historical forces that were industrializing and mechanizing the conflicts of nations now spoke in the friendly voice of the man who embodied America's technological genius.

For decades, Edison had been what today would be called a media event. He used and was used by the press for a variety of purposes. Back in that sensational 1892 *New York World* interview about Edison's purported electric superweapons, the reporter let slip a trade secret: "Edison is the Aladdin's lamp of the newspaper man. The fellow who approaches him has only to think out what he wants to get before taking the lamp in his hand and he gets it."

With the outbreak of the Great War, Edison's cultural status made him a useful instrument in the popularization and organization of what has come to be known as the military-industrial complex. The man who could create the war machines that would win the war to end all wars was now utilized by the press and other forces to promulgate the dogma of industrialized war. Before long, even the government would be calling upon Edison to do for the nation what he had done for the planet in *Edison's Conquest of Mars.*

At first, Edison appeared mainly in his symbolic role. For example, his 1914 inspection of a dreadnought and submarine was reported by the *New York Times* under the headline "Wizard Visits Navy Yard." Two weeks later, the *New York Times* presented him as the very spirit of American peacefulness—blaming the war on "those military gangs in Europe" who "piled up armament until something had to break," and, apparently forgetting his earlier claims to marvelous weapons, asserting that "Making things which kill men is against my fibre."[18]

In early 1915, Edison began a gradual shift, hinting of a more active role in merging industry with war making. "The present war has taught the world that killing men is a scientific proposition," he declared. Therefore, Edison soon announced, in a characteristic melange of contradictions, "the new methods" of waging war "offer an opportunity to democracy such as it never had before": "They have eliminated the fortress and the secret movements of large armies and have made trench warfare the most effective method. In that one addition to military tactics it is fair to assume that the aeroplane has given to the United States what amounts to an addition of two or three million soldiers."[19]

In May, the *New York Times Magazine* ran what was to prove a most influential interview, entitled "Edison's Plan for Preparedness: The Inventor Tells How We Could Be Made Invincible in War. . . ." Arguing that the European war had already "proved beyond the shadow of a doubt the uselessness of large standing armies" and that "the European plan of readiness for war really provoked war," Edison offered an alternative to the potent militarization or "preparedness" propaganda that was barraging the nation. Rather than establishing a large standing army and navy, Edison advocated training reserve officers and noncommissioned officers, expanding and centralizing the state militia, and manufacturing millions of modern firearms, an armada of airplanes, and a vast fleet of surface vessels and submarines—almost all just to be held in readiness in storage and dry-dock.[20]

Edison's formula for invincibility mixes the citizen-soldier concept of the early Republic with modern technology and industry. This concoction underlies his rather curious notion that trench warfare somehow gives advantages to a democratic industrial society such as the United States. Trench warfare seemed to him to offer opportunities for both independent individual fighters and American technological genius. Since American industry had already developed "trenching machines" to "a state of very high perfection," adapting the "existing machinery to the purposes of military trenching would be a very simple matter." As for the citizen-soldier, "In trench fighting, with our unlimited supply of the most intelligent and independently thinking individual fighters in the world, we would be invincible."

Edison adroitly juggles the old ideology of the citizen-soldier (the invincible fighting man) with the emerging ideology of omnipotent technology (the invincible fighting machine), both central to America's future self-image. Since "modern warfare is more a matter of machines than men," America's technological genius may allow us to become "one of the greatest of the military nations without burdening ourselves with any comparatively great, permanent military expense," though "we never must become a military nation in the old sense of the term." In the next paragraph, he evokes the myth of the peaceful American who can be provoked by evil forces into becoming a mighty warrior: "There is practically no military sentiment in the United States, nor ever has been, but we have proved ourselves to be among the world's most powerful fighters whenever we have had to fight."

The most influential part of this interview turned out to be Edison's proposal that "the Government should maintain a great research laboratory, jointly under military and naval and civilian control":

> In this could be developed the continually increasing possibilities of great guns, the minutiae of new explosives, all the techniques of military and naval progression, without any vast expense.
>
> When the time came, if it ever did, we could take advantage of the knowledge gained through this research work and quickly manufacture in large quantities the very latest and most efficient instruments of warfare.

Upon reading this suggestion, Secretary of the Navy Josephus Daniels on July 7 invited Edison to be the principal advisor to a newly organized Naval Consulting Board. Daniels' letter, reprinted on the front page of the July 13 *New York Times* and many other newspapers, called on Edison, as "the one man above all others who can turn dreams into realities," to lead the board into providing the navy with "new devices that will assure peace to our country." Edison's acceptance sent a wave of joy across the nation, as though the superweapons of *Edison's Conquest of Mars* would soon be available.

Amid the hoopla, a few dissenting voices tried to warn of the dangers in looking to technology for peace and security. Under the ironic headline "Wanted—A Device to Insure Peace," the *Harrisburg Star-Independent* pointed to the underlying fallacy in the Navy Secretary's claim that Edison and his technological colleagues would be able to invent " 'new devices that will assure peace to our country' ":

> The international rivalry has resulted in the development of one device after another, each designed not only to be destructive in itself but to render less dangerous devices that had preceded it. Just now the submarine is the most effective new device for which no practical antidote has been discovered. . . . the present conflict, which is the most terrible the world has ever seen, is so largely because of rivalry in the past in inventing more and more destructive devices.

While granting the need to invent a "submarine destroyer," the article ridicules the belief that "even Mr. Edison, with his unequalled inventive genius, can devise a machine that will 'insure peace to our country.' "[21]

The *San Francisco Chronicle* of July 17 attacked the emerging cult of the superweapon even more directly: "There are no mechanical devices . . . which can assure peace. From the first crude stone weapons to the latest submarine or siege gun there has been no form of arms or armament which has preserved a day's peace for any power."

But these voices were drowned out by the exuberant cheers going up from hundreds of newspapers around the country. Many now began to speculate on what form of superweapons the Wizard might have up his sleeve. The *Portland Journal* of July 11 thought that it might be long-range heat beams (a science-fiction favorite popularized by H. G. Wells' 1897 Martians). Backing up this hypothesis, it quoted an interview with Edison then appearing in journals from coast to coast: "Science can find much more effective ways of destroying life than by artillery and rifle fire, or the use of high explosives. The possibilities of chemistry and electricity have hardly yet been touched upon in modern warfare. They can do a lot better." Though expressing his reluctance to devise weapons, Edison announced that he personally could invent something more deadly than gas bombs and that if the United States were pushed into the war he would indeed use his science not "to make the world a better place to live in" but to "help make it worse." Foreign powers were on the verge of transforming this peaceful inventor, merely trying to make fame and fortune by benevolently applying technology to daily life, into an awesome destroyer. The lone inventive genius of nineteenth-century lore was about to animate the full powers of modern industry. Edison here speaks as a voice of America during a moment of transition.

On October 13, 1915, a fleet of German zeppelins bombed London, foreshadowing the role of advanced technology in a Second World War that would make the First look like a medieval tournament. The day after reporting on this raid, the *New York Times* ran a startling interview with the new chairman of the Naval Consulting Board under the headlines "Machine Fighting Is Edison's Idea" and "War on a Business Basis." In this interview, immediately reprinted across the country, Edison boldly defines his philosophy of modern war: "The war of the future, that is, if the United States engages in it, will be a war in which machines, not soldiers, fight." He urges "the gentlemen of the press" to promote this doctrine to the public: "Tell them that preparation for war is not military work, but should be done by shrewd business men in an economical way. Machines should be invented to save the waste in men."[22]

In "actual fighting," Edison declares, "I would rather use machines than men. . . . A machine can be easily as good as twenty men. Then one man, using it, is as good as twenty men. He should be at least that good if he is an American." "The soldier of the future will not be a sabre-bearing, blood-thirsty savage," he explains, "but a machinist; he will not shed his blood, but will perspire in the factory of death at the battle line."

The broad appeal of Edison's images, like so much of the rhetoric crucial to the support of modern warfare, comes not from their consistency but rather from their power to merge and obscure contradictions. Though this technological warfare is supposedly to be less horrendous, Edison advocates that the arsenal of "defense" should incorporate such weapons as "liquid fire and asphyxiating gases." And if modern combat is to be "war in which machines, not soldiers, fight," and these machines will "save the waste in men," then who will be slaughtered in "the factory of death"?

It is the destiny of America, according to Edison, to consummate this industrialization of war: "America is the greatest machine country in the world, and its people are the greatest machinists. They can, moreover, invent machinery faster and have it more efficient than any other two countries. It is a machine nation; its battle preparation should be with machinery."

Edison here becomes the spokesman for a new military age. As he predicted, "shrewd business men" could help convert America, with its prodigious capacities for technological innovation and production, into the world's greatest "factory of death." America's fighting machines would soon become central images in the nation's culture and central facts in the world's history.

But, meanwhile, for over three years the nation waited expectantly for Edison and the Naval Consulting Board to produce their technological miracles. Stories swept the newspapers about Edison working "night and day" on secret projects that would bring the war to a speedy and happy conclusion. Yet the war ended without either the board or Edison having had any noticeable effect on its outcome. Edison himself, who sometimes had a naval vessel at the disposal of his experiments, later complained, "I made about forty-five inventions during the war, all perfectly good ones," but the Navy "pigeon-holed every one of them."[23] His own papers list forty-one inventions during the war, mostly consisting of such items as "Steamship Decoy," "Device for Look-out Men," "Extinguishing Fires in Coal Bunkers," and "Smudging Periscopes," as shown in the following box.

Mr. Edison's War Work

Sound Ranging
Detecting Submarines by Sound from Moving Vessels
Collision mats
Quick Turning of Ships
Taking Merchant Ships out of Mined Harbors
Oleum Cloud Shells
Camoflaging Ships and Burning Anthracite
More Power for Torpedoes
Coast Patrol by Submarine Buoys
Destroying Periscopes with Machine Guns
Cartridge for Taking Soundings
Sailing Lights for Convoys
Saving Cargo Boats from Submarines
Smudging Sky Line
Obstructing Torpedoes with Nets
Under Water Searchlights
High Speed Signalling with Searchlights
Water Penetrating Missile
Aeroplane Detection

Observing Periscopes in Silhouette
Steamship Decoy
Zigzagging
Reducing Rolling of War Ships
Obtaining Nitrogen from the Air
Stability of Submerged Submarines
Induction Balance for Submarine Detection
Turbine Head for Projectile
Protecting Observers from Smoke Stack Gas
Mining Zeebrugge Harbor
Blinding Smudging Submarines and Periscopes
Mirror Reflection System for War Ships
Device for Look-out Men
Extinguishing Fires in Coal Bunkers
Telephone System on Ships
Extension Ladder for Spotting Top
Preserving Submarine & other Guns from Rust
Freeing Range Finder from Spray
Smudging Periscopes
Night Glass
Re-acting Shell

Source: Typescript, Naval Consulting Board Papers, Box 21, Edison National Historic Site.

In fact, 110,000 military inventions were submitted to the board.[24] These proposals came from local tinkerers, successful and unsuccessful inventors, avaricious or patriotic patent attorneys, leading and unknown scientists and engineers, and companies of all shapes and sizes, from one-man shops to prominent corporations. There were fantastic schemes, such as a device for projecting an electric death ray. There were innumerable designs, practical and impractical, for helicopters, airplanes, submarines, motors, artillery, depth charges, torpedoes, and defenses against torpedoes. Many inventions, such as bomb sights, navigational aids, petroleum bombs, and chemical bombs, were designed to make airplanes more deadly. Edison's staff routinely thanked the senders but explained that neither Edison nor the board had the facilities to develop the suggestions. Some were forwarded to the board; others apparently went to Edison's own laboratory and factory, where more than four thousand people were employed. An examination of the twenty-eight boxes of the board's papers in the Edison archives disclosed that most of the inventions listed as Edison's derived from these unsolicited proposals.

Although the Naval Consulting Board hardly seemed to live up to expectations, it was a critical link in the chain of events leading to the fulfillment of Edison's military dreams. Secretary of the Navy Daniels originally had intended to name just two or three famous inventors to the board. Instead, following Edison's recommendation, he asked each of the nation's eleven leading scientific and engineering societies to designate two members. Some of the same corporate and scientific institutions represented by these twenty-two men still interpenetrate each other and the military. Among the members were the director of research for General Electric; Elmer Sperry, founder and head of the Sperry Electric Company; the chairman of the board of the Ingersoll-Rand Corporation; the chief engineer of Westinghouse; the president of the Carnegie Institution; Hudson Maxim, inventor and producer of high explosives; Frank J. Sprague, founder of the Sprague Company; eminent professors from several universities; and leading engineers from various corporations.

Thus it was the Naval Consulting Board, under Edison's guidance, that first brought together leaders of industry, science, and the military into a prototype for the modern American system of weapons development and procurement. Ironically, the most mythologized of all lone inventive geniuses played midwife at the birth of the faceless giant now called the military-industrial complex. The board's centralization of more than a hundred thousand proposed advances in military hardware (even though few, if any, at first materialized in new weaponry) represented a crucial transition from both the bygone days of lone inventors such as Fulton personally peddling their wares and the contemporaneous hodgepodge of hit-and-miss free-enterprise dealings, such as the Sims-Edison Electric Torpedo Company.

Edison's exhortations for America to use its industrial might in war helped initiate a process that would culminate in the truly staggering feats of military production during World War II. The documentary films of that war project the triumph of Edison's vision, with the thrilling sights and sounds of industrial production: fiery cascades of molten steel, machines spinning and humming, sparks flying from armies of welders, countless thousands of planes and bombs and tanks endlessly clanking off the assembly lines of America. This stupendous armada, ever multiplying, ever advancing in designs of deadliness—here indeed was Edison's factory of death stretching from the battle front to the "home front," where children with their little red wagons collected scrap metal from the neighbors to feed the machines of war.

We who grew up on the home front believed that this mighty productivity of America is what won the war (although we also knew that the war had been won by the American fighting unit—with its slow-talking Georgian, lanky, sharp-shooting Texan, passionate Italian from the Bronx, and comical but brave Brooklyn Jew). We knew that the conflict had been decided by the never-ending rain of bombs from thousands of Flying Fortresses and Superfortresses on the cities of Germany and Japan. And then, of course, we also found out that the war had finally been won by the atom bomb. In any case, we were certain that victory was guaranteed by American know-how, American industrial might, and the American superweapon.

We didn't think much about the opposite side of this triumphant Edisonian vision. If warfare were to be mainly a contest of industrial might, decided by who could produce the most devastating technological weaponry, then the "home front"—maybe even including children with little red wagons—would be not just a legitimate target but the main objective to be destroyed.

Weapons to End All War

The colossal European arms race that culminated in World War I had induced a false complacency. Surely, no government would be so irrational as to plunge into a war that would unleash such arsenals. After all, these dreadful weapons had been created for *deterrence,* and deterrence worked, as proven by the fact that no major European war had taken place since the early 1870s.

When the Great War came, industrialized warfare proved to be even more appalling than the nations had imagined. Its technological wonders were terrifying enough to convince many observers and participants that this, indeed, was the war to end all wars. Beyond this catastrophic conflict seemed to lie nothing but Armageddon. In reaction to the governments and ideologies that had led the world into this nightmare many kinds of opposition arose, including pacifism, anarchism, internationalism, and revolution. Disarmament, or at least arms control, seemed to be on the world's agenda.

But the historical forces that had generated the war were still very much in operation, guaranteeing a resurgence of the weapons cult and the dogma of military "preparedness." There were frank advocates of militarization and preparation for war—in the defeated nations to reverse their defeat; in the victorious nations to safeguard their victory; in both camps to "contain Bolshevism." There were those who subscribed to the old argument from future-war fiction between the Franco-Prussian War and 1914 that another major conflict could be deterred only by adequate armaments and forces—for their own nation, of course. And then there were those who favored unrestrained development of terrible new weapons as the surest path to universal peace.

Among the latter was Edison, who after the war became a prominent spokesman for the theory that science would end war by developing superweapons too horrific to be used. This he foresaw as the ultimate consequence of the scientific industrialization of warfare, in which he had already played such an emblematic role.

The aged inventor offered his prescription for universal peace in a 1921 interview published by the *New York American* under the title "How to Make War Impossible: By Thomas A. Edison." Implicitly rejecting his own 1914 national program, Edison now opposed all "quantity production of war materials in peace time,'' because rapid technological obsolescence made this ''sheer waste.'' Instead, governments should ''go on experimenting with death-dealing devices, ceaselessly, inexorably'' until the world became so ''full of such death-dealing devices as would make war utterly impossible.'' This was advice not just for his own country:

> No competent government in the world ever ought to cease experimentation with war-making machinery and substances. This, I think, is especially true with regard to aviation and asphyxiating and other gases. That may sound bloodthirsty. As a matter of fact it is the common sense of a true man of peace. Experimentally every terrible way of war-making should be developed without pause or hindrance.

Edison could not see ''any limit to the future possibilities of instruments of war-making,'' possibly including ''atomic energy.'' So he urged all governments to ''produce instruments of death so terrible that presently all men and every nation would well know that war would mean the end of civilization.''[25]

A few weeks later, still playing the role of his own image in *Edison's Conquest of Mars,* Edison suggested that he single-handedly could arm America with such terrible superweapons. After ''a few more experiments,'' he announced in an interview with the *Springfield* (Massachusetts) *News,* he would be able to '' 'kill everybody in a great city in five minutes.' '' He assured the *News* reporter that ''if the United States ever got into a tight pinch he would put the government in possession of a destructive agency so terrible that nothing short of impending national disaster could cause him to disclose it.''[26]

Edison's cultural stature no longer needed such boasts of technological wizardry, for he was now venerated as the grand philosopher of the new world wrought by technology. So in 1922, when Shaw Desmond decided to interview him for the *Strand Magazine,* it was because ''the one intellect in the world which might conceivably be able to abolish war from the earth is that of Edison.''

As expected, Edison did have the final solution. ''There will one day,'' he told Desmond, ''spring from the brain of science a machine or force so fearful in its potentialities, so absolutely terrifying, that even man, the fighter, who will dare torture and death in order to inflict torture and death, will be appalled, and so will abandon war forever.'' Edison went on, developing his theory until it bears an eerie likeness to the world in which we now live:

> My solution for war . . . is preparation. . . . this preparation or preparedness may one day involve the discovery of some terrific force, some engine of war the employment of which would mean annihilation for the opposing forces. The way to make war impossible is for the nations to go on experimenting, and to keep up to date with their inventions, so that war will be unthinkable, and therefore impossible.[27]

Governments, as Edison advised, have indeed managed to ''produce instruments of death so terrible that presently all men and every nation would well know that war would mean the end of civilization.'' We now have that ''engine of war the employment of which would mean annihilation for the opposing forces." And, in fact, the two principal owners of these suicidal arsenals did once agree that war was "unthinkable, and therefore impossible" since their weapons guaranteed what came to be called "Mutually Assured Destruction."

But would Edison have imagined that the political-economic-military forces he helped develop in America would attempt, at that very point in the history of the species, to devise ways to make war once again "fightable and winnable"? Or thinkable, and therefore possible?

As late as 1927, Edison was insisting that "if wars are ever done away with" it will be because "science and invention may make war so dangerous to everyone concerned that the sheer patriotism of educated people in all nations, plus their common sense, will be universally against the stupid war-idea."[28] But by then a new demigod of war was using Edison's motion picture technology to project on screens across the country the gospel of the latest superweapon. Amid the post–World War I disarmament conferences, air-war hero Billy Mitchell was heralding the epoch of total war from the skies. The world would soon see the fulfillment of Edison's exhortation that "every terrible way of war-making should be developed without pause or hindrance."

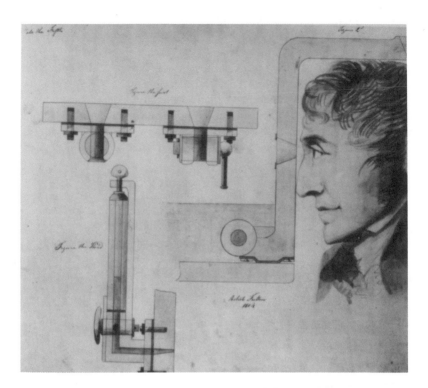

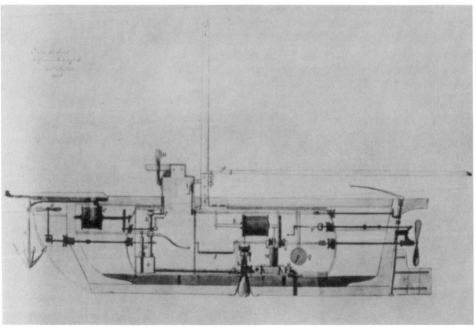

Above: Robert Fulton's 1804 self-portrait at the periscope of his submarine, *Nautilus,* which he built for France in 1800, together with his drawings of the vessel's viewing windows and depth gauge. *Below:* Fulton's engineering sketch of the *Nautilus.* Conceived as a weapon that would eliminate standing navies, his submarine was intended to bring about a "millennium" of reason and progress. Fulton argued that the submarine would be purely defensive and would have no offensive potential. *The New York Public Library, Astor, Lenox and Tilden Foundations.*

A direct descendant of Fulton's purely defensive weapon, the USS *Georgia,* typical Ohio-class Trident nuclear submarine, can hit 192 separate targets with thermonuclear warheads, each eight times as powerful as the atomic bomb that destroyed Hiroshima. *Official U.S. Navy photograph by General Dynamics Corporation.*

Original 1910 magazine illustration for Jack London's "The Unparalleled Invasion" shows the Chinese people as pests being exterminated by airplanes dropping bacteriological bombs. According to London, only such aerial superweapons could eliminate the Yellow Peril. *McClure's Magazine, July 1910.*

The President of the United States christens one of the secret radiation-powered planes in *The Vanishing Fleets*. The President decides that it is his "duty" to use this "most deadly machine ever conceived" to defeat Japan and establish a global Pax Americana, "thereby ending wars for all time." *Illustration from The Vanishing Fleets (1908).*

Published Weekly

The Curtis Publishing Company
Independence Square
Philadelphia

London 6, Henrietta Street
Covent Garden W.C.

Copyright 1914
by The Curtis Publishing Company in
the United States and Great Britain

Entered at the Philadelphia Post Office
as Second Class Matter

Entered as Second Class Matter at the
Post Office Department
Ottawa, Canada

THE SATURDAY EVENING POST

Founded A°D¹ 1728 by Benj. Franklin

Volume 187   PHILADELPHIA, NOVEMBER 14, 1914   Number 20

THE MAN WHO ROCKED THE EARTH

By ARTHUR TRAIN

ILLUSTRATED BY HENRY RALEIGH

The Earth Blew Up Like a Cannon—Up Into the Air, a Thousand Miles Up

When PAX destroys the Atlas Mountains with his atomic beam weapon in *The Man Who Rocked the Earth* (1914), the crewmen of a nearby fishing vessel die excruciating deaths from radiation sickness, with symptoms remarkably similar to those among survivors at Hiroshima and Nagasaki. *Reprinted by permission of the Saturday Evening Post Society.*

THE FRIGHTFUL BATTLE UNDER THE MARTIAN CLOUD.

EDISON'S CONQUEST of MARS.

by GARRETT P. SERVISS.

(Copyright, 1898, by Garrett P. Serviss.)

poured their resistless vibrations in every direction through the quivering air.

The airships of the Martians were destroyed by the score, but yet they flocked often as thicker and faster.

We dropped tower and our blows fell upon the forts, and upon the wide-spread city bordering the Lake of the Sun. We almost entirely silenced the fire of one of the forts, but there were forty more in full action within reach of our eyes.

Some of the metallic buildings were partly unroofed by the disintegrators, and some had their walls riddled and fell with thundering crashes, whose sound rose to our ears above the hellish din of battle. I caught glimpses of giant forms struggling in the ruins and rushing wildly through the streets, but there was no time to see anything clearly.

The Flagship Charmed!

Our flagship seemed charmed. A crowd of airships hung upon it like a swarm of

But the latter was not to be thought of. It was no mere question of self-pride, however, and no consideration of the tremendous interests at stake, which would compel us to continue our apparently vain attempt.

No Hope in Sight.

Our provisions could last only a few days longer. The supply would not carry us one-quarter of the way back to the earth, and we must therefore remain here and literally conquer or die.

In this extremity a consultation of the principal officers was called upon the deck of the flagship.

Here the suggestion was made that we should attempt to effect by strategy what we had failed to do by force.

An old army officer who had served in many wars against the running Indians of the West, Colonel Alonzo Jefferson Smith, was the author of this suggestion.

"Let us circumvent them," he said. "We can do it in this way. The chances are

---

nations, Grimini and English, besides another for the honor of giving the largest sum. Ten thousand millions were raised. The King of Spain threw the great Mogul diamond into the contribution. Amid immense enthusiasm Mr. Edison was elected to lead the expedition against Mars, and given unlimited powers.

Many of the keenest men of science in the world accompanied the expedition when, six months after the raising of the fund, a fleet of one hundred electrical ships started from New York for the attack on Mars. It was determined to rendezvous at the moon. On the way a meteor, travelling eighteen miles a second, and came within an arm's reach of the electrical ships, killing several of its inmates.

Moving again for Mars, the squadron falls into the electrical attraction of a passing comet and has a fearful adventure, which threatens to end by precipitating it into the sun. Suddenly the adventurers find themselves near the earth again.

The Terrestrial chaps must raise them, however, as it extracts them from the power of the comet, and after a short stay on the earth the expedition makes another start and soon reaches a 'wonder-world' where a speed of ten miles a second is acquired, and the fleet bears Mars in a few days. An asteroid is encountered and Martians are discovered upon it. They attack the airship with their terrible gold blasts and knock one into nothingness.

A terrible duel between our Martians and our people ensues, the result of which is that several of the former are killed, their war captors discovered, and the Terrestrians effect a landing upon the asteroid.

A gigantic Martian is captured after much difficulty, and an exploring party on the asteroid makes a marvellous discovery that it is a solid mass of gold. Evidences of a frightful battle are found, and it is plain that the Martians have had a civil war, evidently over the gold.

The asteroid runs, but he's a man of pure gold, as the Martians have been working in an inexhaustible mine. While our people are exploring in another possible route from Mars approaches and lands upon the little globe. A terrific battle ensues, in the middle of which Mr. Edison's disintegrator refuses to work and the terrestrians stand in peril of instant destruction.

The disintegrators are repaired and turned upon the Martians, who return the assault with such energy that the electricians of the air blow how their courage and fire into pieces. The fleet, however, wages the fight and destroys many of the enemy. After the battle many interesting curiosities are noted about the asteroid, and the preparations are made for the next flight to Mars.

Mars is finally at hand, and its beauties and wonders are disclosed to the earth's warriors. Suddenly an immense air ship, manned by Martians,

The ships course in immediate conflict by rising to a greater altitude, and, the Martian ships disappearing over mountains make a circuit of the planet and hide in tantalizing... Remember it now soon, however, a black cloud suddenly arose, and our warriors are choked and blinded, as the airship confounds the immense air pressure and the substance over the water. Mars are mastered for the moment.

The earth's warriors make a circumstance of the atmosphere above Mars, and prepare for an attack upon the planet. Their disintegrators are employed in blowing a hole through the Martians' immense smoke cloud.

A TERRIBLE ENCOUNTER

The Martians and Our Warriors Fight a Battle to the Death.

Instantly there opened beneath us a huge well-shaped hole, from which the black clouds rolled violently back in every direction.

Through this opening we saw the gleam of brilliant lights beneath.

We had made a hit.

"It is the Lake of th... un!" shouted the astronomer who furnished the calculation by means of which its position had been discovered.

And, indeed, it was the Lake of the Sun. While the opening in the clouds made by the discharge was not wide, yet it sufficed to give us a glimpse of a portion of the curving shore of the lake, which was ablaze with electric lights.

Whether the shot had done any damage beyond making the circular opening in the cloud curtain, we could not tell, for almost immediately the surrounding black smoke masses billowed in to fill up the hole.

But in the brief glimpse we had caught sight of two or three large air ships hovering in space above that part of the Lake of the Sun and its bordering city which we had beheld. It seemed to me that they also had been touched by the discharge and was wandering in an erratic manner. But the clouds closed in so rapidly that I could not be certain.

Penetrating the Cloud.

Anyhow, we had demonstrated one thing, and that was that we could penetrate the Martians in their hiding place.

It had been arranged that the first discharge from the flagship should be a signal for the concentration of the fire of all the other ships upon the same spot.

A terrific spitting... however, occurred, and a half a minute had elapsed before the disintegrators from the other members of the squadron were got into play.

them was not so serious, although they were evidently hors de combat for the present.

Our fighting blood was now boiling and we did not stop long to count our losses.

"Into the smoke!" was the signal, and the ninety and more electric ships which still remained in condition for action immediately shot downward.

A Dash Into the Smoke.

Looking back it seemed the very mouth of hell that we had escaped from.

The Martians did not for an instant cease their fire, even when we were far beyond their reach. With furious persistence they blazed away through the cloud curtains, and the vivid spikes of lightning shuddered so swiftly on one another's track that they were like a flaming halo of electric fingers around the frowning helmet of the War Planet.

OUR DISINTEGRATOR DOES AWFUL DAMAGE.

"Four or five of the airships tumbled headlong toward the ground, and it was evident that fearful execution had been done among the crowded structures along the shore of the lake."

Accordingly it was resolved that about twenty ships should be told off for this movement, and Colonel Smith himself was placed in command.

At my desire I accompanied the new commander in his flagship.

Flank Movements

Rising to a considerable elevation in order that there might be no risk of being seen, we began our flank movement while the remaining ships, in accordance with the understanding, dropped nearer the curtain of cloud and commenced a bombardment with the disintegrators, which caused a tremendous commotion in the clouds, opening vast gaps in them, and occasionally revealing a glimpse of the electric lights on the planet, although it was evident that the vibratory currents did not reach the ground. The Martians immediately replied to this renewed attack, and again the cloud-covered globe bristled with lightning, which flashed out in the blackness far below the feet of the stoutest hearts among us quailed. Although we were situated way beyond the danger.

But there was no opportunity supplied us withdrew from our eyes when, having attained a proper elevation, we began our course toward the opposite hemisphere of the planet.

We guided our flight by the stars, and from our knowledge of the rotation period of Mars, and the position which the principal points on its surface must occupy at certain hours, we were able to tell what part of the planet lay beneath us.

Having completed our semi-circuit we found ourselves on the night side of Mars, and determined to lose no time in recovering our camp. But it was deemed best that all exploration should first be made by a single electrical ship, and Colonel Smith naturally wished to undertake the adventure with his own vessel.

Dropping to the Planet.

We dropped rapidly through the black cloud curtain, which proved to be at least half a mile in thickness, and then suddenly emerged as if suspended at the apex of an enormous dome arching above the surface of the planet a mile beneath us, which sparkled on all sides with innumerable lights.

These lights were so numerous and so brilliant as to produce a faint imitation of daylight, even at our immense height above the ground, and the dome of cloud out of which we had emerged assumed a soft fawn color that produced an indescribably beautiful effect.

For a moment we recoiled from our in...

In the above illustration looking from left to right, the following members of the new Naval Advisory Board appear as indicated: 1. Josephus Daniels, Secretary of the Navy; 2. Thomas A. Edison, Chairman of the Board; 3. Elmer Ambrose Sperry; 4. Arthur Gordon Webster; 5. Hudson Maxim; 6. Lawrence Saunders; 7. Benjamin G. Lamme; 8. Frank Julian Sprague; 9. Henry Alexander Wise Wood; 10. Lawrence Addicks; 11. Howard E. Coffin; 12. Spencer Miller; 13. Thomas Robins; 14. Wm. L. Emmett; 15. L. H. Baekeland; 16. Benj. B. Thayer; 17. W. R. Whitney; 18. Peter Cooper Hewitt; 19. Joseph W. Richards; 20. Alfred Craven; 21. Andrew Murray Hunt; 22. Andrew L. Riker; 23. Robert Simpson Woodward; 24. Matthew B. Sellers.

A 1915 publicity photo of Thomas Edison and the Naval Consulting Board exhibits the embryo of the military-industrial complex. *International Film Service.*

OPPOSITE:
During early 1898, while its front pages agitated for war with Spain, the *New York Evening Journal* serialized *Edison's Conquest of Mars,* an imperialist sequel to H. G. Wells' anti-imperialist *The War of the Worlds.* Here Edison's disintegrator beam begins the destruction of the Martian space fleet in preparation for exterminating the Martian species. *New York Evening Journal, January 31, 1898.*

Billy Mitchell trails his command pennant as he flies over the 1921 demonstration bombing of unmanned warships. Thanks to his media team, newsreels of Mitchell and of his bombers dispatching the battleship *Ostfriesland* (sequence below) were shown within twenty-four hours to millions of Americans from coast to coast. *Official U.S. Air Force photographs.*

Are We Ready for War with Japan?

An Authoritative Picture of the Situation in Manchuria, and What It Means to the United States

By GENERAL WILLIAM MITCHELL

Former Commander, Air Forces, A. E. F., and Director, Military Aëronautics, U. S. Army

(Reading time: 21 minutes 30 seconds.)

THE situation in Asia should be a matter of profound interest if not apprehension to all American citizens. The Japanese, finding the rest of the civilized world either too busily occupied or grown too fat and impotent to take any action against them, have jumped on to the mainland of Asia and begun to dismember China in earnest.

Since the Russo-Japanese War the Japanese have felt

[CONTINUED ON NEXT PAGE]

WHAT Japan is in deadly fear of is our air force. Her islands offer an ideal target for air operations.

Picture by
ROBERT A.
CAMERON

In the early 1930s, Mitchell was already agitating for annihilating the cities of Japan. According to this article in the January 30, 1932 issue of *Liberty,* one of the nation's three top-selling general weekly magazines, these cities offered "an ideal target for air operations." *Copyright 1932 Liberty Publishing Corporation; reprinted by permission of Liberty Library Corporation.*

Ronald Reagan, as heroic Secret Service agent Brass Bancroft in Warner Bros.' 1940 *Murder in the Air,* secures the "new superweapon," a defensive beam destined to "make America invincible." *Copyright 1940 Warner Bros. Pictures, Inc., Ren. 1968 United Artists Television, Inc.*

OPPOSITE:

Above: Ronald Reagan, as American RAF volunteer, prepares to give his life attacking a Nazi ammunition dump in Warner Bros.' 1941 *International Squadron.* Like other Hollywood productions, the movie projects images of the Nazis bombing civilians while the British bomb only military targets. *Courtesy Warner Bros. and the Museum of Modern Art/Film Stills Archive.*

Below: The erotic star of the 1943 Warner Bros. hit *Air Force* is the B-17 Flying Fortress "Mary Ann." Here she is being protected from the hordes of "stinkin' Nips" to prepare for the day when she can help "blast 'em off the map." *Courtesy Warner Bros. and the Museum of Modern Art/Film Stills Archive.*

Major Alexander de Seversky and Walt Disney study the story board for *Victory through Air Power*, which in 1943 projected for family audiences the vision of Japan being annihilated by aerial superweapons. *Courtesy The Museum of Modern Art/Film Stills Archive.*

Japan is devastated without any human image of suffering or death in the animated orgy of destruction that climaxes Disney's *Victory through Air Power. Courtesy Disney Productions and the Museum of Modern Art/Film Stills Archive.*

More than two-thirds of the Japanese city of Shizuoka was incinerated in the night raid of July 12, 1945. By the end of July, every major Japanese city had been cremated except for four reserved for trying out a secret weapon. *Official U.S. Air Force photograph.*

On August 6, 1945, Hiroshima became the first city to experience atomic war. The decision to drop the bomb and U.S. policy in the ensuing years were based on the assumption that America possessed the ultimate weapon. *Official U.S. Air Force photograph.*

In the tradition of earlier American fiction that reversed the roles of victim and victimizer, *Life* magazine of November 19, 1945, imagines Washington being attacked by atomic missiles in "The 36-Hour War." *Illustration by A. Leydenfrost.*

In *Life*'s 1945 vision of global nuclear conflict, anti-missile defenses fail to live up to expectations. But U.S. offensive missiles win the war anyhow. *Illustration by Noel Sickles.*

On June 14, 1946, the United States proposed the Baruch Plan, under which an international commission would enforce the U.S. nuclear monopoly until some unspecified future period when the threat of war had been abolished. Two weeks later, the United States began the first peacetime nuclear testing with this atomic bombing of Bikini. *Official U.S. Air Force photograph.*

A popular cold war drink from the same period. *Courtesy The Museum of Modern Art/Film Stills Archive.*

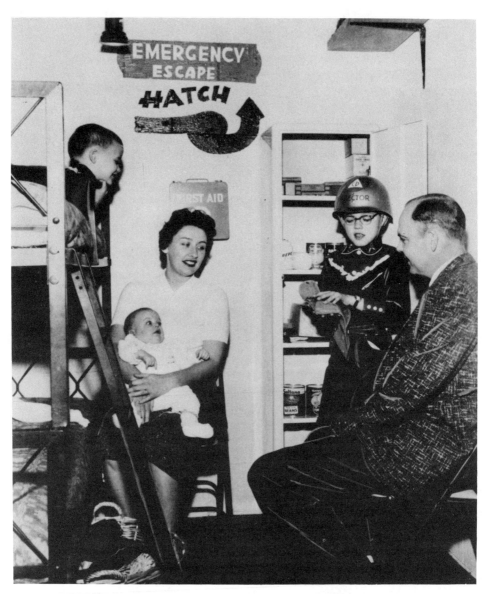

An official civil defense photograph displays a model American family of the 1950s happily living in their fallout shelter. *Courtesy The Museum of Modern Art/Film Stills Archive.*

Rita Hayworth as Gilda adorns this atomic-equipped B-29 in a 1946 Air Force photo. *Official U.S. Air Force photograph.*

The 1952 movie *Above and Beyond* showed that the real tragedy of Hiroshima was the upset it caused in the model American family of Colonel Paul Tibbets (Robert Taylor) because he couldn't tell his wife (Eleanor Parker) about "the billion dollar secret." *Courtesy the New York Public Library at Lincoln Center, Performing Arts Research Center.*

Baseball can't compete with the glorious mission of the B-36 that roars over the head of former B-29 pilot "Dutch" Holland (Jimmy Stewart) in the 1955 film *Strategic Air Command. Courtesy Paramount Pictures and the Museum of Modern Art/Film Stills Archive.*

"She's the most beautiful thing I've ever seen in my life," blurts Jimmy Stewart when he first beholds this B-47 in *Strategic Air Command. Courtesy Paramount Pictures and the Museum of Modern Art/Film Stills Archive.*

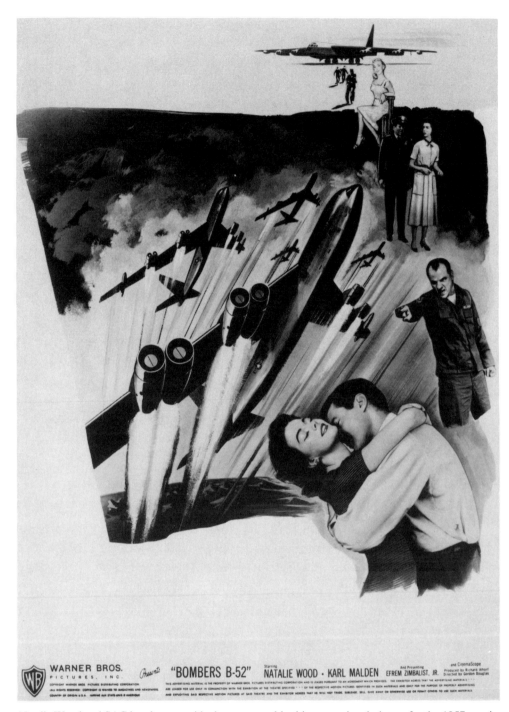

Natalie Wood and SAC bombers provide the sex appeal in this promotional picture for the 1957 movie *Bombers B-52*. *Courtesy the Museum of Modern Art/Film Stills Archive.*

Dr. Strangelove, Or How I Learned to Stop Worrying and Love the Bomb (1964). In a final triumph of American ingenuity, B-52 commander Major Kong (Slim Pickens) repairs the jammed bomb-bay mechanism that has been interfering with Armageddon. *Courtesy Columbia Pictures.*

The *New York Times Magazine* article of January 27, 1985, entitled "Defense in Space Is Not 'Star Wars'" by Zbigniew Brzesinski, Robert Jastrow, and Max Kampelman used this Department of Defense picture to illustrate its argument that SDI "Should Not Be Seen as Science Fiction." *Courtesy Department of Defense.*

The Force Is With Them

Star Wars Defense Will Benefit a Slew of Companies

By JAYE SCHOLL

LOS ANGELES—"Let the force multiplier be with you." That's not a line likely to be spoken in the next *Star Wars* sequel. It's too long. But Obi Won Kenobi could give worse advice to Luke Skywalker. A broad term, force terred by such warnings, the Administration has requested $1.8 billion for research into a Star Wars defense system as part of the DOD budget now being considered by Congress. Regardless of whether the program is approved, bil- electromagnetic battle management, and now abbreviated to "C-cubed"—another; and the field of simulation—creating battle situations on computers, also known as video war games, the third. And, in fact, Wall Street was pretty much in the dark about defense electronics until about five years ago. Once it discovered the industry, though, it couldn't say enough good things about laser guns and the spook houses where

The true value of Star Wars is suggested by this investment advice offered in *Barron's* of April 30, 1984. *By permission of Scott Pollack.*

VICTORY THROUGH AIR POWER

Should a nation . . . attain complete control of the air, it could more nearly master the earth than has ever been the case in the past. Just as power can be exerted through the air, so can good be done, because there is no place on the earth's surface that air power cannot reach and carry with it the elements of civilization and good that comes from rapid communications.

—GENERAL BILLY MITCHELL, *Winged Defense*, 1925

We read our mail and counted up our missions—
In bombers named for girls, we burned
The cities we had learned about in school—

—RANDALL JARRELL, "LOSSES," 1945

"Your objective is to destroy Tokyo."

—Opening instruction for "1942,"
a 1984 arcade video game

4

*Peace Is Our Profession**

After ages of fantasizing about flight, people took to the air in 1783. When the first human ascended in a balloon that autumn in prerevolutionary France, one of the most enthusiastic witnesses was Benjamin Franklin. Reflecting on this "discovery of great importance" that may "possibly give a new turn to human Affairs," Franklin began to speculate about the potential for air power to end all wars. "Convincing Sovereigns of the Folly of Wars may perhaps be one effect" of manned flight, he wrote, "since it will be impracticable for the most potent of them to guard his Dominions." In a burst of optimism, he calculated that "five thousand Balloons, capable of raising two Men each, could not cost more than Five ships of the Line," and certainly no monarch could "afford to so cover his Country with Troops" as to defend against "Ten Thousand Men descending from the Clouds."[1]

Franklin's response measures the historical and ideological distance from Samuel Johnson's satiric "A Dissertation on the Art of Flying," included in *Rasselas* a quarter of a century earlier. Johnson had not only ridiculed the practicality of flight, but also warned of dire consequences if it were feasible. "If men were all virtuous," explains Johnson's would-be inventor of manned flight, "I should with great alacrity teach them all to fly. But what would be the security of the good, if the bad could at pleasure invade them from the sky? Against an army sailing through the clouds neither walls, nor mountains, nor seas, could afford any security."[2] Such warnings about technological progress were rejected by Franklin, with his faith in the industrial and political revolutions that were in fact vehicles of power for men of his class. His conception of flying machines as world-transforming peacemakers was an expression of the same ideology that obsessed Robert Fulton, spurring him to invent the submarine, torpedo, and steam warship as instruments for achieving permanent peace.

*Motto of the Strategic Air Command, United States Air Force.

81

This was of course the ideology that became dominant as part of the triumph of industrial capitalism. A classic rendition of this creed in its heyday appeared in the futuristic vision of Tennyson's 1842 dramatic poem "Locksley Hall." Aircraft make "the heavens fill with commerce" and then bring about a final war in the air that leads to universal peace and unity:

> . . .and there rain'd a ghastly dew
> From the nations' airy navies grappling in the central blue. . . .
>
> Till the war-drum throbb'd no longer, and the battle-flags were furl'd
> In the Parliament of man, the Federation of the world.

In the next few decades, the contributions of flying vehicles to war and peace fell far short of the extravagant expectations they had aroused. Austria used balloons to drop bombs on revolutionary Venice in 1849, but with little effect. Balloons played only an auxiliary role, as aerial observation posts, during the U.S. Civil War and the Franco-Prussian War. And these early lighter-than-air machines had little discernible effect on peacetime human affairs.

But as powered and controlled dirigibles and airplanes began to seem feasible, American fiction and nonfiction ecstatically heralded their advent. Aircraft were now pictured as agents of the kind of millennium that Fulton had promised his seacraft would inaugurate.

Meanwhile, during this dawn of powered flight—the decades between the Franco-Prussian War and the Great War—Fulton's maritime weapons of progress were the main vessels of worldwide war and terror. While his steam warship was extending and fortifying colonial rule around the globe, bringing innovative nightmares to most of the world's nonwhite peoples, his submarine and torpedo were about to transform the crews and passengers on the ships of the world's greatest powers into ducks crossing ocean-wide shooting galleries. Yet American enthusiasm for the latest ultimate weapon kept growing.

In an 1879 essay entitled "Aerial Navigation (A Priori)," the respected poet, influential literary critic, and successful Wall Street broker Edmund C. Stedman envisioned the age of flight as the reign of free trade, global unification, prosperity, and peace. This post–Civil War utopia adds one significant feature, however, to Fulton's ideal world of late-eighteenth-century mercantile capitalism: unbridled rule by the "fittest" over the "ruder, less civilized" peoples:

> Boundaries will be practically obliterated when mountain-chains, rivers, even seas, can be crossed as readily as a level border-line, and oppose no obstacle to the passage of travelers, merchants, or men-at-arms. Laws and customs must assimilate when races and languages shall be mingled as never before. The fittest, of course, will survive and become the dominant types. The great peoples of Christendom soon will arrive at a common understanding; the Congress of Nations no longer will be an ideal scheme, but a necessity, maintaining order among its constituents and exercising supervision over the ruder, less civilized portions of the globe. Free-trade will become absolute, and everywhere reciprocal: no power on earth could enforce an import tariff. War between enlightened nations soon will be unknown. Men will see "the heaven fill with commerce," but after a few destructive experiments there will rain no "—ghastly dew / From the nations' airy navies grappling in the central blue."[3]

Stedman concludes that "troops, aerial squadrons, death-dealing armaments" will be maintained only "for instantly enforcing the judicial decrees of the world's international court" and "for police surveillance over barbarous races" (581).

On the heels of this 1879 panegyric appeared the first American future-war fiction. As seen in Chapter 2, airships and airplanes of all shapes and sizes soared through this literature in the decades leading up to the Great War, materializing fantasies of foreign aggression, American world rule, apocalyptic terrors, and the abolition of warfare. If aerial weapons were feasible, their supremacy seemed almost beyond debate. This premise was common to both the pessimistic conservatism of Samuel Johnson and the revolutionary zeal of Benjamin Franklin. Indeed, echoes of both Johnson and Franklin reverberate throughout these fictions about aerial superweapons, as one can hear clearly in Simon Newcomb's *His Wisdom, the Defender,* published in 1900, three years before Kitty Hawk: "No defence of person or property against an army flying through the air where it chose, and pouncing down at any moment, was possible" (296). As the era of actual aerial combat approached, American fiction welcomed it as a liberating apocalypse.

In the United States, this faith in aircraft as the ultimate weapon was part of the new religion evoked by the first successful experiments with heavier-than-air flight. As Joseph Corn has shown in *The Winged Gospel,* to Americans "the airplane's coming portended a fundamental and marvelous change in human affairs," promising "a virtual millennium." This would be a new epoch in human history, an "air age," when utterly free, democratic movement in three dimensions would bring enlightenment, prosperity, and peace to all.[4] These ecstatic fantasies were updated versions of Robert Fulton's utopian projections of the age to be wrought by his small canals, steamships, or maritime superweapons.

In American future-war fiction, aircraft sometimes end war merely by exhibiting their terrible potential. For example, in Walt McDougall's brief 1914 story in *American Magazine,* "The Last Conflict, the Horror That Awoke the Nation," the world comes to its senses when one British and one German airplane each sinks an enemy battleship: "Within ten days, in the parliaments of seven nations, measures had been taken to forever prohibit the building of war vessels and putting out of commission the navies of the entire world, for the horrible occurrence had suddenly opened the eyes of mankind to the utter futility of such methods of offense and defense." The Hague Museum later displays models of "the two aeroplanes that abolished war" (36). The fleets of airships and spaceships devised, built, and commanded by the title character of *His Wisdom, the Defender* easily drive "the armies of Europe into caves and dens" (306) and inaugurate the Golden Age of perpetual peace under the Defender's beneficent dictatorship. In some fantasies, the flying machines provide platforms for other superweapons that end war, such as the atomic beam weapon fired from the atomic-powered flying ring in *The Man Who Rocked the Earth* (1914).

Tennyson's vision of aerial superweapons forcing an end to war recurred throughout these stories and novels. For example, when "His Wisdom" orders the representatives of the world's nations to attend his declaration that he is about to "put an end to war and assure to all nations and peoples the blessings of security and peace forever," he seats them facing a scroll emblazoned with Tennyson's lines (199, 202). In 1910, a twenty-six-year-old Missouri farmer copied those ten lines prophesying that the rain of "ghastly" weapons from airships would bring peace and inaugurate "the Federation of the world"; he carried this copy with him for decades. In July 1945, on his way to Potsdam, President Harry S. Truman pulled the yellowed slip of paper from his wallet and recited those lines.[5]

The European response to the airplane was much closer to Samuel Johnson's forebodings. Indeed, the American glorification of a new "age" of flight was often ridiculed as romantic and utopian,[6] and very few British and European works shared the American notion that airplanes would prove the great deterrent to war. In fact, British journals were rife with articles, such as "The Aerial Peril," "The Wings of War," "The Airship Menace," and "The Command of the Air," portraying the new aircraft as a menace to national and perhaps even human security.[7] R. P. Hearne's 1909 *Aerial Warfare* painted a terrifying picture of England under attack from the skies. Like the early European future-war fiction, these writings mainly preached military preparedness, advocating that Great Britain "obtain and keep Command of the Air."[8]

Amalgamating elements of the typical British and American responses was Sir Hiram Maxim, an American expatriate and an early pioneer of airplane experiments. The inventor of the Maxim machine gun, often referred to as the deadliest weapon of the age, Sir Hiram asserted unequivocally in a 1911 *Collier's* article, "The Newest Engine of War," that "the aeroplane flying machine is the most potent instrument of destruction ever invented." He agreed that it did indeed introduce "a totally new epoch," but not the blissful age heralded by Americans: "The flying machine is essentially a military weapon. It is difficult to see how it can be used profitably for any other purpose." Sir Hiram believed, however, that since there could be no defense against the airplane and government leaders themselves could be subject to bombing within a few hours, perhaps they would replace war with arbitration. So he shared with many American contemporaries the wish that the airplane would prove the ultimate peacekeeper, though he added a typical imperialist qualification: "Therefore let us hope that the most potent instrument of destruction ever invented will ultimately lead to universal peace and prevent war altogether, at least among the highly civilized nations."[9]

H. G. Wells more accurately foresaw flying machines as neither a successful deterrent to war nor a guarantee of victory. Instead, aircraft, whether the fleets of dirigible bombers in *The War in the Air* (1908) or the small planes dropping "atomic bombs" in *The World Set Free* (1913), merely raised human powers of destruction to apocalyptic levels. Air power could thus inaugurate warfare capable of destroying human civilization. Wells did, however, maintain the Tennysonian vision of a united global civilization emerging from the ruins. His brave new world would be ruled by the technological elite, eventually appearing as the beneficent airmen of his 1936 film *Things To Come*, who call themselves "Wings Over the World."

In America, a third version of flying machines as the ultimate weapon arose and became dominant. Aircraft would neither deter war nor bring about a Wellsian apocalypse. Instead, they would guarantee a painless triumph of the world's technological and national vanguard: American inventive and productive genius. Tennyson's vision of a last dreadful war in the skies bringing about the unification of the human species was twisted in these scenarios into the gospel of victory through air power.

This doctrine, which would eventually determine national policy, was first introduced into the cultural imagination by those future-war novels discussed in Chapter 2. In *Armageddon*, an aluminum dirigible with fearsome bombing capacities brings world peace under the invulnerable military domination of America and its ally, Britain. The steam-powered, forty-thousand-ton airship invented and commanded by the suggestively named John Fulton Edestone in *L.P.M.: The End of the Great War* achieves not only Anglo-Saxon global domination, but also the elimination of socialism and feminism.

Aerial bombing of cities in this fiction is usually the identifying mark of a ruthless, savage enemy, beyond the pale of civilized conduct. In *The Conquest of America*, the Germans display their inhumanity by subjecting American cities to air raids with both high explosives and firebombs, forcing Thomas Edison and another wizard to come to the rescue with a deadly air-launched missile. It is the Japanese who prove their barbarism by bombing American cities from the air in *All for His Country*, but they are finally defeated by a gigantic American flying Superfortress.

The Yellow Peril evokes the most spectacular scenes of aerial war. In *The Vanishing Fleets*, America defeats Japan with an armada of planes powered by radioactivity and capable of flying six hundred miles per hour. Vast fleets of aerial warships meet in apocalyptic showdowns with the Yellow Peril in such futuristic novels as *Under the Flag of the Cross* and *The War of the Worlds; A Tale of the Year 2,000 A.D.* Indeed, the imagined Chinese or Japanese hordes are so menacing that they legitimize the genocidal air war condemned as inhuman when conducted by an imagined enemy. Thus, for example, Jack London offers his final solution for the Yellow Peril in "The Unparalleled Invasion": send waves of aerial bombers to wipe out every man, woman and child in China by raining "infectious death" across their land.

The first actual use of airplanes as bombers came in 1911. As in London's story, published just one year earlier, the target was a nonwhite people. Attempting to gain an empire from the crumbling Turkish dominion in 1911–1912, Italy used airplanes both to bomb Arab troops and to create terror among the civilian population.[10] So the distinction of being the first country to be bombed from airplanes belongs to Libya, attacked by Italian aviators at Benghazi and Tripoli, the sites of U.S. air raids seventy-five years later.

In 1912, France employed airplanes to help put down an anticolonial rebellion in French Morocco. In raids specifically intended to sow terror, French pilots dropped newly invented aerial bombs on villages, markets, and flocks of sheep, and set fire to grainfields with incendiary flechettes originally designed for combat against zeppelins.[11]

Such attacks upon civilian populations clearly violated the restrictions on aerial bombardment that had been tentatively adopted by the Hague Conference of 1907. But the restrictions were designed to govern warfare among the "civilized" nations, not their campaigns against nonwhite colonial subjects or would-be subjects. The Italian and French raids in North Africa set the pattern for the use of airplanes as counterinsurgency weapons by all the great colonial powers when national independence movements spread throughout their empires in the wake of World War I.

During that war, bombing was at first limited to battlefields and to cities being defended against ground assaults, as allowed under the 1907 Hague regulations. But since these ill-conceived agreements also permitted the bombing of targets defined as legitimate military objectives within any city, they inadvertently opened the gates to widespread attacks on civilian populations. In a pattern to be repeated on a far more devastating scale during World War II, attacks from the air led inevitably to escalation upon escalation, limited only by the technical capacity of the belligerents' aircraft. Due to the inherent inaccuracy of dropping projectiles from a moving aircraft through the winds beneath to a target far below, most bombs intended for military targets within a city would fall in nearby or even remote civilian areas. The nation subjected to such an attack would interpret it as terror bombing of the civilian population, and sooner or later begin reprisal raids. These, in turn, would remove the last moral and political barriers to unrestrained bombing of cities.

Early in the war, German, British, and French air raids on cities all more or less conformed to the 1907 Hague guidelines.[12] But the stumbling escalation of the raids then unleashed the only air force already capable of large-scale attacks on cities: the German zeppelin fleet, which began sporadic bombing of Britain in January 1915. The zeppelins were at first forbidden to bomb London, but this restriction was removed after British reprisal raids. While attacking London, the zeppelins were still officially restricted to military targets; but of course they loosed their bombs promiscuously when subjected to fierce air and ground defenses. After the last significant zeppelin raids were defeated in 1916, Germany switched to airplanes. In 1917, attacking from high altitude, often at night, these raids strewed bombs even more indiscriminately. Then the British bomber counterattack in late 1917 and 1918 dropped more than twice the tonnage on German cities that Britain had received.

By modern standards, neither the 300 tons dropped on Britain throughout the war nor the 660 tons unloaded on Germany in the final British air offensive seem significant. But at the end of World War I, the airplane's potential for devastation and slaughter was growing dramatically. All the major belligerents were developing fleets of large bombers equipped with more and more technologically advanced ordnance. When the war began, bombs were so small they could be carried in the aviator's pockets; by the end, both Germany and Britain had developed bombs weighing a ton. Both sides used and kept perfecting incendiary bombs. Both were preparing poison-gas bombs for use on cities.[13] The terrorizing effects of actual air attacks, amplified by this ever more menacing technology, now translated the images of aerial superweapons from science fiction into practical—and fateful—military theory.

As they first appeared in science fiction, conceptions of victory through air power were usually dramatized in relatively innocuous scenarios, especially when picturing war between white nations. The imaginary aerial superweapons typically incapacitate the enemy's military forces, often with minimal bloodletting, sometimes merely by displaying their powers. Victory through air power in these early works does not come from either a surprise massive air assault aimed at knocking out a nation with a single blow or a prolonged bombardment intended to reduce the enemy's cities to rubble. Such strategies, when attempted, fail (unless they are used on nonwhites). Deliberate attacks on cities or other civilian targets are almost never carried out by Americans, and when inflicted on the United States such barbarous actions typically goad the nation into developing precisely the new weapons that bring victory and peace.

The fiction expressed the range of conduct acceptable to the American public and considered permissible by the U.S. government. When single-engine German Taube monoplanes dropped a mere fifty tiny bombs on Paris during several months in late 1914—attacks whose legality was questioned by no European nation—President Wilson informed Berlin that these raids were tarnishing Germany's image among the American people. After the United States entered the war, Secretary of War Newton D. Baker sent instructions to the Air Service that U.S. forces were not to participate in any bombardment plan that "has as its objective promiscuous bombing upon industry, commerce or population."[14]

This remained the official U.S. policy through the 1920s and 1930s, even after the outbreak of World War II in Europe. President Hoover actually proposed to the Disarmament Conference of 1932 the total elimination of all airplanes designed for bombing. In September 1939, President Roosevelt pleaded with the belligerents to refrain from bombing civilian populations and unfortified cities, a plea that, at first, all vowed to heed.[15] Throughout the period between the two world wars, American fiction and also film—as we shall see—tolerated no notion of U.S. participation in such barbarous conduct.

Meanwhile, however, interwoven material forces—including the ineluctable pressures of industrial development, a deepening worldwide economic crisis, and resulting collisions among the major powers—were preparing the machinery for total war on an unprecedented scale. In the realm of ideas, one manifestation of things to come was the growing conception of the bombing airplane as a superweapon destined for a transcendent role in warfare.

Three men are credited with shaping the doctrine of "strategic" air power, a euphemism for the annihilation of cities by an "independent" air force as the way to victory in modern war: British Chief Air Marshall Sir Hugh Trenchard, Italian General Guilio Douhet, and American General Billy Mitchell. Questions about who influenced whom among these three are academic. In the wake of World War I, all three sought to overcome the technological and moral constraints that had limited the effectiveness of aerial bombardment. Each developed his theory within a particular national cultural context that both shaped and was influenced by his conception of total war from the air.

Historically, war has often been more or less limited to combat between opposing armed forces. But in various times and places there has also appeared total war, waged without restraint against an entire population. After the U.S. Civil War and the Franco-Prussian War had demonstrated the terrifying potential of industrialized, technological warfare, numerous efforts were made by the industrial states to formalize rules of war to govern combat among each other. The main goal was to exempt all or most of the civilian population, including of course the governments themselves, from the awesome killing potential of modern weaponry, leaving victory to be decided by a contest between armed forces.

Despite some serious breaches, World War I was settled by and large within this framework. German and British cities suffered only minimal damage from air attack. When its armed forces were defeated, Germany gave up.

Restrictions of combat to armed forces benefited not only civilian populations in wartime but also armies and navies during peacetime. These traditional services could lobby for funds as the primary forces for "security" and "defense." The advocates of "independent" and "strategic" air power, however, proposed soaring right over armies and navies to attack the enemy's major cities as the primary target. Thus they posed an explicit threat to the surface forces of their own nations, brazenly seeking to make them irrelevant to the outcome of any future war.

Among the three leading proponents of total air war, only British Air Marshall Trenchard escaped court-martial. Indeed, thanks in no small measure to Trenchard, Great Britain became the first nation to deploy an independent air force capable of sustained long-range massive bombardment of civilian population centers. Trenchard's relatively easy triumph may be attributed to several factors, including Britain's insularity and its experience of receiving and conducting air attacks during World War I. Nothing, however, influenced the role of the bomber in British military theory and practice more directly than its use in colonial conflict.

The rules of so-called civilized warfare tentatively adopted by the industrial nations of course did not apply to the "pacification" of colonial peoples, who posed no apparent threat to turn conflict with their colonizers into mutual suicide. The bombing of nonwhite civilian populations was routine. France referred to it openly as "colonial bombing" and developed an aircraft, the *Type Coloniale,* specifically for bombing and strafing what a proimperialist British author jocularly referred to as "any odd Syrians and Moroccans and Senegalese and Gaboonese and French Equatorial Africans and Indochinese who happened to object to French rule."[16] Under Trenchard's reign as chief of the Air Staff between 1918 and 1930, air power played an even more decisive role in the British Empire.

New ways of targeting the civilian population were developed expressly for "pacifying" the Pathan peoples of India. When systematic bombing of their villages proved ineffective, the Royal Air Force (RAF) began bombing their reservoirs, as described with approval in 1941:

Every Pathan village lives on the irrigation of the hillside. The food is grown on terraces, which are watered by the skies during the rains and from the communal tank during the dry season. When a reservoir is bombed it bursts and washes all the soil off the terraces and there is no more food for that village for that season. And that, I believe, has done more than anything else to produce the peace which has existed on the North-West Frontier for several years past.[17]

Air power proved so successful in maintaining colonial law and order that Trenchard's RAF actually became the dominant military arm in several countries. For the first time in history, an air officer assumed command of all British forces in a foreign country, when Air Marshall Sir John Salmond was sent to Iraq in 1922. Trenchard even initiated in parts of the empire a system of air rule called "Control without Occupation."[18]

When World War II erupted in Europe, Trenchard, now Chief Air Marshall, had no difficulty applying the theory and practice of air power within Britain's far-flung empire to the cities of Germany. It was the RAF that, despite Hitler's repeated threats of reprisal against England, launched the total air assault against civilian populations that was to distinguish this war from all previous human conflict.[19]

Of course Nazi and Italian Fascist planes had already bombarded the civilian populations of Madrid, Guernica, Barcelona, and other unfortified Spanish cities during the Spanish Civil War, and it was hardly humanitarianism that had led Hitler promptly to accept Roosevelt's plea to spare the cities of his European opponents. The Nazis did not want to expose their own cities to air attack, and were also still trying to present themselves as the apex of civilization. In fact, the Luftwaffe, unlike the RAF, had not been designed primarily to conduct and defend against strategic bombing of cities. Its main purpose was support of the ground army in the blitzkrieg strategy of rapid armored attack. In this, Nazi air theory deviated from the true Fascist model of air war promulgated in Italy by General Giulio Douhet.

Douhet, who began enunciating grand visions of the role of military aircraft as early as the first Italian bombing of Libya in 1911, developed the first systematic exposition of the theory of strategic air war. In 1916 his strident criticism of the Italian military resulted in a court-martial and one-year prison sentence, but by 1918 he had been rehabilitated and made head of the Central Aeronautical Bureau. Then in 1921 appeared the first edition of his major treatise, *Il dominio dell'aria* (*The Command of the Air*).

Mussolini, recognizing that this theory fully embodied the ruthless imperial militarism at the center of his Fascist creed, made Douhet his commissioner of aviation shortly after seizing power in 1922. Douhet, however, soon left government to spend full time embellishing and promulgating his doctrine. In 1927, the Instituto Nazional Fascista di Cultura published an enlarged version of *The Command of the Air,* and in 1930 appeared "La guerra del' 19—", Douhet's fantasy of the successful implementation of his theory in a future war between Germany and France. In each of these works, air war is glorified as the perfection of sudden, total terror aimed primarily at the civilian population and leading, if practiced without compromise, to swift, sure, and complete victory.

Douhet's theory is distinguished by its reliance on an "Independent Air Force" consisting mainly of immense fleets of giant, long-range, heavily armed "battleplanes" capable of neutralizing any opposing air force and then raining devastation on the enemy's cities without hindrance. These flying fortresses "will be able to achieve victory regardless of what happens on the surface,"[20] for their main target is the morale of the civilian population. Organized, systematic terrorism, relying on industrialism and technology, is the essence of Douhet's strategy.

Since the goal is "spreading terror and panic," "it is much more important" to destroy "a bakery" than "to strafe or bomb a trench" (126):

> In general, aerial offensives will be directed against such targets as peacetime industrial and commercial establishments; important buildings, private and public; transportation arteries and centers; and certain designated areas of civilian population as well. To destroy these targets three kinds of bombs are needed—explosive, incendiary, and poison gas—apportioned as the situation may require. The explosives will demolish the target, the incendiaries set fire to it, and the poison-gas bombs prevent fire fighters from extinguishing the fires. (20)

Because the effect on civilian morale is more significant than the material effects (57–58), "high explosives will play a minor role" relative to "incendiaries and poison gases," particularly in attacking "civilian objectives such as warehouses, factories, stores, food supplies, and population centers" (40–41). After describing scores of burning cities blanketed with poison gas and "panic-stricken people" fleeing to the countryside "to escape this terror from the air" (58), Douhet offers his assurance that this kind of warfare is actually "more humane" than earlier forms because victory will come so fast (61). In any event, only this total air war can possibly fulfill Italy's "imperial destiny" (141).

Italy never did make serious efforts to build Douhet's strategic air force. Of course its bombers could devastate at will the villages of preindustrial Ethiopia during its 1935 conquest and the weakly defended cities of Spain during the Civil War. But the mammoth bomber fleet envisioned by Douhet was beyond Italy's industrial potential.

Indeed, despite sporadic attempts by France and Germany to build strategic air fleets, among the European powers only Britain succeeded at all, partly because its insularity meant that it did not need large numbers of close-support tactical aircraft. The other nations soon discovered that their industrial base could not support both a tactical air force capable of effective support of surface forces and a major bomber fleet, with its enormous demands on production. Japan, though successful in bombing defenseless Chinese cities during the 1930s, faced a similar choice, and opted for a potent naval arm designed for its imperial offensives and fleets of fighters to protect the home islands. With the possible exception of Britain, only one nation was making serious progress toward the kind of invincible air fleet of long-range battleplanes imagined by Giulio Douhet, the only nation with the industrial potential to support total warfare on such a scale: the United States.

5

Billy Mitchell and the Romance of the Bomber

Billy Mitchell is the central figure in the triumph of air power in America. Combating both entrenched resistance in the army and navy and the powerful antimilitarist forces that before World War II were still a characteristic feature of American life, Mitchell more than any other person led the nation toward a quest for world supremacy through the airplane as a strategic weapon.

Trenchard's primary field of activity was practice, Douhet's was theory, but Mitchell's was culture. Trenchard built an air force capable of exerting British power at great distances against rebellious colonial subjects or hostile foreign powers. Douhet created a systematic theory of bomber fleets as an ultimate terror weapon. Mitchell's main role was within American culture itself. Through his agency, the cultural forces that had been leading toward a religion of the superweapon found their appropriate icons and rituals in the airplane, and their institutional base in America's industrial infrastructure.

Robert Fulton had articulated an American revolutionary ideology of the superweapon and engineered some advanced designs for it. Thomas Edison had personified in American culture the myth of the technological wizard who could create the necessary hardware, and had then embodied the union of this individual genius with the industrial organization capable of producing and deploying it. Billy Mitchell turned the affair with superweapons into an American romance.

The daring lone hero who defies the conservative, bureaucratic, corrupt establishment as coolly as he defies death—that was Billy Mitchell. The visionary apostle of the future—that was Billy Mitchell. The handsome, dashing image on the giant screen of the neighborhood movie theater—that too was Billy Mitchell, perhaps the first American political figure who successfully turned himself into a matinee idol.

Mitchell's early biography offers almost too pat a symbol for the conversion of America's nineteenth-century antimilitarist republicanism into its twentieth-century global imperialism. Young William's father, Senator John Mitchell of

Wisconsin, was one of the staunchest congressional opponents of the rising tide of militarism and imperialism that was leading toward war with Spain in 1898. On the eve of that war, he published the pamphlet *Against the Annexation of the Hawaiian Islands* and declared on the Senate floor, "No soldier should be mustered in for the purpose of shooting our ideas of liberty and justice into an alien people."[1] Yet one of the very first enlistees was his eighteen-year-old son, who left his studies at George Washington University to fight for three years in Cuba and the Philippines, first against Spain, then against the Cubans and Filipinos resisting U.S. colonization. From Senator Mitchell's halcyon nest somehow emerged a fledgling imperial hawk.

In these colonial wars, the U.S. army began its transformation from a force designed mainly for the conquest and suppression of the indigenous population at home into an arm of a global Manifest Destiny. The Confederates had been shocked by the foreshadowing of twentieth-century total war waged against them in Georgia by General William Tecumseh Sherman. But they experienced only a semblance of the devastation routinely meted out to Native Americans since the late seventeenth century. The warfare waged against the Cuban and Philippine nationalists, for whose ostensible benefit we had defeated Spain, was an export of the genocidal campaigns against the "savages" and "redskins" who had inhabited North America. General Jacob H. Smith, one of the leaders of U.S. counterinsurgency in the Philippines, expressed the guiding philosophy: "Neutrality must not be tolerated on the part of any native . . . if not an active friend he is an open enemy. . . . I want no prisoners! I want you to kill and burn! The more you kill and burn the better you will please me! I want the interior of Samar made a howling wilderness!"[2]

Young William Mitchell, then the youngest officer in the U.S. Army, was no passive observer in this fighting. Writing to his father from Cuba, he explained that the United States had to annex the island and fight the Cubans "until they dare fight no more," for they "are no more capable of self-government than a lot of ten-year-old children." In the Philippines, Mitchell remorselessly and resourcefully applied the principles of colonial warfare: burn, kill, and terrorize.[3]

In early 1917, amid the rising campaign for the country to join the war against Germany, the War Department sent an aeronautical observer to Europe. The choice was Major William Mitchell, now a career officer and one of the handful of U.S. servicemen with flight training. Fortuitously arriving in Paris only four days after the U.S. declaration of war on April 6, Mitchell soon was chosen to be the Aviation Officer of the American Expeditionary Force.

In the battle of Saint-Mihiel, Mitchell commanded the war's greatest concentration of air power—fifteen hundred Allied planes—and in the prolonged Meuse-Argonne campaign, he experimented with raids by massed bombers. But throughout this combat, he found himself chafing at the reins on U.S. air power, which was limited by matériel and policy to support of the ground forces. Flying over the static lines of the gigantic armies locked in grinding slaughter below, Mitchell, like many other airmen, began to conceive of the airplane as the ultimate weapon that would make ground combat peripheral to the outcome of war.

Hardly had the war ended when Mitchell, now a brigadier general, began to wage a campaign against new opponents: the leaders of the United States Army and Navy. With an increasingly messianic zeal, he threw himself into what became a lifelong struggle to create an independent air force with a transcendent military mission. His main target shortly became the surface fleet of the Navy, which he claimed airplanes could "put out of business in a very short time."[4]

In 1920, Mitchell took his fight to the public, authoring provocative articles in national magazines and issuing declarations accurately calculated to win newspaper headlines. Soon he was openly demanding that the Navy provide battleships for him to bomb so he could demonstrate that they were anachronisms. (Today's readers may have trouble comprehending the ferocity of interservice rivalry in the decades between the two World Wars; in those days, military spending was not sacrosanct, and each branch, rather than expecting almost every weapon requested, had to compete for limited funds.) By January 1921, resolutions had been introduced in the Senate and House ordering the Navy to give Mitchell the ships. Although during these early skirmishes Mitchell emphasized the defensive role of the bomber, he and many of his supporters, unlike Robert Fulton, knew that a weapon that could sink warships had capabilities for other kinds of destruction. The *New York Times* enthusiastically editorialized in favor of his demand so that Congress could "be made to understand the offensive power of aircraft on land and sea."[5]

When, in Feburary, Secretary of the Navy Josephus Daniels capitulated to the pressure and invited the Air Service to participate with the Navy in bombing tests on ships, Mitchell responded by revving up his public campaign in order to excite interest in the tests. A sympathetic biographer describes how he cultivated "friendly congressmen and newspaper publishers," who, together with "active and reserve Air Service officers throughout the United States and abroad," served as his public-relations "network":

> His staff prepared editorials, letters to editors, and articles for newspapers which would use them. Mitchell's own magazine article production increased markedly after he had won approval for the bombing tests. Many of the mass-audience publications of the day, including that of the newly formed and politically potent American Legion, carried articles by Mitchell.[6]

So far, Mitchell was merely using methods of propaganda campaigns like those that had helped sweep the nation into war with Spain in 1898 and Germany in 1917, combined with Edison's flair for capturing headlines about purported superweapons and industrialized warfare. But when the actual tests came, he showed his mastery of more advanced techniques for manipulating the media.

In mid-June, a great flotilla of Navy ships, carrying military brass, cabinet officers, senators, congressmen, delegations from foreign countries (including Japan), and reporters from around the world, gathered off the Virginia coast. A month of war games was about to commence, featuring aerial bombardment of assorted war vessels surrendered by Germany. Even during the first phases, conducted by the Navy's own planes, many eyes were on Mitchell's orbiting twin-seat De Havilland, which sported his personal insignia and trailed a long blue command pennant.

On June 21, the Navy fliers speedily dispatched the German submarine *U-117*, a feat dismissed as insignificant by the partisans of capital ships. Mitchell's first opportunity did not come until July 13, when his Martin bombers sank the German torpedo destroyer *G-102* in nineteen minutes. Next came the German light cruiser *Frankfurt,* which succumbed to heavy bombs after surviving most of a day of bombardment with lighter ordnance. All the players in this war game—as well as much of the world's press—now anticipated the grand climax, the final showdown between two superweapons: the battleship and the bomber.

The Navy confidently placed its bets on the German dreadnought, the *Ostfriesland,* whose twenty-seven-thousand-ton bulk was protected by four layers of advanced alloys and whose hull, an intricate honeycomb of independent watertight compartments, was considered unsinkable. Indeed, the *Ostfriesland* survived a day of attack from light and medium bombs dropped by Navy, Marine, and Army planes, and several direct hits the following morning by half-ton bombs specially prepared under Mitchell's direction. That afternoon, however, Mitchell's heavy bombers, in direct violation of the Navy's rules for the game, which called for intermittent damage inspection, pummeled the battleship with one-ton behemoths until the unmanned hulk rolled over and sank.

The Navy now argued that attacking an undefended stationary ship proved little. But Mitchell had a better argument: media. His goal was to hit the American public with immediate, nationwide images of the airplane's triumph over the warship. The audacity of this enterprise in 1921 was remarkable. There were no satellites to relay images, and no television; in fact, the first experimental radio broadcast station had begun operation only in November of 1920.

Back in 1919, Mitchell had given the young photographer George Goddard his own laboratory, where, with assistance from Eastman Kodak, Goddard developed high-resolution aerial photography. As soon as Mitchell had won the opportunity to bomb ships, he put Goddard in command of a key unit in his plan of attack on the Navy: a team of aerial photographers provided with eighteen airplanes and a dirigible. Mitchell's instructions to Goddard were unambiguous: "I want newsreels of those sinking ships in every theater in the country, just as soon as we can get 'em there." This necessitated more than mere picture taking. With his true flair for public-relations work, Mitchell explained to Goddard: "Most of all I need you to handle the newsreel and movie people. They're temperamental, and we've got to get all we can out of them."[7] Goddard also solved unprecedented logistical problems, flying the film first to Langley Field and thence to Bolling Field for pickup by the newsreel men who would take it to New York for development and national distribution. Each sinking, artfully photographed by relays of Goddard's planes, was screened the very next day in big-city theaters across the country.

This spectacular media coup implanted potent images of both Mitchell and the airplane in the public mind. Mitchell became an overnight national hero and a figure of American folklore as millions watched the deaths of great warships on newsreel screens. He was a prophet. The battleship was doomed. The airplane would rule the world.[8]

Mitchell's innovative command of American media did not stop there. On July 29, eight days after the sinking of the *Ostfriesland,* his heavy bombers flew in formation from Virginia to New York City, which they proceeded to "pulverize" in a simulated raid. Mitchell then landed on Long Island to conduct a press conference. His sensationalistic description of New York in ruins provided material for bannered newspaper articles the next day that read like passages from the pre–World War I novels of future war. Wrote the *New York Herald:*

> The sun rose today on a city whose tallest tower lay scattered in crumbled bits of stone. . . . Bridges did not exist. . . .
>
> The sun saw, when its light penetrated the ruins, hordes of people on foot, working their way slowly and painfully up the island. A few started with automobiles but the masses of stone buildings barricading the avenues soon halted their vehicles. Rich and poor alike, welded together in a real democracy of misery, headed northward. They carried babies, jewel cases, bits of furniture, bags, joints of meat and canned goods made into rough packs.
>
> Always they looked fearfully upward at the sky. . . . bodies lay like revelers overcome in grotesque attitudes. . . . The majority had died swiftly of poison gas.[9]

Mitchell next led his bombers in similar raids on Philadelphia, Wilmington, Baltimore, and the Naval Academy at Annapolis. The newspapers in each of the "bombed" cities vied to outdo each other in frightful reports on the hypothetical carnage.

Despite all the dreadful images of foreign air raids in newspaper fantasies and future-war novels, the only air weapons ever launched by another nation on the United States mainland were the ineffective Japanese balloon bombs that drifted over during World War II. In fact, the only two American cities ever bombed from the air were attacked by American fliers. The first occurred in May 1921, in the midst of Mitchell's publicity campaign prior to the tests, when a mob of ten thousand whites invaded the black section of Tulsa, Oklahoma, looting and burning as they advanced. Armed blacks defended their homes, but their resistance was overcome with the help of eight airplanes, some manned by police, that rained improvised dynamite bombs on neighborhoods that the ground forces had drenched with oil and gasoline. Most of the ghetto was burned to the ground, and between 150 and 200 black people, mostly women and children, along with fifty of the white invaders, lost their lives.[10] The second bombing of a U.S. city occurred in 1985, when an explosive device dropped by police ignited a conflagration in a black ghetto in Philadelphia.

Mitchell did not publicly advocate "strategic" bombing—the euphemism for bombing civilian populations of cities—until late 1924. Before that, even his most sensationalistic activities were to promote an air force designed for "defense" against hostile surface fleets or air armadas. Having demonstrated his mastery of the newsreel and the daily press, Mitchell next propagated his most radical philosophy of war in what was then perhaps the most formative medium of American national popular culture: the "general-interest" weekly magazine. Three of these magazines—the *Saturday Evening Post, Liberty,* and *Collier's*—maintained weekly readerships of over a million from the mid-1920s through 1941. These were the chosen vessels for the new gospel.

In a series of five articles published in the *Saturday Evening Post* between December 1924 and March 1925, Mitchell began openly challenging the prevailing consensus throughout the government, the armed services, and the American people that the United States was not the kind of society that would intentionally bomb the civilian populations of other civilized nations.[11] Indeed, the great metropolises of industrial nations now overtly became the primary targets of Mitchell's strategy.

His reasoning would later become the rationale for the "strategic" bombing of World War II. As he explained to the readers of the *Saturday Evening Post,* these are the "cities that are producing great quantities of war munitions." And since the people living in these cities are essential to the productive activities, they become legitimate military targets and terrorizing them becomes a legitimate military strategy. Mitchell therefore proposed enormous saturation raids using high explosives and tear gas.

Ignoring any ethical barrier to indiscriminate attacks, Mitchell also proposed "aerial torpedoes, . . . airplanes kept on their course by gyroscopic instruments and wireless telegraphy, with no pilots on board"; these "can be directed for more than 100 miles in a sufficiently accurate way to hit great cities."[12] When a conception like Mitchell's robot planes was actually implemented in the 1944 War Weary Bomber project (see page 106), in retaliation against the Nazi V-1 and V-2 missiles, Chief of Staff William D. Leahy described such indiscriminate attacks as an "inhumane and barbarous type of warfare with which the United States should not be associated."[13] One of Mitchell's main goals, however, was total terror, "So that in the future the mere threat of bombing a town by an air force will cause it to be evacuated and all work in munitions and supply factories to be stopped."[14] The unforeseen but logical extension of his "aerial torpedoes" would be nuclear-armed missiles aimed at cities.

Mitchell's paramount belief was that an air force could win wars all by itself. The nation that achieved air supremacy would be able to annihilate the opponent's air force and devastate "all the places capable of building aircraft." Then the air force would bring "victory, unaided, to the side which was able to control the air" since on the other side "all the important cities would be laid waste, the railroads and bridges destroyed, roadways constantly bombed and torn up so as to prevent automobile transportation, and all seaports demolished."[15]

The *Saturday Evening Post* series, with its open assault on the military and political leaders opposed to Mitchell's aims, led directly to his demotion, in March 1925, from his temporary rank of brigadier general to his permanent grade of colonel, and his reassignment to an obscure post in Texas. In August, Mitchell published *Winged Defense;* incorporating the *Saturday Evening Post* articles, the book sported on its inside covers an assortment of newspaper cartoons lampooning men opposed to his views, such as Secretary of War John Weeks. In September, Mitchell's attacks on the highest military authorities grew yet more vociferous. He was now clearly provoking what he intended to be his biggest media spectacular yet—his own court martial. Ordered by President Coolidge himself, the two-month court martial had two outcomes, neither unanticipated: a guilty verdict and the establishment of an American legend.

The legend of Billy Mitchell glorifies a farsighted air-power crusader crucified by entrenched conservative forces. It was proclaimed by the media in 1921–26, embellished in later books, and mythologized by Hollywood in the 1955 movie *The Court Martial of Billy Mitchell,* starring Gary Cooper. The legend tends to obscure the issues. Many of Mitchell's opponents were just as dedicated to air power. The airplane was rapidly becoming a principal tool of U.S. counterinsurgency, as demonstrated by the U.S. bombing of Ocotal and other Nicaraguan towns in 1927. (Frank Capra's 1929 film *Flight* celebrated the U.S. Marine pilots who helped defeat the Nicaraguan revolution.) And at the very time Mitchell was charging the government and Navy with neglecting air power, the United States was becoming the world's leader in naval aviation with the construction of the aircraft carriers *Saratoga* and *Lexington,* both completed in 1927. Mitchell, however, scoffed at aircraft carriers as "a delusion and a snare" that would be "quickly swept off the sea in case of war against a great air power."[16] Mitchell championed not all forms of air power, just one destined to transform human history.

Mitchell's opponents saw air power as only one component of the military forces combating the military forces of an enemy (at least when that enemy was another "civilized" nation). Indeed, Mitchell himself had started out with this view. But as his vision became increasingly messianic, he saw the airplane as a superweapon with its own mission. The key words in his vocabulary, like those of his counterparts in Italy and England, were *independent* and *strategic.* As he explained in an article published in *Liberty* the month after his court-martial, the strategic mission of the independent air force was to defeat the enemy single-handedly by flying "straight over the armies to the vital centers," "paralyzing them and making it impossible for the population to carry on war or live in peace."[17] In "When the Air Raiders Come," published three months later in *Collier's,* Mitchell became more lurid in painting the features of the apocalypse awaiting the losing side in the next war. "One side or the other will be completely victorious," he predicted. If the United States does not strike first at "the vital centers" of the enemy, each major city of America would quickly be reduced to "a heap of dead and smoldering ashes" in which "the fortunate ones" would be those who died at once.[18]

Less than twenty years after Mitchell's court-martial, his was to become the dominant strategy for waging war. It would then prove to be a formula for a new world in which the human species lives in perpetual danger of annihilation.

In his final years, Mitchell focused this strategy more and more on one particular people: the Japanese. His writings about war with Japan have been called prophetic, but they may have been a self-fulfilling prophecy.

Mitchell had been writing secret reports about eventual war with Japan as early as 1911.[19] In the early 1930s, he added his public voice to the growing concern about Japanese imperial conquests on the mainland of Asia, using the opportunity to promote strategic air power on a grand scale. In articles published in 1932, he provided a precise picture of what he advocated for Japan:

> What Japan is in deadly fear of is our air force. . . . Japan offers an ideal target for air operations. The islands are not only very unprolific but there is little fresh water on them. Most of the towns are found along the water courses, usually hemmed in by mountains on each side. These towns, built largely of wood and paper, form the greatest aerial targets the world has ever seen. . . . Incendiary projectiles would burn the cities to the ground in short order.[20]

In a few years, these images would help propel the leaders of both Japan and the United States into fatal acts.

Although Mitchell himself was no longer the popular idol he had been a decade earlier, his triumph was already beginning to materialize. In 1933, the year after his enthusiastic description of how to burn the cities of Japan to the ground, the Boeing Company won the Army Air Corps competition for the design of a long-range, multi-engine bomber. Its XB-17 was tested in 1935, and the Air Corps placed its first orders. Mass production got under way in 1938, two years after Mitchell's death.[21] A vast fleet of Boeing B-17 Flying Fortresses soon would be coming off the assembly lines, capable of carrying out wholesale aerial bombardment. One of their first proposed missions was the incendiary holocaust that Mitchell had advocated.

When the cities of Japan were indeed incinerated by American strategic bombers, the rationale then advanced and still largely accepted was that this was in retaliation for the December 1941 Japanese attack on Pearl Harbor. But plans for the U.S. firebombing of Japanese cities came prior to Pearl Harbor, and even may have helped precipitate the Japanese first strike.

In late 1940, retired Air Corps General Claire Chennault, then advisor to the Chinese air force, suggested a scheme for basing hundreds of U.S. bombers in China so that they could "burn out the industrial heart of the Empire with fire-bomb attacks on the teeming bamboo ant heaps" of Japanese cities. President Roosevelt was "simply delighted" by this proposal. When it proved not feasible, other plans were developed. In February 1941, Chief of Naval Operations Admiral Harold Stark, apparently acting under orders from the White House, asked Admiral Husband Kimmel in Hawaii to prepare for "making aircraft raids on the inflammable Japanese cities (ostensibly on military objectives)." As the B-17s became operational in sizable numbers, they seemed to offer the practical means to carry out this strategy. In mid-1941, the first contingents of B-17s were deployed to the American Pacific colonies of Hawaii and the Philippines. Incendiary bombs were then rushed to the bomber bases. The plan was to have 340 of the bombers ready to bomb Japan by February or March of 1942.[22]

Ostensibly a closely guarded secret, the U.S. strategy was widely discussed in the press. For example, in October 1941, *United States News* ran a two-page map showing the various routes and flying times for U.S. bomber raids on cities that constituted "the head and the heart of industrial Japan," particularly Tokyo, that "city of rice-paper and wood houses."[23]

Confidence in the aerial superweapon seemed boundless. If the Japanese were rational, the B-17s would be a supreme deterrent to their imperial ambitions. If not, the Flying Fortresses would speedily finish them off. On November 15, Army Chief of Staff General George C. Marshall flatly informed an extraordinary secret gathering of press bureau chiefs and senior correspondents: "We are preparing an offensive war against Japan. . . ." To deter war, Marshall confided, Japanese officials were going to be officially informed of the U.S. bomber buildup and intention to bomb their cities. If this deterrence does not work, Marshall explained, we will wage war "mercilessly": "Flying fortresses will be dispatched immediately to set the paper cities of Japan on fire."[24]

Billy Mitchell's vision was now beginning to shape history, in Tokyo as well as in Washington. The Japanese had been paying close attention to Mitchell's spectacular advertisements for air power ever since they had been invited guests at his 1921 demonstration of how to sink ships. As the Japanese learned about the movement of the Flying Fortresses to bases from which they might be able to bomb the cities of Japan, they remembered what Mitchell had advocated and what now seemed imminent. For example, Japanese naval officer Kinoaki Matsuo seemed to be echoing Mitchell:

> Needless to say, Uncle Sam is aiming at none other than Japan. . . . As a matter of fact it is feared that the United States air force may carry out a record-breaking air raid upon Japan. . . .
>
> Once the enemy succeeds in making an air raid on a large scale upon the Japanese cities, which are mostly of wooden houses, it is feared that all the cities will be reduced to ashes.[25]

When Japan struck first in December 1941—showing they had learned well how to sink ships and wipe out planes from the air—one of their principal objects was to neutralize this threat, and they swiftly succeeded in knocking out the B-17 bases in Hawaii and the Philippines. It is ironic that Billy Mitchell and others who publicly trumpeted plans for an all-out terror bombardment of Japan would be retrospectively credited with predicting the war they helped trigger.

Of course rampaging Japanese imperialism was already on a collision course with the United States, for reasons Mitchell had outlined in the 1932 article in which he proposed firebombing Japan's cities. As "the only great white power whose shores are washed by the Pacific," the United States, according to Mitchell, blocked the path of Japanese expansion with its own colonial possessions. Therefore it was essential that the Philippines and other U.S. colonies in the Pacific not be given independence, for this merely would invite Japanese encroachment. Instead, he argued for giving full play to the fact that "America is one of the most warlike nations on earth."[26]

Mitchell's ultimate goal for air power was even more grandly imperialist than the Greater East Asia Co-Prosperity Sphere, Japan's rationalization for the conquest of Asia:

> Should a nation . . . attain complete control of the air, it could more nearly master the earth than has ever been the case in the past.

Just as power can be exerted through the air, so can good be done, because there is no place on the earth's surface that air power cannot reach and carry with it the elements of civilization and good that comes [*sic*] from rapid communications.[27]

Mitchell's colonial outlook, shaped by his experience in Cuba and the Philippines, eventually expanded into his vision of a world mastered by one product of advanced technology and industry. His life personifies the evolution of American military goals for almost a century: from pacification, to colonialism, to fighting wars to end all war, to victory through air power, and then to the motto of the Strategic Air Command: "Peace Is Our Profession." The rationale for genocidal warfare against the native peoples of North and Central America, Cuba, Puerto Rico, and the Philippines in the name of "civilization"—literally "citifying"—evolved into the rationale for warfare aimed at the total destruction of the cities of particular industrialized nations and, ultimately, of civilization itself.

6

The Triumph of the Bombers

On September 1, 1939, the day the Nazi invasion of Poland started World War II in Europe, President Roosevelt communicated an "urgent appeal to every Government which may be engaged in hostilities publicly to affirm its determination that its armed forces shall in no event, and under no circumstances, undertake the bombardment from the air of civilian populations or of unfortified cities." Alluding to the air raids on Ethiopia, China, and Spain, he declared:

> The ruthless bombing from the air of civilians in unfortified centers of population during the course of the hostilities which have raged in various quarters of the earth during the past few years, which has resulted in the maiming and in the death of thousands of defenseless men, women and children, has sickened the hearts of every civilized man and woman, and has profoundly shocked the conscience of humanity.

He warned that "hundreds of thousands of innocent human beings" would be victimized if the warring nations were to sink to "this form of inhuman barbarism."[1]

The president seemed to express the collective will of the nation. Secretary of State Cordell Hull had claimed to be "speaking for the whole American people" when he denounced the Fascist bombing of Barcelona the previous year, declaring, "No theory of war can justify such conduct."[2] That same year, the Senate had voted its condemnation of the "inhuman bombing of civilian populations," and the government had proclaimed a "moral embargo" on shipments of airplane manufacturing materials to any nation engaged in the bombing of civilians.[3]

Rejection of the bombing of cities was virtually universal in American fiction and film. Before the nation entered World War II, American movies, to the best of my knowledge, contained not a single image of the United States engaging in such conduct. This is true even in science-fiction films calculated to arouse hostility to the Yellow Peril, from D. W. Griffith's 1916 film, *The Flying Torpedo,* in which radio-controlled flying weapons are used to repel an invasion of "yellow men from the East," to the 1938 serial *Fighting Devil Dogs*, in which U.S. Marines defending democracy in a Far Eastern protectorate are subjected to artificial thunderbolts from an evil Asian dictator.

Only a handful of the most bizarre future-war novels and stories project scenes of American bombing of cities. Even among these few, perhaps only London's "The Unparalleled Invasion" condones unrestrained bombing of the civilian population. Take W. D. Gann's outlandish 1927 novel, *The Tunnel Thru the Air; or, Looking Back from 1940*. Gann prophesies wholesale bombing of American cities by Japan and its allies Spain, Mexico, Britain, Germany, Italy, Austria, and Russia, followed by the invention of American superstealth bombers able to attack without being seen or heard. Yet the U.S. retaliation is limited to destroying buildings and neutralizing the civilian population with sleeping gas.

Few novels or films even mention the possibility of the United States devastating cities from the air. One of the rare exceptions is Hector Bywater's *The Great Pacific War: A History of the American-Japanese Campaign of 1931–1933*, a popular 1925 novel that chronicles in detail the military history of an imagined conflict with Japan. The seven-volume official history, *The Army Air Forces in World War II*, in 1953 cited a passage from this novel as "an uncannily shrewd forecast" of the actual course of the war.[4] The passage quoted from Bywater is a forecast that "when the Americans had secured a base within striking distance," "the war would then be carried to the coasts of Japan proper" by "the gigantic fleet of aircraft which was building for the express purpose of laying waste to Tokyo and other great Japanese cities."

In the novel, however, the passage is offered not as an accurate prediction but as a preposterous piece of disinformation planted in a newspaper by U.S. intelligence in order to mislead the Japanese authorities. The only bombs dropped on Japanese cities during Bywater's imagined war are filled with leaflets advocating surrender. The novelist could assume that his American readers in 1925 would recognize that of course the United States would not engage in such atrocities as laying waste whole cities. But by 1953, the authors of the official history of the U.S. air war could misinterpret the novel as stating precisely the opposite because they accepted the devastation of cities by bombers as normal behavior. Their misreading suggests just how thoroughly American moral consciousness about warfare had been transformed.

Bywater, an ardent advocate of naval power, aims other parts of the novel directly at the partisans of independent air power.[5] Indeed, it sometimes seem that they, rather than the Japanese, are his antagonists. When American planes attempt to bomb a Japanese fleet, they are annihilated by antiaircraft fire and interceptors. Bywater alludes directly to the great propaganda triumph carried off in 1921 by General Billy Mitchell's successful air attack on undefended battleships: "Less than half the American machines broke through and made for the Japanese fleet, which received them with a hurricane of fire. How different now were the circumstances from those which aviation enthusiasts, deceived by artificial peace tests against helpless targets, had pictured!" (280)

Bywater's 1925 broadside against strategic air power appeared at the peak of Billy Mitchell's public campaign, conducted in the *Saturday Evening Post*,

Liberty, Collier's, the Hearst newspapers, and other popular forums, to sell the American people on air power as an invincible weapon, and massive bombing of cities as its most efficient use. Mitchell's 1925 court-martial, and the gradual fading of apparent interest in his media antics, seemed to suggest that he had failed. But forces that he had helped set in motion were inexorably, albeit quietly, working to insure the triumph of the strategic bomber in American culture and history.

In the boom times of the 1920s, American industry had no need for huge military contracts. Indeed, there was no major component of U.S. industrial capacity not in use. With the onset of the Depression, however, rearmament became useful not only as a political strategy designed to augment U.S. power amid growing global conflict but also as part of an economic strategy aimed at revitalizing American industry. So the 1935 award to the Boeing Company of its first production orders for what was then still called the XB-17 presaged a new feature of American political economy, a dependence on what would later be called the "aerospace" industry. And although strategic bombing was still anathema to both official U.S. policy and the culture, the material means for conducting it were being created in the form of the B-17 Flying Fortress.

The senators from Washington State, known even then as "the gentlemen from Boeing," were enthusiastic advocates of expanding the bomber fleet. While the Nazi and Italian Fascist bombings in Spain, according to President Roosevelt, "sickened the hearts of every civilized man and woman" and "profoundly shocked the conscience of humanity," Senator Lewis Schwellen-bach of Washington argued on the Senate floor that they actually proved the bomber's effectiveness.[6] In 1939, while the government and the media were so eloquently denouncing the bombing of cities, Boeing's assembly lines had already begun mass production of B-17s.

At first, the belligerents in Europe all accepted President Roosevelt's appeal to refrain from bombing civilian populations and unfortified cities. But even with the tight restriction of air power to battle zones, restraints were almost certain to be compromised. When the Polish army resisted the Nazi advance on Warsaw, for example, the city thereby became an allowable target for German aerial assault; in mid-September, the Luftwaffe eagerly seized the opportunity to demonstrate the havoc its planes could wreak on a major urban center. The British Air Ministry then secretly decided it would no longer be bound by its acceptance of Roosevelt's appeal. Those terms were still honored in practice, however, until May 1940, when two accidents led to the predictable breaching of any sustainable restraints.[7] On May 10, the German town of Freiburg was mistakenly bombed by German bombers that had lost their bearings on the way to military targets in France. The Nazi government publicly blamed the Allies, who in turn assumed the incident had been engineered to justify raids on British and French cities. On May 14, while Rotterdam's defenders were negotiating their surrender, the Germans called off their scheduled air attack, but fifty German bombers missed the cancellation message and carried out a devastating raid.[8]

Already familiar with the spectacle of Warsaw in flames after Nazi bombardment, Americans now heard of the gutting of Rotterdam after the city had already surrendered. So when the war's first intentional bombing of cities outside a ground combat zone was initiated the next day by the British Bomber Command, it appeared to most Americans to be nothing more than retaliation.

The British air onslaught against the cities of Germany was to go on for five years. Its only limits were the availability of crews and bombers—many of which were furnished by the United States—and the defenses they encountered. Their strategy was based squarely on the Trenchard-Douhet-Mitchell doctrine of neutralizing urban centers and industrial production by wholesale devastation and terror. Their operations were from the outset conducted primarily at night, decreasing the vulnerability of the bombers, which, in any case, were none too accurate. Soon the center of a city became their customary point for defining the target of their "area" bombardment.

Hitler's warnings about retaliation went unheeded by the RAF, but caused great consternation among the French. Germany lacked the long-range bomber force that had put its own cities within reach of the British Bomber Command, so it was not Britain but France that was vulnerable. Even after the fall of France, when German medium-range bombers based in Norway, the Low Countries, and northeastern France were well within range, Germany still hesitated before launching an aerial offensive against Britain. When the Battle of Britain finally began in August 1940, nearly two months later, the German raiders at first restricted themselves to daylight bombing of specific military targets deemed relevant to the planned invasion of Britain.

The intentional bombing campaign against British cities—known as "the Blitz"—did not begin until mid-September. So for the first year of the war, the Nazis had essentially stayed within Roosevelt's guidelines. Of course, given their previous bombing of undefended Spanish cities, they were no doubt restrained by expediency and fear of retaliation, not moral qualms. Several factors induced them to switch to blatant attacks on British cities. The very effectiveness of British defense broke up attempts at precision bombing of military targets, causing bombs to fall at random and pushing the Lutfwaffe away from daylight raids toward night attacks, when precise bombing was even less feasible. Given the difficulty of hitting small, well-defended, often camouflaged military targets, large cities became more and more tempting. The escalation of British attacks on German cities encouraged Hitler to shift from raids with military objectives to pure reprisal. Some analysts have argued that in fact Churchill's main objective in bombing Berlin was to bait Hitler into shifting the Luftwaffe attack onto London, away from the embattled RAF fighter bases.[9] Targeting cities also allowed fuller use of incendiary bombs, which were relatively unsuitable for reinforced or isolated military sites but proved capable of igniting large-scale conflagrations in crowded urban areas.

By the spring of 1941, the German air offensive against Britain had been decisively defeated, frustrating the planned invasion, and most German bombers were being shifted to the east to prepare for the June attack on the Soviet Union. Together, the Battle of Britain, the Blitz, the sporadic air raids during the ensuing years, and the final desperation attacks with V-1 and V-2 missiles killed approximately sixty thousand Britons. Before long, there would be instances of

Anglo-American air raids in which about that many Germans would be killed in a single city in a single night.

One of the most remarkable transformations in American cultural history took place between 1938 and 1942. From almost universal condemnation of the bombing of cities, American society from top to bottom shifted to almost universal approval—when America and its allies were doing the bombing. The moral outrage against the Fascists' use of airplanes on civilian populations transmuted into a craving to use airplanes on the enemy's civilian population. Thirst for vengeance went far beyond an eye for an eye. For example, even in their surprise attacks on Pearl Harbor and the Philippines, the Japanese, who had shown no moral qualms about attacking the cities of China, prudently avoided bombing civilians. Nevertheless, calls for retaliation clamored not for sinking the Japanese fleet or destroying Japanese military airfields but for bombing and burning Japanese cities.[10]

Yet throughout the war in Europe and until the last few months of the war against Japan, the United States remained officially committed to a policy of bombing only military targets, defined to include industrial sites involved with military production or transportation. Such "precision" bombing was to be conducted by daylight, when clouds did not obscure the targets.

In the European theater, this doctrine facilitated a convenient partnership in the great Anglo-American Combined Bombing Offensive against the Axis powers: the British would continue to conduct their nighttime "area" attacks against whole cities, while the Americans would attack well-defined military and industrial targets during the day. These roles were maintained through the end of 1943. While the growing force of B-17 Flying Fortresses struck deeper and deeper into vulnerable parts of the Nazi war machine, the RAF stepped up its systematic obliteration of cities, with an official goal of destroying the "morale" of the population.

Against Japan, the stated commitment to precision bombing was a moot point until mid-1944, when B-29 Superfortresses attained bases within range of the home islands. Although the celebrated April 1942 raid on Tokyo by sixteen carrier-borne medium bombers led by Colonel Jimmy Doolittle ostensibly aimed at military targets, it was hailed in America as exactly what it was intended to be: a display of American capacity for revenge and a direct attack on the morale of the Japanese people. In fact, most of the bombs from these B-25 Mitchells fell into thickly populated districts of Tokyo, Kobe, Yokohama, and Nagoya.[11]

Even in the earliest aerial campaigns in Europe, revealing gaps yawned between the theory and the practice of American bombing. No matter how narrowly intended for military targets, the inherent inaccuracy of high-altitude bombardment, especially from large formations of planes under attack by antiaircraft batteries and enemy fighters, guaranteed that most bombs would be strewn randomly, resulting in many civilian casualties. This was true even for the first major assaults, which, because they targeted military sites in occupied France, demanded special care for accuracy. Indeed, raids on such venerable French cities as Rouen and Lille killed and maimed numerous civilians, thereby undermining cooperation with the French Resistance.[12] Concentrated attacks on the virtually impregnable Nazi submarine pens actually wiped out the French port cities around them without penetrating the targets. Commenting on the annihilation of Saint-Nazaire and Lorient, German Grand Admiral Doenitz put it this way in his official report: "No dog or cat is left in these towns. Nothing but the submarine shelters remain."[13]

Coordination with the British terror bombing of major German cities also tended to blur moral distinctions. As early as March 1942, the RAF had developed massive firebombing techniques that were used to incinerate the historic town of Lübeck and the Baltic port of Rostok. In May, a thousand-plane raid devastated Cologne. On three nights between July 24 and July 30, almost eight hundred British heavy bombers saturated Hamburg with a rain of high explosives and incendiaries, generating unextinguishable firestorms and killing at least forty-eight thousand people. In the daytime, while the fires were still raging, waves of American B-17s aimed their "precision" attacks at a submarine yard and aircraft engine factory obscured by a blanket of smoke.[14] The U.S. bombers had by now become de facto collaborators in terror bombing of civilian populations.

By November 1943, the Combined Bombing Offensive virtually annihilated at least nineteen German cities and towns.[15] Yet, despite the inexorable buildup of the bombing fleets, Nazi air defenses continued to stiffen. At the close of 1943, a virtual stalemate had been reached in the battle for control of the German skies. Then, in early 1944, the offensive against Germany was interrupted as the bombers shifted their objectives to occupied France in preparation for the June 1 landing at Normandy. The invasion stripped away the Nazi early-warning system, and the battle for air supremacy over the western battlefields ravaged the Luftwaffe. In July the strategic bombing of Germany could resume with a vengeance.

From then until Germany's defeat the following May, vast armadas of British and American bombers, now accompanied by swarms of long-range escort fighters, pulverized the German heartland by night and day. As German air defense and the ability to retaliate weakened, so did U.S. restraints against intentional attacks on the civilian population. In late 1944, the Army Air Forces initiated the "War Weary Bomber" project, in which explosive-laden heavy bombers were aimed at cities and abandoned by their crews many miles from their targets; this was halted mainly because of British fears of German retaliation. In February 1945, operation CLARION dispatched thousands of bombers and fighters to roam above the German countryside attacking small towns and villages. Formations sometimes including a thousand "heavies" routinely bombed major cities through cloud cover, guided only by unreliable radar. One Eighth Air Force raid on Berlin, according to conservative estimates, killed at least twenty-five thousand people. By the spring of 1945, fifty of Germany's principal cities were reduced to rubble and ashes.[16]

In January 1945, as the Soviet army launched the offensive that drove the Nazi forces back through eastern Europe and captured Berlin, the Anglo-American Combined Bombing Offensive began to concentrate on annihilating cities and industrial facilities in the path of the Soviet advance. Whether these raids on east Germany were intended primarily to assist the Soviets or to guarantee that they would capture nothing but ruins remains a matter of debate. Whatever their motivation, the attacks dispelled any lingering ethical illusions about the doctrine of victory through air power.

The historic city of Dresden, which contained no important military targets, had so far suffered no major raid. Attacks on Dresden were considered so unlikely that it lacked the extensive air-raid shelters common in other German cities, and was used to house numerous Allied prisoners of war. Tens of thousands of refugees poured into Dresden as Nazi defenses collapsed before the Soviet advance.

On the night of February 13 came the first of several catastrophic air raids. Hundreds of RAF heavy bombers filled the skies with firebombs. A vast fire storm, fed by raging winds sucked in by its own updrafts, swept across the city. Daylight brought hundreds of American heavy bombers, using radar to bomb through the clouds of smoke. As the survivors, including refugees and Allied prisoners of war, escaped from the flaming and collapsing buildings to bare ground, they were strafed at low levels by American fighters. Most of those killed were so consumed by the fire or so inaccessible amid the devastation and chaos that no accurate casualty toll has ever been made. Some estimates suggest at least one hundred thousand civilian deaths from the first day's attacks alone.[17] The number killed probably exceeded all civilian deaths in England for the entire war and rivaled the carnage in Hiroshima a few months later.

A month before the Anglo-American destruction of Dresden, the United States had already begun the first stage of the systematic cremation of the cities of Japan. Billy Mitchell's 1932 dream of American strategic bombers raining incendiaries on "the greatest aerial targets the world has ever seen"—Japanese cities "built largely of wood and paper"—was about to come true. His prediction that such attacks "would burn the cities to the ground in short order" proved accurate.

By 1943, Mitchell's vision of airpower had become dominant in American culture. Public enthusiasm for bombing German cities outran the capabilities of the air force, and the frenzy to burn Japan now matched the earlier genocidal fantasies of the Yellow Peril future-war fiction. Article after article echoed Mitchell in emphasizing the flammability of Japanese cities.[18] Typically, they justified mass attacks on Japanese civilians as a means to save American lives. A piece in the January 1943 *Harper's* entitled "Japan's Nightmare: A Reminder to Our High Command" put it in terms that would become even more familiar in August 1945: "It seems brutal to be talking about burning homes. But we are engaged in a life and death struggle for national survival, and we are therefore justified in taking any action which will save the lives of American soldiers and sailors."[19] A professor of geology at Colgate University proposed in the January 1944 *Popular Science* that we wipe out not just the Japanese people but also all the islands of Japan by using high-explosive bombs to initiate volcanic eruptions. After the firebombing of Japan's cities began, even more voices clamored for the extermination of the Japanese. For example, in March 1945, the U.S. Marine monthly *Leatherneck* acknowledged that the "flame throwers, mortars, grenades and bayonets'' of the Marines could not complete "the gigantic task of extermination'' of the Japanese—drawn as a hideous buck-toothed louse, "Louseous Japanicas'': ". . . the origin of the plague, the breeding ground around the Tokyo area, must be completely annihilated.'' The following month, the chairman of the War Manpower Commission, Paul V. McNutt, announced that he favored "the extermination of the Japanese people in toto.''[20]

The most effective propagandist for implementing Mitchell's total air war on Japan was his long-time associate, Russian émigré Alexander de Seversky, then president of Republic Aviation Corporation. In 1942, de Seversky gathered his articles from *American Mercury*, *Atlantic Monthly*, *American Magazine*, *Flying and Popular Aviation*, *Look*, *Coronet*, *Reader's Digest*, and *Town and Country* into his exceptionally influential book *Victory through Air Power*, which sold hundreds of thousands of copies through the Book-of-the-Month Club alone. In the first chapter, he uses the rhetorical strategy that Mitchell had adopted from future-war fiction, painting a lurid picture of what enemy bombers will do to America's cities if America doesn't do it to theirs first. The airplane has brought the age of "total war,'' and "the very ease with which a machine-age country can be blasted into chaos from on high is an invitation to the war of annihilation.'' In this war of "total destruction'' and "extermination,'' the enemy's goal is to "destroy our civilization.'' Our goal then must be "to eliminate rather than take over Japan,'' and so we must base our strategy on "a war of elimination.''[21]

Walt Disney was so impressed by de Seversky's book that he decided to turn it into an animated feature film. His 1943 *Victory through Air Power* opens with an old newsreel clip of Billy Mitchell enunciating his doctrine of the bomber. Disney's animation turned out to be an ideal medium for this message. Gigantic American bombers blacken the sky as omnipotent machines, then magically transmute into an American eagle that claws the Japanese octopus to death. In what James Agee's review called "the gay dreams of holocaust at the end,'' Japan is bombed and burned into ruins with no human image of suffering, terror, or death on the screen. As the audience beholds the smoldering remains of this make-believe nation, it hears the swelling strains of "America the Beautiful.'' Then across the screen is emblazoned "VICTORY THROUGH AIR POWER.''[22]

Audiences had to be a bit patient, however: the Mitchell-de Seversky-Disney holocaust for Japan was not yet ready to begin. Since Anglo-American strategy in World War II gave first priority to the European theater, no significant strategic bomber force was allocated to Asia until mid-1944. As a result, the U.S. heavy bomber fleet in the Asian theater was composed mainly of the new B-29 Superfortresses, with greater range, heavier defenses, and larger bomb capacity than the B-17s and B-24s committed to the European bombing offensive. The first operational force of B-29s, XX Bomber Command, was based in India and carried out most of its raids from advanced staging bases in areas of China beyond Japanese occupation.

Though at first still officially bound to the doctrine of precision bombing, the B-29s in Asia from the start were guided by ethical standards looser than those technically applying to the U.S. bombardment of Europe. Before launching any attack on Japan itself, XX Bomber Command needed what the official history refers to as a "trial run," "the New Haven tryout before the Broadway opening."²³ The city chosen for this debut was Bangkok, which on June 4, 1944, was bombed by radar from high altitude, and then subjected to repeated raids for months. The first raid on Japan came eleven days after the trial run on Bangkok; ostensibly aimed at the iron and steel works in Yawata, it was a night attack that strewed bombs promiscuously around the area. Sporadic attacks on Japan, rendered fairly ineffective by problems of range and logistics, gradually gave way to intensive bombardment of cities in occupied China and Southeast Asia.

Whereas the Anglo-American raids on occupied cities in western Europe had at least made some effort to avoid the innocent and presumably friendly civilian population, no such consideration was given to the Asian peoples under Japanese occupation. For example, in December General Curtis LeMay experimented with the massive firebombing tactics that he would shortly use on Japan by cremating the Chinese city of Hankow, which was left in flames for three days.²⁴ General Claire Chennault assessed the raid:

> The December 18 attack of the Superforts was the first mass fire-bomb raid they attempted. LeMay was thoroughly impressed by the results of this weapon against an Asiatic city. When he moved on to command the entire B-29 attack on Japan from the Marianas, LeMay switched from high-altitude daylight attacks with high explosives to the devastating mass fire-bomb night raids that burned the guts out of Japan. . . .²⁵

U.S. strategic bombers carried out mass attacks on such Chinese cities as Anshan, Canton, Mukden, and Nanking, as well as repeated raids on urban sites throughout Southeast Asia, including Rangoon, Singapore, and Kuala Lumpur. Occupied French Indochina came under especially severe bombardment, with raids on Phnom Penh and many of the Vietnamese towns and cities revisited by U.S. bombers two decades later. Radar bombing of Saigon by B-29s took place in January and February 1945, with a heavy assault the same week as the Dresden attack. Between April 19 and June 12, a fleet of B-24s attempted a complete "take-out" of Saigon.²⁶

Meanwhile, XX Bomber Command had been disbanded and replaced by XXI Bomber Command, operating from new bases built on Guam, Saipan, and Tinian, islands in the Marianas wrested from Japanese control in 1944. Now the Japanese home islands lay within easy striking range of the Superfortresses.

Back in Hollywood, First Lieutenant Ronald Reagan was taking part in what he refers to in his autobiography as one of the major "secrets of the war, ranking up with the atom bomb project": creating a complete miniature of Tokyo, so authentic in detail that even top Air Corps generals could not distinguish it from reality. Footage of fake bomb runs on the toy city were then used to brief bombing crews, who were taken by Reagan's voice-over narrative all the way to his dramatic "Bombs away." As areas of Tokyo were burned out, Reagan tells how the Hollywood team would "burn out" their counterparts in "our target scene," obliterating, along with the city, the boundaries between illusion and reality.[27]

Meanwhile, scientists from corporations and universities constructed models of typical Japanese dwellings and urban residential neighborhoods to determine how to generate the most relentless fire storms.[28] Between late November and early January, experimental fire raids were carried out on a few Japanese cities. But the commander of XXI Bomber Command, Brigadier General Haywood S. Hansell, Jr., protested when ordered to switch from "precision" bombing of industrial targets to saturation firebombing of whole cities. Hansell was quickly removed, to be replaced, on January 20, 1945, by Major General Curtis LeMay. Dropping all pretenses at precision bombing, LeMay swiftly began his famed campaign to cremate Japan. Within a few months, this strategy would produce more casualties among Japanese civilians than their armed forces suffered throughout the war.[29]

The first stage of LeMay's incendiary raids culminated in the February 25 attack on Tokyo, which left one square mile of the city burned out. These promising results encouraged LeMay to concentrate the full potential of the Superfortresses on incendiary bombardment intended to saturate whole urban environments, thus creating fire storms that would burn the cities to the ground. Accordingly, he removed the one-and-a-half tons of defensive armaments from the B-29s so that they could carry an even heavier payload of firebombs. Like the British raids on Germany, the usual target from this point on would be the center of the city, the attack would occur at night, and the bomb pattern would be designed not for accuracy but for generating the widest possible holocaust.

The new tactics were tried out on Tokyo on the night of March 9–10. The target, marked off by "pathfinder" planes, was a twelve-square-mile rectangle housing one-and-a-quarter million people. For three hours, hundreds of B-29s unloaded their firebombs. The results were indeed spectacular. Instead of mere fire storms, the blazes produced a new and even deadlier phenomenon—the sweep conflagration, a tidal wave of fire igniting every combustible object in its path by radiant heat, melting asphalt streets and metal, leaping over canals, and searing the lungs of anyone within reach of its superheated vapors. The heat was so intense that it generated towering thunderheads with bolts of lightning. The last waves of bombers had difficulty finding anything left to bomb, and were tossed around like leaves by the thermal blasts from the fire below. Sixteen square miles of the city were burned out. More than 267,000 buildings were destroyed, and over a million people were rendered homeless. At least eighty-four thousand (and probably more than one hundred thousand) people died that night in the Tokyo inferno.[30]

The success of the Tokyo raid confirmed LeMay's strategy, which was then implemented against the entire Japanese nation. A city's "crowded districts of highly inflammable houses offered"—as *The Official History of the Army Air Forces* explains—an "ideal incendiary target."[31] Within ten days of the March 9 raid on Tokyo, thirty-one square miles had been burned out in four of Japan's largest cities. By the end of June, every major city in Japan—except a few preserved to measure the effects of a new secret weapon—had been destroyed. In July, the bombers, having run out of bigger targets, were attacking cities with populations between one and two hundred thousand. By early August, the only available targets had populations under fifty thousand.[32] All that was left were the four reserved cities, any of which could be destroyed in a routine incendiary raid. Two of these were Hiroshima and Nagasaki.

7

The Final Catch

When he had exhausted all possibilities in the letters, he began attacking the names and addresses on the envelopes, obliterating whole homes and streets, annihilating entire metropolises with careless flicks of the wrist as though he were God.
—JOSEPH HELLER, *Catch-22*

The grand finale of the global war waged against fascism and nazism—forces epitomized by rampant militarism and wanton slaughter of civilians—turned out to be dropping atomic bombs from flying Superfortresses on the cities of Hiroshima and Nagasaki. American dreams of ultimate weapons seemed finally to have come true, with the United States possessing not one but two—the strategic bomber and the atomic bomb.

Was World War II decided by these superweapons? Whether strategic bombing was a major factor in the military outcome of the war remains a matter of considerable disagreement, and since atomic bombs were dropped only after that outcome had been determined, historians have been debating for over four decades why they were used at all. Nevertheless, the bomber and the bomb were now to become crucial in U.S. military strategy, the postwar economy, the political process, and the world outlook of the nation.

Boeing's preparations for the new era might be detected in its final wartime ads, which assigned a new name to its B-29 Superfortress: the Peacemaker. In March 1946, just seven months after the end of World War II, the Strategic Air Command (SAC) was formed; two months later it was assigned the mission of preparing to carry out nuclear attacks anyplace on the planet.[1] Billy Mitchell's goal of an independent air force was reached in 1947. And preeminent within the Air Force was SAC, whose nuclear-capable B-29s were stationed at forward bases throughout the Northern Hemisphere. Soon these were superseded by true intercontinental bombers, the B-50, B-36, B-47, and B-52, all produced by Boeing.

In the Depression before the war, peacetime markets could not consume the output of American industry's enormous capacity. Then this capacity was bloated far beyond its prewar limits by the herculean tasks of wartime production. So after a brief binge of pent-up consumer demand, only one kind of appetite could gobble up the productive excesses of postwar American heavy industry: the ever-growing, insatiable bulimia induced by war or the threat of war. Hence the permanent conversion of the great war-making capacity envisioned by Edison and achieved during World War II into what President Eisenhower in 1961 labeled the "military-industrial complex." At its core was a network of airplane manufacturers and related enterprises, now styling themselves "the aerospace industry," with an ongoing propaganda campaign and organizational structure designed to guarantee that their most technologically advanced products would seem obsolescent as soon as they began to roll off the assembly lines. From this point on, the cult of the superweapon would be a fact of daily American economic and cultural life.

Mass-produced culture helped create an environment congenial to the needs of the aerospace industry. Space-oriented science fiction moved from the ghetto of fandom's "pulps" to new homes in widely circulated "slicks" like the *Saturday Evening Post* and the *American Legion Magazine*. Movies such as *Destination Moon* (1950) intoned hymns of praise to the marriage of American industry and the military.[2] But before it was superseded by ever more futuristic weapons, the main vehicle to carry the new message was the bomber.

Looking backward at American culture during the World War II bombing offensives, one can easily detect the fantasies that veiled and romanticized the bomber. The virtually unmitigated public approval of these aerial assaults is not surprising, in light of the images and illusions promulgated by newspapers, radio networks, and movies. What was seen and heard of the bombers was almost as innocuous and recreational as their scale models that most young boys (and even a few girls) were gluing together. Missing from the verbal and cinematic pictures of Anglo-American bombers was the world of the bombed, which appeared only in stories about our side, especially Britain, under Axis attack. As in Disney's *Victory through Air Power,* the Japanese cities that Americans wanted to destroy were not inhabited by real people. After all, it was the Nazis who bombed civilians, while the British and Americans bombed military targets. Warner Brothers' 1941 *International Squadron* expressed this perfectly: while holding the body of a young girl killed in a raid on London, the American hero—played by Ronald Reagan—decides to enlist in the RAF; he gives his own life bombing a Nazi ammunition dump.

Hollywood's images of bombers cannot be separated from certain material facts of the movie industry. Making a commercial film about bombers is difficult without two things: large financial resources and the cooperation of the Air Force. This tends to limit the product to what is at least acceptable to both financial and military interests. Of course, during World War II, when there was virtually no dissent about either American purposes or methods, no external pressures were needed to determine the ideology of war movies. But even before the U.S. entered the war, collaboration among the Army Air Corps, airplane manufacturers, and major film studios began to condition the public's responses to the bomber. World War II–era movies forged a crucial link between Billy Mitchell's early use of the silver screen to propagandize the cult of the bomber and postwar films designed to make SAC's bombers the emblem of the nation.

Hollywood's role in glamorizing the bomber was established as early as 1938 in MGM's *Test Pilot,* with Clark Gable and Spencer Tracy flying a wondrous experimental plane, actually the YB-17, prototype of the Boeing Flying Fortress. The Army Air Corps, in the midst of a massive publicity campaign aimed at winning congressional support for full production, assigned all twelve existing YB-17s and their crews, along with dozens of other aircraft, directly to the MGM production crew.[3] *Test Pilot* succeeded splendidly for all the collaborators. The film was one of the biggest box-office attractions of 1938, and was nominated for three Academy Awards, including Best Picture and Best Original Story; the Air Corps and Boeing got their contract.

By 1940, propaganda for war production, which was already booming, was less useful than images designed to recruit the men to fly the bombers and sail the warships. The year of the first peacetime draft in U.S. history, it also produced Paramount's *I Wanted Wings,* archetype of the combat flier recruitment movie. A romantic semidocumentary about the training of air crews, *I Wanted Wings* defined a subgenre that would persist for decades, as evidenced by one of the most popular films of the 1980s, *Top Gun* (1986). Although common ingredients in these films are the camaraderie of the fliers or their love affairs with pining wives and sweethearts, the most intense passions are evoked by the warplane itself. The Air Corps did its part to stimulate these feelings, providing more than a thousand airplanes, including a squadron of B-17s, for the production of *I Wanted Wings,* which became Paramount's top money-maker in 1941.[4]

The paradigmatic movie image of the bomber was established by *Air Force*, a 1943 box-office smash produced by Warner Brothers at the suggestion of Air Corps General "Hap" Arnold and directed by Howard Hawks. The star of the movie is "Mary Ann," a B-17 Flying Fortress, whose tale begins December 6, 1941, as she takes off from California for Hickam Field, Hawaii. "Mary Ann" becomes both the center of erotic interest and a machine endowed with almost godlike destructive power.

Air Force offers a revealing concoction of actual history with fantasies rooted in early American future-war fiction. As "Mary Ann" approaches Hawaii, the Japanese bomb Pearl Harbor and Hickam Field. To the actual attack, the film adds a racist fantasy in the Yellow Peril tradition: ethnic Japanese running amok in Hawaii, sabotaging planes, assaulting American men and women, and treacherously sniping at "Mary Ann." Next her crew has to rescue its beautiful maiden plane from the clutches of Yellow hordes at Wake Island and Clark Field in the Philippines.

So far, "Mary Ann" has been just an intended victim. As her custodians, her crew represents all the American airmen trying to preserve a few bombers from the Japanese onslaught "until we get enough airplanes to blast 'em off the map." Finally, as "Mary Ann" heads for Australia, she gets a chance to demonstrate her awesome powers, leading a flight of B-17s, followed by Navy torpedo bombers, to annihilate a gigantic Japanese invasion flotilla. Some modern film historians have mistaken this imaginary conflict for the Battle of the Coral Sea.[5] But it was filmed, using miniatures, in early 1942, before that battle, which it only vaguely resembles.[6] The four B-17s and other Army bombers involved in the actual Battle of the Coral Sea hit no Japanese ships, narrowly missed sinking a U.S. destroyer they attacked by error, and concentrated their main effort on an ocean reef mistakenly identified as an aircraft carrier.[7] Though land-based bombers did take a toll in other naval battles, *Air Force*'s image of a Japanese fleet being eliminated mainly by Army Air Corps bombers was a fantasy à la Billy Mitchell.

Air Force concludes with a more portentous vision of a future dominated by U.S. bombers. At an unnamed base 862 miles from Tokyo, the briefing of a mammoth assembly of bomber crews opens with thrilling news: "Tonight your target is Tokyo." In the final sequence, B-17 after B-17 takes off to bomb the city, the roar of their engines mingling with a voice-over from President Roosevelt vowing to carry the war to the Japanese homeland.

The first movie to picture an air attack on Japan was *Bombardier,* another 1943 release. Virtually a feature-length advertisement for the Norden Bombsight—whose legendary accuracy was crucial to the strategy of "precision bombing"—*Bombardier* begins by ballyhooing the prewar school where bombardiers learned to use this specimen of U.S. technological wizardry. It ends with a nighttime high-altitude precision raid on a Tokyo aircraft factory by a formation of B-17s (placed at the disposal of RKO studios by the Air Corps).

The first actual air attack on Japan, Doolittle's early-1942 raid by sixteen B-25 medium bombers, each carrying four bombs, was the subject of several films. Though set in 1942, two 1944 movies place Doolittle's mainly symbolic attack in the context of 1944, when the aerial offensive against the Japanese homeland was being prepared and launched.

Some of the American airmen captured after the raid had been put on trial for murdering civilians, and three had been executed. *The Purple Heart* imagines the torture and trial of eight captured fliers; apparently its main intent was to portray the Japanese as fiends and so justify the impending air offensive against their cities. After resolutely resisting torture, one of the downed airmen defiantly tells the court that more American fliers will come to "blacken your skies and burn your cities to the ground and make you get down on your knees and beg for mercy."[8]

Later in 1944 came one of the rare movies to raise ethical questions about the bombing of cities, *Thirty Seconds Over Tokyo,* scripted by Dalton Trumbo from the book by Ted Lawson, one of the survivors of Doolittle's raid, with Lawson, played by Van Johnson, as the central character. Presenting the 1942 bombing purely as a military necessity, the film celebrates the humanitarian aspects of the official U.S. policy of daylight precision bombing, and, quite atypically, studiously avoids racist stereotypes.

The targets are strictly limited to factories producing war material, which (contrary to what actually happened) are hit with pinpoint accuracy. In the final briefing before takeoff, Colonel Doolittle, played by Spencer Tracy as a crusty but humane leader, sternly admonishes the crews: "You are to bomb the military targets assigned to you and nothing else." "Of course," he acknowledges, "on a mission like this you cannot avoid killing civilians because war plants are manned by civilians." Then, in a spirit more characteristic of prewar America than of 1944, he raises the forgotten moral issue: "If you have any moral feelings about these killings, if you feel that you might later think of yourself as a murderer,'' he says fervently, ''I want you to drop out. . . . nobody will think the worse of you.''

The considerable box-office success of *Thirty Seconds Over Tokyo* probably owed less to its philosophizing than to its splendid cinematographic rendering of the planes themselves, including the obligatory training sequences, the incredible launching of B-25s from an overcrowded aircraft carrier, and the simulation of the audacious raid. Whatever its intentions, *Thirty Seconds Over Tokyo,* perhaps even more effectively than many of its more gung ho counterparts, communicates the emotional appeal of the bomber.

In the late 1940s, as the Air Force became preeminent in the Cold War strategy of maintaining the permanent threat of a U.S. nuclear attack, movies and novels about World War II bombers became vehicles for glorifying not aerial combat but top-down social control. The central characters were no longer the men flying the missions but the officers who commanded them. As if by coincidence, the terrific burden borne by high-ranking Air Force officers became the subject of popular and acclaimed works: William Wister Haines' 1947 novel, *Command Decision,* made into a long-running Broadway play and a 1948 film; the 1949 film *Twelve O'Clock High*; and James Gould Cozzens' 1948 novel *Guard of Honor,* awarded the Pulitzer Prize for 1949. Both novels are unabashed apologetics for a ruling military elite; the two movies are a bit less overt. The subtext of all these works, though each ostensibly concerns World War II, celebrates the commanders of the nuclear-armed Air Force of the Cold War.

The movies *Command Decision* and *Twelve O'Clock High* are both set at British bases from which U.S. bombers were conducting the ''daylight precision bombing'' offensive against German forces. Each focuses on the grave responsibilities of a commanding general (Clark Gable in the first, Gregory Peck in the second) who must direct missions that he knows will lead to terrible losses among his men. Each seeks to show how the commanders' vision transcends the limited view of their subordinates, forcing them to assume the anguish of ordering acts that might seem inhuman to less highly positioned observers. The giant World War II bomber offensive here becomes a totem for postwar corporate-military supremacy.[9]

By the mid-1950s, the strategic bomber had become a major icon of American culture. Hollywood paid homage to the primitive church in the 1955 film *The Court Martial of Billy Mitchell,* with the martyred prophet played by Gary Cooper. The persecuted cult of the 1920s was now a state religion. SAC's nuclear-armed bombers no longer needed to be camouflaged as the revered Flying Fortresses that had blasted the cities of the Third Reich. Films like *Strategic Air Command* (1955) and *Bombers B-52* (1957), directed by Gordon Douglas (who five years earlier had directed *I Was a Communist for the FBI*), could openly glorify the Cold War union of the strategic bomber and the nuclear bomb. Gone were the clanking, awkward, propeller-driven craft that had supposedly defeated fascism. In their place were gleaming, streamlined, multijet beauties, each carrying a load of destruction equal to hundreds of World War II bombers. These were superweapons that men could love with an ardor surpassing anything inspired by a woman or even by a Flying Fortress. In *Strategic Air Command,* Jimmy Stewart's commanding general enthuses about the hydrogen bomb and the latest SAC bomber: "With the new family of nuclear weapons, one B-47 and a crew of three carries the destructive power of the entire B-29 force we used against Japan." Reverentially ushered into the presence of the B-47 he will pilot, Stewart ecstatically blurts out, "She's the most beautiful thing I've ever seen in my life."[10] Within a decade, *Strategic Air Command*'s perverted love affair with this deadly machine, its characteristic and potentially catastrophic confusion between Eros and Thanatos, would inspire the bomber movie to end all bomber movies, *Dr. Strangelove, or How I Learned to Stop Worrying and Love the Bomb.*

The main announced purpose of the thousands of U.S. nuclear-armed bombers was to "deter" the alleged plan of the Soviet Union and its "satellites"—such as China—to overrun the world with hordes of Communist fanatics. These warplanes were supposedly the "peacekeepers" of their era, preventing war, as Edison and dozens of future-war novels promised, by making it too horrific to wage. Yet in the decades after World War II, the United States became involved in almost unending warfare—warfare that disclosed the true meaning of Victory through Air Power.

In World War II, U.S. strategic bombing concentrated on what Billy Mitchell and other theorists referred to as "industrial targets" (cities), "transportation centers" (cities), "communication complexes" (cities), and "nerve centers" or "vital centers" (cities). Using the manned bomber as an ultimate weapon of strategic warfare on predominantly peasant nations such as Korea and Vietnam would reveal new dimensions of its genocidal capacity, but also would expose its very limited military effectiveness.

By September 1950, just three months after the Korean "police action" began, U.S. warplanes had annihilated all the cities of North Korea. Even the smallest villages were targets of saturation attacks with high explosives, rockets, and napalm. Flight crews were complaining, "It's hard to find good targets, for we have burned out almost everything."[11] North Korea was bombed so intensively that, for years afterward, trees that survived were marked with commemorative signs designating them "prewar trees." Moreover, U.S. tactical air supremacy over the battlefield was unchallenged. Nevertheless, despite the numerical superiority of U.S. and allied forces (contrary to contemporaneous propaganda about "hordes" of Chinese Communists), the war ended in a stalemate.

Meanwhile, total rule of the skies by France's U.S.-equipped air force was not saving the French from military defeat in Indochina. Replacing France, the United States unleashed upon Vietnam, Laos, and Cambodia the longest and most devastating air assault in history.

In February 1965, just a few days before the sustained air offensive against North Vietnam began, appeared *Design for Survival* by General Thomas S. Power, recently retired head of the Strategic Air Command. General Power explained how easily air power could bring U.S. victory in Vietnam:

Let us assume that, in the fall of 1964, we would have warned the Communists that unless they ceased supporting the guerrillas in South Vietnam, we would destroy a major military supply depot in North Vietnam. . . . If the Communists failed to heed our warning and continued to support the rebels, we would have gone through with the threatened attack and destroyed the depot. And if this act of "persuasive deterrence" had not sufficed, we would have threatened the destruction of another critical target and, if necessary, would have destroyed it also. We would have continued this strategy until the Communists had found their support of the rebels in South Vietnam too expensive and agreed to stop it. Thus, within a few days and with minimum force, the conflict in South Vietnam would have been ended in our favor.[12]

Echoing superweapon fantasies in the future-war fiction of 1880–1917, this chimera of the bomber's omnipotence was just as far removed from reality.

During all of World War II, the United States dropped a total of about two million tons of bombs in all theaters, including the strategic bombing of Europe and Japan. From 1965, when General Power gave his prescription for painless victory through air power, through 1973, the United States dropped a minimum of eight million tons of bombs on Indochina. According to this conservative estimate, the explosive force of this onslaught was equivalent to approximately 640 Hiroshima-size atomic bombs. In just two years, 1968 and 1969, U.S. planes dropped on South Vietnam alone more than one-and-a-half times the tonnage dropped on Germany throughout World War II. The 1972 Christmas bombing ravaged the cities of Hanoi and Haiphong with more tonnage than Germany dropped on Great Britain from 1940 through 1945. Great swaths of Vietnam came to resemble the lunar landscape, with more than twenty-one million bomb craters just in South Vietnam, which the United States was allegedly defending.

High explosives were not the whole story. Newly engineered incendiary, chemical, and antipersonnel fragmentation weapons were used with unprecedented concentration. The napalm and phosphorus bombs that had burned out the cities of Japan and North Korea were refined into improved incendiaries designed to stick better to human skin and burn more intensely. Cluster bombs were carefully crafted to maximize internal body wounds with plastic flechettes that would escape detection by surgeons' X-rays. The poisoning and defoliation of Vietnamese cropland and forests with Agents Orange, White, and Blue lasted from at least 1961 through 1971; about half of South Vietnam's coastal mangroves were wiped out, over a third of the tropical hardwood forests were destroyed, and six million acres of farmland were inundated with toxic chemicals known to have severe mutagenic effects.[13]

Despite this unprecedented assault from the air, the United States was militarily defeated in Vietnam. And it was through this war that growing numbers of Americans became aware of the significance of America's most advanced weapons.

The cities and villages of Vietnam were perceived by the U.S. military as legitimate military targets because that was where the enemy—that is, the people—could be located. The American people were appalled once they learned that U.S. ground troops were massacring and burning whole villages, such as My Lai. Why was this more reprehensible than massacring and burning countless villages from the air? Because one of the distinguishing features of the warplane as an instrument of genocide is the dissociation it offers from its own effects. The interior of the plane does not even seem to be in the same universe as the victims on the ground. Divorced from the carnage it wreaks, the warplane becomes an icon of power, speed, beauty, cooperation, and technological ritual.

When the dissociation between these icons and the human suffering they inflicted began to break down during the Vietnam War, the most potent symbols of American technological might—such as the carrier-launched supersonic fighter-bomber and the giant eight-jet B-52—became objects of revulsion for many Americans. While many other Americans still responded with thrills and veneration to their roar and flash, the reality of the aerial superweapon was beginning to pierce the fantasy.

In the midst of the changing consciousness about America's wars and weapons appeared the two most popular and influential American works of literature about the bombing offensive of World War II: Joseph Heller's *Catch-22* and Kurt Vonnegut's *Slaughterhouse-Five*. These two novels contributed to the new awareness, which at the same time helped provide their wide audience. It is no coincidence that both books waited until the 1960s to be born, though both authors had been personally involved in the World War II bombing, which had been crucial in shaping the creative vision of each. Both books ask us to confront the world actually wrought by the victory of strategic air power.

Kurt Vonnegut, Jr., experienced this victory as a prisoner of war in Dresden during the Anglo-American raids that turned the city into an inferno, inflicting what he labeled "the greatest massacre in European history."[14] Dresden has never been far from his fictive imagination, which almost obsessively keeps returning to that nightmare and its implications about the weapons we build and use.

Even before Dresden appeared explicitly in his fiction, the superweapon was one of Vonnegut's major themes. His first published story, "Report on the Barnhouse Effect" (1950), concerned "the first superweapon with a conscience": an eccentric professor who accidentally develops so much mental power over the material world that he is able to destroy the world's armaments merely by wishing their doom. This "man who disarms the world," starting with ten V-2 rockets and fifty U.S. jet bombers, is no technological wizard, like his predecessors in early American future-war fiction or the Edison myth: his powers are quite antitechnological and fortuitous. Vonnegut's archetypal scientific genius is Dr. Felix Hoenikker of *Cat's Cradle* (1963), one of those designated a "father" of the atomic bomb. His final gift to the world is *ice-nine,* an ultimate weapon that freezes all water and extinguishes the human species. In *Galápagos* (1985), narrated by the ghost of a U.S. Marine who deserted after initiating the massacre of a Vietnamese village, our species of "big brains" succeeds in annihilating itself with various new weapons, which supposedly prove our fitness to survive. We are replaced by much fitter descendants, creatures with flippers instead of hands, possessing small, harmless brains housed in streamlined skulls suitable for pursuing fish to be caught with their beaks—their only weapons.

Dresden surfaced in Vonnegut's fiction in the mid-1960s. The title character of *God Bless You, Mr. Rosewater* (1965), who is erotically aroused by images of fire, becomes so excited by a graphic description of the Dresden fire storms in Hans Rumpf's *The Bombing of Germany* (an actual book) that he hallucinates a gigantic phallus-shaped fire storm consuming Indianapolis. But it was not until 1969, twenty-four years after the Dresden raids, that Vonnegut finally was able to finish a novel expressing the inferno at the center of his tormented imagination, *Slaughterhouse-Five; or, The Children's Crusade: A Duty-Dance with Death.*

In the opening chapter, the author tells of wrestling with the Dresden experience ever since he returned from World War II. At first he thought that simply narrating the story would generate a "masterpiece": ". . . all I would have to do would be to report what I had seen." His first attempts were in the late 1940s, before any of his fiction was published: "Even then I was supposedly writing a book about Dresden. It wasn't a famous air raid back then in America. Not many Americans knew how much worse it had been than Hiroshima, for instance. I didn't know that, either. There hadn't been much publicity" (10). Before *Slaughterhouse-Five* could be composed, much had to happen to America and to Vonnegut's fiction.

The aerial superweapon of America's endless postwar wars had to be desanctified and perceived as a genocidal mechanism. By 1969, Vonnegut was able to assume that many of his readers would respond with not wonder but disgust to the aerial killing machines. Thus, they could comprehend the ironies of a story embedded in *Slaughterhouse-Five*—attributed to Vonnegut's mythical science-fiction writer Kilgore Trout—about a robot with bad breath, "who became popular after his halitosis was cured":

> . . . what made the story remarkable, since it was written in 1932, was that it predicted the widespread use of burning jellied gasoline on human beings.
>
> It was dropped on them from airplanes. Robots did the dropping. They had no conscience, and no circuits which would allow them to imagine what was happening to the people on the ground.
>
> Trout's leading robot looked like a human being, and could talk and dance and so on, and go out with girls. And nobody held it against him that he dropped jellied gasoline on people. But they found his halitosis unforgivable. But then he cleared that up, and he was welcomed to the human race. (168)

Abandoning the straightforward narrative that had proved so inadequate for Dresden, Vonnegut constructed a fiction composed of wildly diverse modes of narrative reality. The autobiographical opening introduces what seems at first to be more or less realistic fiction about Billy Pilgrim, an oafish American prisoner of war at Dresden, who eventually becomes a well-to-do, apparently complacent optometrist—that is, a person whose job is to make people see things more clearly and accurately. In the novel, actual and imaginary books about the Anglo-American bombing offensive and the Dresden raids offer other narrative modes. Another layer of reality is composed of fictions by Billy's favorite author, Kilgore Trout, who appears as a character on still another level. Billy's view of time, life, and death is radically altered by experiencing a Trout-like science-fiction adventure: a flying saucer from the planet Tralfamadore kidnaps him, installs him in a zoo, and provides him with a movie-star mate so that the Tralfamadorians can ogle the earthling pair.

Just before being captured and sent to Dresden, Billy becomes "unstuck in time," so that he flips around erratically in his personal history like a dented Ping-Pong ball in a maze of air hoses. Without volition, he bounces around among his capture by the Germans, his bombing by the Allies, his kidnapping by the Tralfamadorians, and his lethargic absorption into middle America. He is thus the perfect protagonist for this novel that strips the romance and glamor from war, that offers no John Wayne, no Jimmy Stewart, no Sylvester Stallone, no character the audience might want to be. As the author explains, since this is a novel about war, "There are almost no characters in this story, and almost no dramatic confrontations, because most of the people in it are so sick and so much the listless playthings of enormous forces" (164).

Billy is thus also an ideal potential disciple of the Tralfamadorians, who believe that all moments in time occur simultaneously and that every creature in the universe is a mere machine without volition. So he ends up preaching with messianic zeal the determinist gospel of Tralfamadore in the 1976 America of the future: "The United States of America has been balkanized, has been divided into twenty petty nations so that it will never again be a threat to world peace. Chicago has been hydrogen-bombed by angry Chinamen" (142).

The Tralfamadorians advise Billy to forget about war and think only about pleasant events, a message opposed to the danse macabre of the novel. In his usual listless, almost autistic way, Billy goes along far enough with this Pollyanna philosophy to blind himself to the reenactment in the late 1960s of the massacre he had witnessed in 1944.

Whereas novels and movies of the late 1940s drew upon the popular support of the Second World War to legitimize the Cold War, *Slaughterhouse-Five* draws upon the popular revulsion against the Vietnam War to expose its continuity with the triumph of the superweapon in the Second World War. Shuttling back and forth in history, Billy merges Dresden into Vietnam. As a prelude to the ghastly scene in which Billy and the other prisoners of war dig for corpses in the bombed-out shell of Dresden, the author reminds the readers that "every day my Government gives me a count of corpses created by military science in Vietnam" (210).

In 1967, Billy drives through a burned-out black ghetto, which reminds him of Dresden, on his way to a luncheon at the local Lions Club, of which he is a past president. The speaker, a major in the Marines, regales his audience with anecdotes about the Vietnam War and advocates "bombing North Vietnam back into the Stone Age" (60)—the famous words of General Curtis LeMay, who became commander of SAC and Air Force Chief of Staff after directing the incineration of Japan. The optometrist fails to see any connections: "Billy was not moved to protest the bombing of North Vietnam, did not shudder about the hideous things he himself had seen bombing do" (60). Billy agrees with the major that he is very proud of his son, a Green Beret sergeant in Vietnam, who is undoubtedly doing "a great job" (61). He thus ignores the one clear lesson the author had drawn from Dresden: "I have told my sons that they are not under any circumstances to take part in massacres. . . . I have also told them not to work for companies which make massacre machinery, and to express contempt for people who think we need machinery like that" (19).

Yet Billy's "unstuck" perspective on time offers unique visions of the aerial superweapons that superimpose Dresden on Vietnam. He watches on television an archetypal World War II bomber movie glorifying the entire military-industrial process that ends in burning down cities. Played backward in Billy's mind, the movie turns into its opposite: the bombers put out the fires and suck up the bombs, the German fighters repair the bombers and heal the wounded crewmen, the bombs are unloaded at the base and disassembled in factories, from which their mineral contents are shipped to specialists in remote areas whose job is "to put them into the ground, to hide them cleverly, so they would never hurt anybody ever again" (75).

Juxtaposed with the fantasy is the reality of being bombed, pieced together in kaleidoscopic chips of the Dresden experience. The British officer who briefs the American POWs tells them: "You needn't worry about bombs, by the way. Dresden is an open city. It is undefended, and contains no war industries or troop concentrations of any importance." "The loveliest city that most of the Americans had ever seen," Dresden was "jammed with refugees." "About one hundred and thirty thousand people" die when this "voluptuous and enchanted and absurd" city, which looks like "a Sunday school picture of Heaven," is consumed by "one big flame" that "ate everything organic, everything that would burn." U.S. fighter planes fly under the smoke to strafe any survivors, firing at but missing Billy and the other American POWs. Like the genocidal bombing of Vietnam, "The idea was to hasten the end of the war" (146–80).

This was also the rationale for dropping atomic bombs on Japanese cities, as explained in President Truman's August 1945 announcement. Truman's words are inserted here amid extended quotations from David Irving's *The Destruction of Dresden* (185–88), including the forewords by two key figures in the Anglo-American bombing offensive, U.S. Air Force General Ira C. Eaker and British Air Marshal Sir Robert Saundby. Eaker offers a calculus and an ideology to justify the bombing: "I deeply regret that British and U.S. bombers killed 135,000 in the attack on Dresden, but I remember who started the last war and I regret even more the loss of more than 5,000,000 allied lives in the necessary effort to completely defeat and utterly destroy nazism" (187). Saundby (who engineered the saturation firebombing of Hamburg) admits that the bombing of Dresden was "a great tragedy" for which there was no military necessity: "Those who approved it were neither wicked nor cruel, though it may well be that they were too remote from the harsh realities of war to understand fully the appalling destructive power of air bombardment in the spring of 1945" (187–88). Then Saundby uses Dresden to argue in favor of nuclear weapons: "The advocates of nuclear disarmament seem to believe that, if they could achieve their aim, war would become tolerable and decent. They would do well to read this book and ponder the fate of Dresden, where 135,000 people died as the result of an air attack with conventional weapons'' (188).

Billy encounters these published statements while sharing a hospital room with Harvard Professor Rumfoord, a retired brigadier general, multimillionaire since birth, apostle of Theodore Roosevelt's cult of strenuous manhood, and now ''the official Air Force Historian'' working on a history of the U.S. Army Air Corps in World War II. At first Rumfoord refuses to believe that Billy, whom he considers a ''repulsive non-person,'' was actually in Dresden during the raids. When finally forced to confront this unpleasant fact, Rumfoord insists on a deterministic—that is, perfectly Tralfamadorian—rationalization, to which Billy blandly assents:

> ''It *had* to be done,'' Rumfoord told Billy, speaking of the destruction of Dresden.
> ''I know,'' said Billy.
> ''That's war.''
> ''I know. I'm not complaining.'' (198)

Rumfoord concedes, '' 'It must have been hell on the ground,' '' which Billy verifies in two words: '' 'It was.' '' Then the official Air Force historian offers his only words of sympathy: '' 'Pity the men who had to *do* it.' ''

One of the men who did do the bombing of Europe was Joseph Heller, a bombardier with sixty combat missions over France and Italy. It took sixteen years of postwar personal and American history for Heller to publish his first book, *Catch-22*, the most widely read work of American literature to come out of the war, and arguably the single most influential American novel of the twentieth century.

Catch-22 goes much further than the historians who deride the military effectiveness of the Anglo-American bombing offensive in defeating the Nazi-Fascist Axis. The novel suggests that the bombers helped the enemy win the war.

The enemy in *Catch-22* is ''they,'' all those who embody forces inimical to life, love, and other values for which the American people thought they were fighting. The ultimate interpretation of the novel's title comes from an old woman in Rome: '' 'Catch-22 says they have a right to do anything we can't stop them from doing' '' (416). By this point in the story, and in the history it interprets, ''they'' have very few constraints on their power.

World War II, everybody knew back then, was to be the final victory over fascism. American bombs were defeating the would-be conquerors of the world, with their storm troopers, secret police, armed gangs of thugs, torturers, and rapists, war profiteers, militaristic madmen, demagogues, big lies, anti-Communist crusades, fiendish scientists creating diabolical new weapons for giant cartels, and corporate states determined to make their war economy and culture permanent and universal. Yossarian discovers in 1944 what many more Americans later suspected: that these forces might achieve their global empire draped in red, white, and blue. Even President Eisenhower, several months before the publication of *Catch-22*, had warned the American people in his January 1961 Farewell Address:

Now this conjunction of an immense military establishment and a large arms industry is new in the American experience. The total influence—economic, political, even spiritual—is felt in every city, every state house, every office of the federal government. . . . In the councils of government, we must guard against the acquisition of unwarranted influence, whether sought or unsought, by the Military/Industrial Complex.

At the end of *Catch-22,* power is being consolidated by all the forces that at first seemed to be just bad jokes. The competing war profiteers and military bureaucrats have all merged into one giant cartel: M & M Enterprises. Like the "M.M.," the storm troopers who take over America in Sinclair Lewis's 1935 *It Can't Happen Here!,* M & M Enterprises is as American as apple pie. The syndicate's slogan, "What's good for M & M Enterprises is good for the country," echoes the famous declaration by Charles E. Wilson, president of GM: "What's good for General Motors is good for the country." M & M's most implausible and outrageous acts, such as contracting with both sides to maximize war profits, are no more outlandish than GM's construction of weapons for both sides throughout the war.[15]

Gestapo-like secret police, embodied by the CID men, have become ubiquitous and indistinguishable from their Nazi and Fascist counterparts. This is one of the meanings hidden in the metaphor of déjà vu, as the chaplain learns when he is arrested and interrogated by American officers in a torture room replete with rubber hose and blinding spotlight. Those who, like Dunbar, oppose these fascists are "disappeared," a transitive verb coined in the novel and soon absorbed into everyday American usage. Frenzied "loyalty" campaigns foreshadow the repression of the late 1940s and 1950s, part of the background of the novel.

Yossarian gets part of the message during his walk in Rome, which has fallen again, this time to a mob of "vandals," the hordes of U.S. military police terrorizing the civilian population. Yossarian realizes that the cry "Help! Police!" is no longer meant "as a call for police but as a heroic warning" to "everyone who was *not* a policeman with a club and a gun and a mob of other policemen with clubs and guns to back him up" (425). He sees the outcome of the war: "Mobs of policemen were in control everywhere" (426). And then he too is seized by MPs with clubs, "icy eyes," and "unsmiling jaws," each apparently "powerful enough to bash him to death with a single blow" (429), who haul him back to submit to Colonels Cathcart and Korn.

Colonel Cathcart, whose murderous ambitions are directly responsible for the deaths of almost all of Yossarian's friends, is manipulated by Colonel Korn, whose "brown face with its heavy-bearded cheeks," bearing "deep black grooves isolating his square chin from his jowls" (436, 441) gives him more than a coincidental resemblance to Mussolini. In Yossarian's crucial confrontation with the colonels, they announce that they have become America:

"Won't you fight for your country?" Colonel Korn demanded. . . . "Won't you give up your life for Colonel Cathcart and me?"

Yossarian tensed with alert astonishment when he heard Colonel Korn's concluding words. "What's that?" he exclaimed. "What have you and Colonel Cathcart got
to do with my country? You're not the same."

"How can you separate us?" Colonel Korn inquired with ironical tranquillity.

"That's right," Colonel Cathcart cried emphatically. "You're either for us or
against us. There's no two ways about it."

"I'm afraid he's got you," added Colonel Korn. "You're either *for* us or against
your country. It's as simple as that." (433)

The logic of the colonels is as flawless as the corporate slogan of M & M
Enterprises, for they are in command. But where did they and Milo Minder-
binder get their power, and how did a maniac named Scheisskopf get put "in
charge of everything" (400)?

The answer lies in the temporal and verbal structure of *Catch-22*, a labyrinth
that rearranges both fictive and historical chronology into intricate patterns of
meaning. Charting this labyrinth leads to the discovery that one crucial
episode, not mentioned until three-fourths of the way through the novel, is
central to the novel's structure and its message about the bombing of Europe.
After this, the narrative becomes qualitatively more nightmarish and surreal.
Before this (in the actual rather than the narrative sequence of events), hardly any
of Yossarian's friends die, but afterward all but a very few are killed or
"disappeared." This event is the mission against an undefended Italian mountain
village.[16]

We first hear about the mission in chapter 29, which begins with "no word
about Orr," whose escape to Sweden will later enlighten Yossarian. Next comes
news of the arrival in Italy of Colonel Scheisskopf. When last heard of,
Scheisskopf was a mere lieutenant obsessed with parades and frustrated because
he couldn't turn all the men into precision marching machines. Now he has been
assigned to the command of General Peckem, whose "objective is to capture
every bomber group in the U.S. Air Force" (331). Since Scheisskopf embodies
the aesthetics of Nazi and Fascist militarism, it is appropriate that General
Peckem considers placing him in charge of bomb patterns (334).

Among the most notorious tactics responsible for prewar American outrage
against the bombing of civilians was the Italian bombardment of undefended
Ethiopian villages. Especially infamous was the aesthetic ecstasy expressed
about these raids by one of the pilots, the dictator's son Vittorio Mussolini, who
enthused about the beauty of groups of tribesmen "bursting out like a rose after
I had landed a bomb in the middle of them."[17] For the Fascists, beautiful images
were only a bonus. But in *Catch-22* the sole purpose of the American annihilation
of the undefended Italian village is to produce publicity photos of tight bomb
patterns.

General Peckem privately explains to Colonel Scheisskopf that although
bombing this " 'tiny undefended village, reducing the whole community to
rubble' " is " 'entirely unnecessary,' " it is a fine opportunity to extend his
power over the bombing squadrons. For he has convinced them that he will
measure their success by "a neat aerial photograph" of their "*bomb pattern*"—
" 'a term I dreamed up,' " he confides, that " 'means nothing.' "

The ostensible purpose of destroying the village is to create a roadblock to delay German reinforcements, although Headquarters has no reason to believe that any reinforcements will be traveling on the road. All the previous missions had applied daylight precision bombing to targets with arguable military significance. But the perversion of this doctrine to legitimize the annihilation of a village filled with friendly civilians disgusts the sensibilities of all the fliers, except outright sadists like Havermeyer.

"'What the hell difference will it make?'" demands Dunbar, pointing out that it would take the Germans only a couple of days to clear the rubble. McWatt and Yossarian want to know why the people in the village can't be warned. "'They won't even take shelter,'" Dunbar argues, "They'll pour out into the streets to wave when they see our planes coming. . . .'" (335–36). Major Danby, the well-intentioned briefing officer, can justify the raid only in terms of obedience to orders and confidence in higher authority. When Colonel Korn demands to know why the men don't want to bomb the village, Dunbar replies, "'It's cruel, that's why.'" Colonel Korn responds with the argument used to legitimize the firebombing of Dresden and Tokyo, the nuclear attacks on Hiroshima and Nagasaki, and the aerial slaughter of Asian peasants:

> "Cruel?" asked Colonel Korn. . . . "Would it be any less cruel to let those two German divisions down to fight with our troops? American lives are at stake, too, you know. Would you rather see American blood spilled?" (336)

Korn quells the mutiny with two additional arguments: a threat to send the men on a more dangerous mission and a reminder that "'we didn't start the war and Italy did.'" He of course does not mean this to remind anyone that air war began with Italy's 1911 imperial war in Libya, that the Anglo-American bombing offensive put into practice the theory of the Italian Fascist General Douhet, or that the raid on this undefended village is a grotesque reenactment of Italy's air war on the villages of Ethiopia, now carried out by B-25 Mitchells, bombers named for Douhet's principal disciple in America.

With the dissent stilled, Major Danby resumes briefing the crews on how to space their bombs to achieve their ostensible military objective, converting the village into a roadblock. He is interrupted by Colonel Korn:

> "We don't care about the roadblock," Colonel Korn informed him. "Colonel Cathcart wants to come out of this mission with a good clean aerial photograph he won't be ashamed to send through channels." (337)

The actual bombing of the village is not described, except for a key paragraph focusing on the roles of Yossarian and Dunbar:

> Yossarian no longer gave a damn where his bombs fell, although he did not go as far as Dunbar, who dropped his bombs hundreds of yards past the village and would face a court-martial if it could ever be shown he had done it deliberately. Without a word even to Yossarian, Dunbar had washed his hands of the mission. The fall in the hospital had either shown him the light or scrambled his brains; it was impossible to say which. (339)

After this, Dunbar seldom laughs, seems to be "wasting away," and soon is "disappeared."

Yossarian is habitually slow in comprehending and acting. He bandages the wrong wound on Snowden, the dying gunner. He literally throws away his one opportunity for redemptive love with a woman, casually tossing out the address of Luciana, not realizing that "his heart cracked, and he fell in love" with her when he learned that it was an air raid by "*Americani*" that had deeply scarred her body (163). Not until the very end does he understand Orr's crafty plan for survival; in fact he furtively connives to avoid being assigned to fly with him. He vacillates until too late before accepting Dobbs' proposal to kill Colonel Cathcart, and fails to fathom or follow Dunbar's revolt. Thus he misses the alternatives offered to him by his comrades before they are dead or gone: escape (Orr), direct action (Dobbs), and resistance (Dunbar). All that is left for him is his existential, absurdist final flight from what his nation has become.

When we first meet Yossarian, in the hospital, he is amusing himself by censoring mail as perversely as possible, "attacking the names and addresses on the envelopes, obliterating whole homes and streets, annihilating entire metropolises with careless flicks of his wrist as though he were God" (8). The joke of course is that he is annihilating only verbal symbols of homes, streets, and metropolises. But the joke turns inside out when we discover that Yossarian is a bombardier, whose duty requires him to obliterate actual towns and people. Yossarian escapes from the Air Force before he becomes part of the machinery designed to turn the joke into the central fact of human life. *Catch-22* shows the bombers of World War II helping to forge a world in which we all live in terror of those who may annihilate entire metropolises with careless flicks of the wrist.

This universal terror is one subtext of the novel, which appeared in 1961, during the period in which the threat of global nuclear holocaust first became the subject of widely popular novels and movies, such as *On the Beach* (novel, 1957; film, 1959), *Red Alert* (novel, 1958; filmed as *Dr. Strangelove*, 1964), *A Canticle for Leibowitz* (1959), *Level 7* (1959), *Alas, Babylon* (1959), and *Fail-Safe* (novel, 1962; film, 1964). A later chapter will explore why these particular years were the first in which such visions became part of mass culture, though they had been common in science fiction for decades.

Six years after *Catch-22*, Yossarian's joke expanded to become Heller's play *We Bombed in New Haven*: actors playing Air Force crews wait to learn which cities their script has them annihilate. Tired of this piecemeal approach, some of them wish to blow up the whole globe: "That's what I would do if I were in charge, instead of picking it apart so slowly, piece by piece and person by person. Why don't we just smash the whole fucking thing to bits once and for all and get it over with?" (34–35). Some of the actors keep insisting that this is all make-believe, just a verbal construct. The play ends with the main character insisting to the audience, "There is no war taking place here now!" and "There has been a war. There never will be a war." *We Bombed in New Haven* was first performed in December 1967, to an audience beginning to learn from Vietnam just what kind of victory had been achieved through air power.

III

CHAIN REACTIONS

8

Don't Worry, It's Only Science Fiction

Before nuclear weapons could be used, they had to be designed; before they could be designed, they had to be imagined. Their history involves complex interplay among the insights and illusions of those who conceived of them as fiction, those who actually constructed them, and those who decided to use them. For fifty years, from the first atomic explosion in Robert Cromie's 1895 novel *The Crack of Doom* until 1945, nuclear weapons existed nowhere but in science fiction, and in the imagination of those directly and indirectly influenced by this fiction, including scientists who converted these inventions from fantasy into facts of life.

Radioactive and atomic weapons played an important role in American future-war fiction before World War I (see Chapter 2). In such works as *The Vanishing Fleets* (1907), *The Man Who Ended War* (1908), and *The Man Who Rocked the Earth* (1914), these superweapons contrived by American scientists effortlessly bring the abrupt abolition of war, through either universal disarmament or American global rule. Similar scenarios would determine the thinking of American leaders who actually wielded such weapons at the end of World War II and after (see Chapters 9 and 11).

American readers, however, were also exposed to a less sanguine view, imported from Britain. Just as British culture had viewed the airplane with less euphoria, British future-war fiction before World War I projected atomic weapons in a far more ominous light.

For example, in George Griffith's 1906 *The Lord of Labour* (published posthumously in 1911), a German professor invents a disintegrating ray and proposes, like the scientific geniuses in the American novels, ''to make warfare impossible by making it so awful that no man in his senses would go upon a battlefield.'' But the Kaiser responds: ''My dear Professor, before you make war impossible you will have to make another discovery. You will have to find out how to alter human nature'' (50–51). Meanwhile, a British scientific wizard has discovered a vast deposit of radium and invented a gun to turn it into what he, too, thinks will be the ultimate weapon, rendering "all explosives used by land or sea, and every great gun and rifle . . . obsolete" (111). The resulting world war waged with rival superweapons slaughters millions before ending in Anglo-American global conquest.

In the midst of these contrasting British and American visions of atomic superweapons appeared a novel destined to play a critical role in the development of nuclear bombs: H. G. Wells' *The World Set Free: A Story of Mankind,* serialized in *English Review,* 1913–14, and published as a book in New York and London just prior to the outbreak of the First World War. *The World Set Free* introduces what Wells christens the "atomic bomb," a true nuclear weapon, converting mass into fiery and explosive energy in a chain reaction induced by a triggering device.

Wells, who dedicated his novel "To Frederick Soddy's Interpretation of Radium," comprehended the relations between advanced technology and warfare far more accurately than either the other fiction writers of his time or the devotees of the cult of the superweapon in our time. He shows how superweapons emerge not from the brain of some lone genius but from the current level of technology. He therefore dismisses as an absurdity the notion that they could be monopolized by a single individual or nation. Indeed, Wells' astonishingly accurate projected timetable for nuclear development errs only in predicting that commercial nuclear power would precede nuclear weapons.

Wells imagined the harnessing of "atomic energy" (50) in the early 1950s, precisely when reactors actually began producing electricity in the United States and the Soviet Union. In the novel, the scientist who first discovers the process, realizing that there is no way to suppress such knowledge, nevertheless "felt like an imbecile who has presented a box of loaded revolvers to a crèche" (44). At first, this technological advance merely exacerbates all the contradictions of capitalism. Tremendous wealth piles up alongside deepening poverty, while masses of the unemployed turn inevitably to crime. International conflicts become ever more ominous, with governments "spending every year vaster and vaster amounts of power and energy upon military preparations, and continually expanding the debt of industry to capital" (286).

By 1956, these socioeconomic contradictions of industrial capitalism, which Wells brands a "barbaric" form of social organization, generate a global nuclear holocaust in which the major cities are destroyed by small "atomic bombs" dropped from airplanes. Civilization virtually collapses, and hordes of survivors, many scarred by radioactivity, are left to wander through desolate landscapes (in scenes that would become familiar in fiction and films after World War II).

From the ruins of industrial capitalism emerges "the Republic of Mankind," directed by a farsighted elite who establish "science" as "the new king of the world" (178). As implausible as this Wellsian technocratic utopia may now appear, it does offer today's readers a disturbing perspective on the resemblances between the fictional and actual history of nuclear weapons. Looking backward from Wells' utopian postatomic world to the mid-twentieth century, it seems "that nothing could have been more obvious" than the fact that "war was becoming impossible." But governments and peoples remained "invincibly blind to the obvious." Then comes a chilling passage, far more relevant today than when it was penned in the days just before World War I:

> They did not see it until the atomic bombs burst in their fumbling hands. Yet the broad facts must have glared upon any intelligent mind. All through the nineteenth and twentieth centuries the amount of energy that men were able to command was continually increasing. Applied to warfare that meant that the power to inflict a blow, the power to destroy, was continually increasing. . . . Every sort of passive defense, armour, fortifications, and so forth, was being outmastered by this tremendous increase on the destructive side. . . . These facts were before the minds of everybody; the children in the streets knew them. And yet the world still, as the Americans used to phrase it, "fooled around" with the paraphernalia and pretensions of war. (117–18)

When this passage appeared, American fiction writers, along with scientists like Thomas Edison, were conjuring up marvelous new weapons—purely defensive, of course—to save us from war.

Wells forecast in *The World Set Free* that artificial induction of atomic disintegration in a minute amount of heavy metal would be achieved in 1933. This was precisely the year in which Irène and Frédéric Joliot-Curie did first induce artificial radioactivity. Meanwhile, Hungarian physicist Leo Szilard had read *The World Set Free* in 1932, while working at the Institute of Theoretical Physics of the University of Berlin.[1] He fled Germany when Hitler came to power, also in 1933. That fall, while ruminating about Wells' novel and Lord Rutherford's assertion that atomic energy was mere "moonshine," Szilard suddenly conceived of a way to sustain a nuclear chain reaction, making it possible to "liberate energy on an industrial scale, and construct atomic bombs." From that moment on, working out this method became what he called his "obsession."[2] By the spring of 1934, Szilard had developed a detailed description of the laws governing such a chain reaction. Influenced by *The World Set Free,* he decided to keep the process secret by patenting it and assigning the patent to the British Admiralty: "This was the first time, I think, that the concept of critical mass was developed and that a chain reaction was seriously discussed. Knowing what this would mean—and I knew it because I had read H. G. Wells—I did not want this patent to become public."[3]

Then in 1938 the uranium atom was split—in Berlin. As soon as Szilard learned, in early 1939, that fission of uranium had been achieved, he at once grasped the fateful implications: neutrons must be emitted during fission, "and if enough neutrons are emitted in this fission process, then it should be, of course, possible to sustain a chain reaction." "All the things which H. G. Wells predicted appeared suddenly real to me," he recalled, and he therefore resolved that this must be "kept secret from the Germans."[4] Szilard met with two other Hungarian émigré physicists, Edward Teller and Eugene Wigner, and the three enlisted Albert Einstein in their plan to counter the potential menace posed by atomic bombs in the hands of Nazi Germany. The outcome was the letter, composed by Szilard and Einstein, and sent over Einstein's signature to President

Roosevelt on August 2, 1939, that initiated the chain of events leading to the Manhattan Project and the epoch of nuclear superweapons.[5]

Szilard and the other physicists who led the United States into atomic weaponry had two motives: to deter the Nazis from using atomic weapons, and to bring a final end to war. As Eugene Wigner explained:

> We realized that, should atomic weapons be developed, no two nations would be able to live in peace with each other unless their military forces were controlled by a common higher authority. We expected that these controls, if they were effective enough to abolish atomic warfare, would be effective enough to abolish also all other forms of war. This hope was almost as strong a spur to our endeavors as was our fear of becoming victims of the enemy's atomic bombings.[6]

One of Szilard's earliest preparations for the atomic scientists was ordering two books about atomic bombs for their library: *The World Set Free* and Harold Nicolson's 1932 *Public Faces*.[7] Nicolson's novel was not at all typical of other British future-war fiction between the two world wars. Most of these novels and stories were either apocalyptic or preparedness messages about catastrophic raids by bomber fleets, along the lines of the theories of Trenchard, Douhet, and Mitchell. Often the bombers carried deadly new gas, as well as advanced types of high-explosive and incendiary bombs.[8] Nicolson, an experienced diplomat, imagined a different scenario.

Public Faces is set in 1939, when Britain secures a monopoly on an ore, found exclusively on an island belonging to Persia (Iran), from which can be refined one metal essential to the production of rocket planes and another that is sufficiently unstable to yield "atomic bombs." In this curious comedy of manners, petty human foibles, vanity, and bungling permit the inertia of these weapons to lead to the brink of world war.

The Liberal foreign minister convinces the Cabinet that "it would be impious, for us to dabble in the Satanic possibilities of the atomic bomb" (18). But, aware that the Conservative Opposition would look upon it as a "weapon of world dominion" (14) and frightened by the catastrophic potential of their secret, the Cabinet vacillates. This allows the Air Ministry to produce rocket planes, thereby impelling the other powers into war preparations, and to test a gigantic nuclear bomb that engulfs Charleston, South Carolina, in a fatal tidal wave. The war that threatens to overwhelm the planet is prevented only when the British government, "recognizing that these new and potent engines of destruction are inimical to existing civilisation," pledges to destroy all its atomic bombs and to manufacture none in the future (299). This brings disarmament and peace.

Nicolson's main theme is that "the unavowed weaknesses of human character" (153) can permit our own weapons to lead to our destruction, and that our survival depends upon our confronting the responsibility imposed by these weapons. In a novel published a decade before the Manhattan Project, in the same year that Szilard was reading *The World Set Free*, Nicolson labeled "the atomic bomb" as "this Manhattan of responsibility" (21).

Between World War I and the bombing of Hiroshima, the consequences of splitting the atom became a commonplace theme in American science fiction. Most of the stories and novels were concerned not with atomic weapons but with atomic energy. Some reveled in its thrilling potential, imagining virtually free electricity, atomic-powered spaceships (as in "Doc" Smith's 1928 *The Skylark of Space* and its sequels), an occasional atomic-powered time machine (Victor Rousseau's 1930 "The Atom Smasher"), or even the artificial creation of a new sun when ours dies (Raymond Gallun's 1931 "Atomic Fire"). Others warned about the deadly dangers of atomic experimentation, even hypothesizing a chain reaction that could make the planet uninhabitable, as in Isaac Nathanson's "World Aflame" (1935). But more common than either of these farfetched extremes were thoughtful stories—such as Harl Vincent's "Power Plant" (1939), Robert A. Heinlein's "Blowups Happen" (1940), Lester Del Rey's "Nerves" (1942), and Clifford Simak's "Lobby" (1944)—blending enthusiasm about atomic energy with warnings about its associated financial and environmental hazards.[9]

There was far less enthusiasm about atomic weapons. Gone was the naive celebration of them that characterized the pre–World War I tales. Such fiendish inventions might be used by "Asiatics" (as in "The Atom Smasher," a 1938 tale by "Gordon A. Giles" [Otto Binder]) or the Nazis (as in Cleve Cartmill's 1944 "Deadline")—but not by Americans. In only two stories of which I am aware does the United States use, or even threaten to use, atomic weapons. Both appeared in the 1940–41 period when America was slipping into World War II, and each, as will be seen, expressed the fateful transition in American attitudes toward superweapons also manifested in the growing approval of strategic air power.

Long before the scientists of the Manhattan Project began their feverish quest for an atomic deterrent, American science fiction was probing the fallacies of deterrence theory. In *The Pallid Giant,* published in 1927, Pierrepont B. Noyes tells of how the development of atomic weapons by advanced civilizations in remote prehistory led to the virtual extinction of the human species. The deadliest aspect of these weapons, he writes, is the *fear* they inspire, the overpowering motive called "the pallid giant." Supposed to deter attack, fear instead leads to preemptive strikes, for each nation finds itself unable to continue living in "fear of the other's fear." In the words of one who urges such a first strike: " 'I fear not their desire to kill . . . *I fear their fear.* They dare not let us live, knowing or even fearing that we have a power so terrible. . . .' "[10]

Another remarkable early novel that exposed the fallacies of atomic deterrence, *The Final War,* was serialized in 1932 in *Wonder Stories* and became quite popular among science-fiction fans. Its author, Carl W. Spohr, was an artillery captain in the German army during World War I; his experience seems to have helped induce a vision of atomic weapons quite different from that of American artillery captain Harry Truman.

The Final War opens in the twenty-first century, when two unnamed superpowers have divided the world. Each one's enormous arsenal supposedly deters the other from war. One side launches its bomber fleets in a massive surprise first strike, "a bold attempt to start and end the war with one crushing blow." But "the strategic value of the air raids" turns out to be "surprisingly small, and the slaughter of the non-combatants appalling" (1117).

After full-scale retaliation, a ghastly stalemate ensues. Year after year, each side introduces fiendishly ingenious new weapons only to see them copied or outdone by the other side. All large cities are laid waste; armies on the battlefield and civilian populations alike are forced to burrow deep into the ground, where they share squalid concrete tunnels with hordes of huge rats and roaches. Drone aircraft guide more and more deadly missiles into these underground warrens, while the secret police—known as the SS—enslave all those too weak for combat into war production. Women are turned into breeding animals to produce the next generation of combatants, and children are forced to toil in the weapons factories.

In the midst of this nightmarish vision of scientific war, a little hunchbacked scientist strains his "gigantic brain" to find an explosive powerful enough to end the war that was destroying civilization. What he discovers is a way to produce atomic explosions. When he witnesses one, his response foreshadows those of some of the atomic physicists witnessing the Trinity test at Alamogordo: "'And the power of the atoms is free. I wanted an explosive, I did not want to do this. It is awful'" (1277). Realizing the dreadful implications of his invention, he tries to keep it from the military. But a spy in his laboratory turns it over to the enemy, and the SS of his own country torture him until he reveals the secret.

Both sides now frantically produce great quantities of various atomic weapons: tactical nuclear shells for battlefield use, short- and intermediate-range missiles, and atomic bombs. Realizing, however, the apocalyptic destructiveness of the two atomic arsenals, each side tentatively adopts a policy of no first use. But they equip their armies in the field with the tactical shells, instructing the field commanders not to be the first to use these secret weapons, labeled by one side with a green dart, by the other with a red dot.

The inevitable happens. With his unit about to be overrun and annihilated by an armored offensive, a young field officer decides to try a red dot shell. The results are impressive: "... a black crater yawned, where a tank battalion had been" (1281). A battlefield barrage of these tactical atomic shells quickly escalates into a massive exchange of short- and intermediate-range atomic missiles. Both sides then launch their fleets of long-range bombers.

Spohr's analysis of the ensuing global nuclear holocaust is even more startling than his description. Since the logic of deterrence has led each side into planning the total annihilation of the other, the first use of atomic weapons triggers an unstoppable mutual devastation amounting to joint suicide. Most remarkable is Spohr's realization that the deadliest term in what we now call the policy of "mutually assured destruction" is the middle one. To be effective, the threat of full-scale nuclear retaliation requires inflexible, flawless, absolute certainty. There *must* be no way to stop the onslaught, even if the political and military leaders who have planned it are wiped out in the other side's first strike, even if continuing the attack means the end of human civilization:

Tossed and whipped by the tempest, flights and flights of airships fought their way to their destinations. Their commanders thought of orders, that had been given them, before they left the ground. In years of spy work and patient observation all life on earth had been mapped carefully as "enemy activity." . . . plans had been worked out on either side, to make the annihilation of the enemy complete. And now these plans were carried out by men, who had been learning through years and years, that the only human virtue was unscrupulous, unrelenting discipline.

These plans were carried out, after the ground commanders, that had sent the ships, were dead. They were completed, after the men, in whose power-mad brains the plans had originated, were crushed in their deep concrete dugouts. There were no staffs, no governments, only these orders, that had to be carried out. (1283)

Each side has deployed defensive weapons beyond the most extravagant promises of Star Wars. But the offensive nuclear arms, with their carefully planned redundancy, overwhelm these massed batteries of beam weapons and energy shields. The plans are successful, "and the world was swallowed in black, raging darkness": "Earth and debris rained from overhead. Dust clouds blotted out the world" (1282–83).

For three days after the nuclear tempest, "rain fell in sheets from the dark sky" (1283). Tiny clusters of pitiful survivors emerge from the planet's ruin. Foraging for food, sometimes resorting to cannibalism, they form straggly little tribes that in the end try to establish a truly human world society.

In 1932, the year *The Final War* was published, Billy Mitchell's articles advocating incendiary bombardment of Japan appeared in *Liberty,* one of America's top three general-interest magazines, which had just been acquired by publication magnate Bernarr MacFadden, then an ardent admirer of Benito Mussolini. The next year, Adolf Hitler came to power in Germany, Franklin Delano Roosevelt assumed the presidency of the United States, and Leo Szilard conceived of a method to initiate a nuclear chain reaction. By the end of 1939, World War II had begun in Europe and physicists in Germany, Great Britain, the United States, the Soviet Union, Japan, and several other nations were working on atomic energy and atomic weapons. During that year, scientists from these nations published almost a hundred articles on nuclear fission.

By the summer of 1940, American newspapers and magazines—from *Popular Mechanics* to *Time* and *Newsweek*—were exciting their mass audience with the potential wonders to come from splitting the atom. Since atomic research had not yet been enveloped in the cult of "security," millions of Americans learned that the nuclear chain reaction essential for atomic energy, and possibly even atomic bombs, now depended mainly on developing practical means for producing significant quantities of the unstable isotope uranium 235. All this common knowledge would soon become transformed into the "atomic secret," legitimizing the most fundamental censorship in the history of the American press.

Typical of the popular articles in mid-1940 were John J. O'Neill's "Enter Atomic Power" in *Harper's,* California Institute of Technology physicist R. M. Langer's "Fast New World" in *Collier's,* and William L. Laurence's "The Atom Gives Up" in the *Saturday Evening Post,* which all foresaw atomic energy leading to the kind of futuristic wonderland projected by the New York World's

Fair of 1939–1940. Far more blandly optimistic than the typical science-fiction story of the period, these articles dismissed the use of atomic energy for superweapons as "unthinkable," except by "eccentrics and criminals."[11] But this very same year, *Liberty* exposed millions of Americans to quite a different view of atomic weapons.

Between 1932 and 1940, *Liberty* remained an ardent and influential advocate of Mitchell's independent air force capable of devastating strategic bombardment. In 1939, while sponsoring a national contest for the official Air Force song, the magazine commissioned veteran journalist Fred Allhoff to write a novel for serialization depicting a future invasion of the United States. The result was the exceptionally revealing *Lightning in the Night,* which ran in thirteen installments from August to November 1940. According to the blurbs accompanying the serial, Allhoff had consulted extensively with Lieutenant General Robert Lee Bullard (who had headed the National Security League since his retirement in 1925), Rear Admiral Yates Sterling, George Sokolosky (the right-wing columnist), and "many others" in preparing his scenarios.

Appearing in the midst of President Roosevelt's campaign for a full-scale military buildup, *Lightning in the Night* was blatant propaganda in the hackneyed preparedness mode of future-war fiction. But it did not merely rework stale materials, for the forms of preparedness it advocated in 1940 included not only Billy Mitchell's dogma of victory through strategic air power but also an all-out scientific-industrial effort to develop atomic bombs.

The action of the novel begins five years in the future. Having conquered the British and French empires, the Greater United German Reich has established menacing bases in the American hemisphere. In July 1945, Japan and the Soviet Union carry out a sneak attack on Hawaii, while Germany seizes the Panama Canal. Massed formations of Soviet, Japanese, and Nazi bombers, using incendiary, high-explosive, and gas bombs, devastate the cities of the East and West coasts. Hordes of Reds, Japanese, Mexicans, and Germans invade on three fronts, inflicting on an unprepared United States the "macabre nightmare of modern warfare."[12]

The military lesson is spelled out repeatedly: "Billy Mitchell told us years ago that the nation controlling the air in the next war would win. Hitler listened to him. We court-martialed him" (Installment V). "The conquest of Europe had shown unmistakably that mastery of the air over a theater of operations was vital in the new modern warfare. Yet the United States had stubbornly rejected the idea of a separate air force" (VI).

The social message of *Lightning in the Night* is virtually identical to that of the 1987 television miniseries *Amerika,* which showed a flabby, self-centered America forced by foreign occupation to reassert patriotism and nationalism. In the novel, America wakes up and returns to its noble frontier ethic only after its capital has been moved from Nazi-occupied Washington into its midwestern heartland: "A great country built by sweat and hardship and sacrifice was returning to the precepts of its pioneers; was abandoning the good things of the present for the better things of the future—and being proud of the right, jealously guarded for centuries, to make its sacrifices voluntarily!" (IX). Factory workers relinquish their selfish demands for a forty-hour week (IX). They build new bombers, designed with American technological ingenuity and equipped with "America's bombsights, the finest in the world," to sink Nazi ships in Billy Mitchell style (X). The alien invaders now confront a fierce, revitalized nation: "With a newfound faith both in itself and its leaders, America promised to be invincible. That there were dark days ahead was calmly accepted. A common danger had bludgeoned panic, defeatism, and differences of all kinds out of a now firmly united people" (IX). But all this heroism, unity, and pioneer greatness could be nullified by the atomic superweapon—if the Nazis get it first.

Sure enough, Hitler calls for a conference with the American president in the neutral city of Cincinnati, with leading physicists from both sides to be in attendance. Ostensibly motivated by the "humanitarian desire to save lives," Hitler begins by explaining the theory of atomic energy and summarizing the state of atomic research at the time *Lightning in the Night* was published. Rather surprisingly to modern readers, who have grown accustomed to the secrecy surrounding nuclear weapons, the Nazi leader reminds everybody that " 'all of us knew the overwhelming implications,' " so " 'by the year 1939 the physicists of the Reich, of Denmark, and of America were frantically at work attempting to free and harness atomic energy' " (XI).

" 'The secret of world mastery,' " Hitler continues, of course would go to the nation that " 'first could produce great quantities of pure U-235,' " the uranium isotope sufficiently unstable to sustain an explosive chain reaction. The Reich, he announces, has discovered this " 'key to atomic energy,' " and has begun production of pure U-235, with its " 'destructive power beyond present-day comprehension, . . . the power to blow entire cities off the face of the earth.' " " 'Within one month, that devastating power can be unleashed against your cities, your people,' " Hitler boasts, so " 'further resistance becomes utterly foolhardy.' " After this " 'one month of grace,' " he will unleash " 'literal and total annihilation.' "

The President readily concedes that it would be hopeless for a nation without atomic weapons to wage war against a nation with them. And so the November 9 installment of *Lightning in the Night* concludes with the United States apparently ready to surrender to the Nazis.

Thus the millions of readers of *Liberty* in 1940 were confronted with a picture of the future that lies in wait for them if the United States does not build a separate air force capable of strategic bombardment and does not win a nuclear arms race with the Nazis. One wonders whether this effort to build popular support for financing the atomic weapons research that was already under way might have had some tacit semiofficial sponsorship. In any case, having seen what could happen to the United States if it were to lose the race for atomic bombs, the readers were treated in the next chapter to a vision of the future if the United States were to win.

This final installment reveals that the United States had secretly been working on its own atomic weapons. Great cyclotrons had been moved from the coasts to the interior, and marvelous new equipment was built for the nation's "most ingenious and resourceful scientists." The President expounds America's vision of atomic energy, a vision like that of those 1940 articles in *Harper's, Collier's,* and the *Saturday Evening Post,* a vision that would reappear after Hiroshima and Nagasaki under the slogan Atoms for Peace:

> "We saw its potentialities as a weapon of war, but even more clearly as an unlimited source of heat, of light, of power for peaceful production and transportation—all this at an almost incredibly low cost.
> "We saw a new world in which the most densely populated country would have ample room for all its citizens to live in well and cheaply; a world in which this new wealth of energy would be shared by the people of every land and race and creed.
> "International boundaries, money as we know it today, and poverty would vanish from the earth. So would war itself; for the economic causes of war would no longer exist. That, gentlemen—that Utopia, if you like—was what we envisioned: a free world of free peoples living in peace and prosperity, facing a future of unlimited richness."

The President acknowledges that the Nazis have acquired " 'the secret of producing a weapon that must inevitably overwhelm and subdue any nation on earth,' " but they are too late. At this very moment, " '50,000 feet over the Atlantic, great United States stratospheric bombers,' " specially modified for intercontinental flight and carrying atomic bombs, are " 'heading for every great city in Germany.' "

Next, the President presents the American terms for peace. These are precisely those that would be central to the only proposal for nuclear disarmament ever actually offered by the United States, the Baruch Plan of June 1946: a body dominated by American scientists would control both the world's supply of uranium and the licensing of nuclear-energy facilities to other nations; the United States would maintain its monopoly on nuclear weaponry until some unspecified date in the future when it would be turned over to an international agency. In this 1940 novel, the American proposal of May 1946 puts it this way:

> . . . that the desire of the American nation was for true peace and prosperity for all the world; that to preserve that peace for future generations it was necessary that the world's supply of uranium should fall only into the hands of those who would use it not for destruction but for construction. Temporarily, American engineers would limit the amount of uranium obtained by any and every other country, while American physicists would keep watch of the uses made of it.
> "We have no wish," the President said, "to assume for long the task of policing the world. When the world is restored and made free, a Council of Nations shall take over the task we inaugurate now."

Hitler remains obdurate, but is assassinated by the German generals, who surrender on America's terms. Japan and the Soviet Union capitulate a day later, after "an American bomber, dropping just one 500-pound bomb of the new explosive on the deserted Russian steppes, had blasted a hole in the earth several hundred feet deep and fifteen miles in diameter." The two American super-weapons—the atomic bomb and the strategic bomber—have brought the utopian Pax Americana to the planet.

The nation's motives for developing atomic weapons in the novel are precisely those of the Manhattan Project: to forestall Nazi use and to achieve a lasting peace. Like those who later were to make the decision to use atomic bombs, *Lightning in the Night* assumes that the first nation to deploy atomic weapons wins and ends the race. The actual American president in 1945–46, like his fictional counterpart, would fail to realize that this might just accelerate the race for superweapons and open an epoch increasingly dominated by them.

No such failure of imagination appears in the most distinctively *American* work about atomic weapons prior to Hiroshima, Robert A. Heinlein's aptly titled "Solution Unsatisfactory." Published in 1941, this prescriptive tale about the prevention of full-scale nuclear war is in some ways more ominous than the fiction that foresaw a global holocaust.

By 1941, only two years after publishing his first story, Heinlein was already being acclaimed as the most popular living writer of science fiction. He soon became a major figure in American culture. By the mid-1960s, he had an audience of millions. His works have now been translated into about thirty languages, and all but one of his thirty-nine published books are currently in print in mass-market paperbacks. Perhaps more than any other person, he helped shape the modern science-fiction movie, television serial, and juvenile novel. Several words he coined have become part of our language.

The varied uses of Heinlein's coinages suggest the diversity of his appeal. Engineers call a type of servomechanism a *waldo,* a term he invented in 1942. His 1961 verb *grok* became part of the youth movement of that decade. *TANSTAAFL* ("There Ain't No Such Thing As A Free Lunch"), coined in 1966, has become a shibboleth for the Libertarian movement, which has made him one of its standard-bearers. His 1961 novel *Stranger in a Strange Land* was a bible for some participants in the counterculture of the 1960s. Yet Heinlein has also long been revered in militaristic circles: he presented a major address to the United States Naval Academy, flew as a distinguished guest in one of the first models of the B-1 bomber, and became a founding father of the campaign for Star Wars.

In *Robert A. Heinlein: America as Science Fiction,* I tried to show how the Heinlein cultural phenomenon presents crucial aspects of modern American ideology and imagination. Nowhere is this more evident and significant than in "Solution Unsatisfactory," a story of how the ultimate American atomic superweapon will change the world.

Heinlein accurately describes the state of atomic research at the time the story was published in 1941, including this description of the secrecy that was actually decreed in mid-1940:

> Someone in the United States government had realized the terrific potentialities of uranium 235 quite early and, as far back as the summer of 1940, had rounded up every atomic research man in the country and had sworn them to silence. Atomic power, if ever developed, was planned to be a government monopoly, at least till the war was over. (59)

Then he extrapolates forward until 1945, when Hitler has gained the upper hand in Europe. Britain's main ally, America, is still officially neutral, and the American atomic researchers, running out of time in their efforts to produce deliverable atomic bombs, opt for the much simpler production of deadly radioactive dust to be scattered by bombs dropped from airplanes.

In fact, Heinlein's analysis of the probable form and chronology of atomic weapons was closely paralleled by the original National Academy of Sciences report, delivered by Arthur Compton on May 17, 1941, a month after Heinlein's story appeared in the nominally dated May issue of *Astounding Science-Fiction.* Compton argued that the "production of violently radioactive materials . . . carried by airplanes to be scattered as bombs over enemy territory" might be achieved a year after a successful chain reaction, which meant "not earlier than 1943," while "atomic bombs can hardly be anticipated before 1945." In December 1941, Eugene Wigner and Henry D. Smyth "concluded that the fission products produced in one day's run of a 100,000 kw chain-reacting pile might be sufficient to make a large area uninhabitable." By 1942, scientists working on the Manhattan Project were discussing how airplanes might be used to drop such fission products. In 1943, the idea resurfaced in Enrico Fermi's proposal, supported by J. Robert Oppenheimer, to poison hundreds of thousands of people with lethal radioactive strontium.[13] Heinlein's tale may have directly influenced these schemes, for, as his editor, John Campbell, attested, it "was read, and widely discussed, among the physicists and engineers working on the Manhattan Project."[14]

In "Solution Unsatisfactory," America's radioactive dust is the ultimate weapon, against which there can be no defense. Heinlein argues that the very possession of such a weapon immediately nullifies democracy and most political freedom, confronting the nation with a limited number of options, all "unsatisfactory."

The story is told by "an ordinary sort of man" who suddenly finds himself thrust into the center of history as the gofer of "liberal" but "tough-minded" Clyde C. Manning, western congressman, U.S. Army colonel, overseer of American atomic research, and later the much-hated savior of the world from its own follies. Both the narrator and the hero use an American vernacular to comprehend the issues and choices before them, giving the story its peculiarly American tone and attitudes. This is typical Heinlein, whose language and metaphors, like those of fellow Missourian Harry Truman, express America's handling of atomic weapons in terms of a romanticized frontier past and mythologized future destiny.

Heinlein was familiar with earlier future-war fiction. His narrator, serving as a kind of American Everyman and surrogate for the reader, voices the typical thinking about superweapons dating from the late 1880s, when the United States was preparing to transform from a republic into a colonial power, through *Lightning in the Night,* published just a few months earlier: " 'It's our secret, and we've got the upper hand. The United States can put a stop to this war, and any other war. We can declare a *Pax Americana,* and enforce it' " (65). But Colonel Manning, unlike the men who were to make the decision to explode atomic weapons on Japan, understands the naïveté and superficiality of such thinking. He explains that " 'it won't remain our secret; you can count on that.' " The atomic weapon will soon become " 'a loaded gun held at the head of every man, woman, and child on the globe!' "

The narrator sees no possible solution other than American hegemony, though even he has a few misgivings:

> It seemed to me that a peace enforced by us was the only way out, with precautions taken to see that we controlled the sources of uranium. I had the usual American subconscious conviction that our country would never use power in sheer aggression. Later, I thought about the Mexican War and the Spanish-American War and some of the things we did in Central America, and I was not so sure—. (66)

At this stage, Colonel Manning also accepts the *"Pax Americana"* as the only option. Indeed, he foresees that the superweapon will determine the next sequence of events, forcing the United States to use atomic weapons and accept its destiny as global ruler:

> . . . it would not do to wait, to refrain from using the grisly power, until someone else perfected it and used it. The only possible chance to keep the world from being turned into one huge morgue was for us to use the power first and drastically—get the upper hand and keep it. . . .
>
> The United States was having power thrust on it, willy-nilly. We had to accept it and enforce a world-wide peace, ruthlessly or drastically, or it would be seized by some other nation. There could not be coequals in the possession of this weapon. (67–68)

So Manning arranges for the United States to supply British bombers with enough radioactive dust to exterminate the population of Berlin. But even this, the only American fiction prior to August 1945 in which the United States actually uses atomic weapons on human beings, recognizes grave moral obligations that were utterly ignored by those who ordered the atomic bombing of Japan without any demonstration or warning, and with the intent to kill as many residents of Hiroshima and Nagasaki as possible. Before authorizing the massacre, the United States provides a demonstration for the German ambassador and has German blanketed with three waves of leaflets, calling for peace on terms that would not be "vindictive," showing photographs of a herd of cattle slaughtered by the atomic weapon, and finally warning people to evacuate the cities: "As Manning put it, we were calling 'Halt!' three times before firing. I do not think that he or the President expected it to work, but we were morally obligated to try" (70).

When the bombs with their radioactive dust are dropped on Berlin, every person and animal in the city dies. After seeing films of the carnage, the narrator loses "what soul I had" (73). The story, however, is less concerned with the moral issues that might deter first use than with the inescapable future consequences.

Manning had expressed the problem in one of his characteristically American metaphors:

"Once this secret is out—and it will be out if we ever use the stuff!—the whole world will be comparable to a room full of men, each armed with a loaded .45. They can't get out of the room and each one is dependent on the good will of every other one to stay alive." (65)

So now that the weapon has been used, what can be done? Allowing other nations to possess it would mean, according to Manning, inevitable global atomic war. When a cabinet member proposes turning over the dust to a "world-wide democratic commonwealth" of nations, Manning replies with the argument that no nation other than the United States is capable of democracy:

"Four hundred million Chinese with no more concept of voting and citizen responsibility than a flea. Three hundred million Hindus who aren't much better indoctrinated. God knows how many in the Eurasian Union who believe in God knows what. The entire continent of Africa only semicivilized. Eighty million Japanese who really believe that they are Heaven-ordained to rule. Our Spanish-American friends who might trail along with us and might not, but who don't understand the Bill of Rights the way we think of it. A quarter of a billion people of two dozen different nationalities in Europe, all with revenge and black hatred in their hearts." (76)

The only solution, according to this "wise and benevolent" American in whom "ordinary hard sense had been raised to the level of genius," is "'a military dictatorship imposed by force on the whole world'" (75).

To maintain its monopoly, the United States must implement world rule with no delay, which means ignoring not only world law but the U.S. Constitution. So the President declares "a state of full emergency internally" and sends his "Peace Proclamation" to every sovereign state:

Divested of its diplomatic surplusage, it said: The United States is prepared to defeat any power, or combination of powers, in jig time. Accordingly, we are outlawing war and are calling on every nation to disarm completely at once. In other words, "Throw down your guns, boys; we've got the drop on you!" (77)

This quaint frontier metaphor is to be implemented by all nations turning over to the United States all their airplanes, that other superweapon necessary to deliver the radioactive bombs.

However, there is one nation so uncivilized, unreasonable, antidemocratic, and dastardly as to dispute the American global rule—the "Eurasian Union," now under the control of the "Fifth Internationalists," who have paralleled our atomic research. In the fictional 1945, unlike the actual history of that year, the sneak atomic air attack is delivered not by but upon the United States. We retaliate by wiping out Vladivostok, Irkutsk, and Moscow, and sending "the

American Pacification Expedition'' to occupy the conquered land. Now the United States has the job of "policing the world.''

The President of the United States is a good man, so he and Colonel Manning wish to prevent the atomic weapon from being used "to turn the globe into an empire, our empire,'' for "imperialism degrades both oppressor and oppressed.'' They decide that the power "must not be used to protect American investments abroad, to coerce trade agreements, for any purpose but the simple abolition of mass killing.'' In characteristic American and Heinlein style, "Manning and the President played by ear'' and used "horse sense,'' establishing treaties "to commit future governments of the United States to an irrevocable benevolent policy.''

Manning becomes Commissioner of World Safety, forming the international Peace Patrol, whose pilots, armed with the atomic weapon, are never to be assigned to their own country. The Peace Patrol is welded together by "esprit de corps,'' while the main check on its new recruits is "the President's feeling for character.''

But in 1951 the good President is killed in a plane crash, and his office is assumed by the isolationist Vice President, allied with a senator who had tried to use the Peace Patrol to recover expropriated holdings in South America and Rhodesia. They attempt to arrest Manning, but the pilots of the Peace Patrol intervene and make Manning "the undisputed military dictator of the world.''

Though Manning dislikes this role, he embodies the true national hubris. Heinlein, who may be said to represent America as science fiction, could imagine no solution to the dilemma of superweapons more satisfactory than the one proposed in such early-twentieth-century tales as Simon Newcomb's 1900 *His Wisdom, the Defender* and John Stewart Barney's 1915 *L.P.M.*: an American mastermind would have to rule the world for its own good.

When Groff Conklin reprinted "Solution Unsatisfactory'' in his influential 1946 anthology *The Best of Science Fiction,* he noted that he did so "against my better judgment'' for the story "seems quite dangerous to me.''[15] He prefaced it with a quotation from Harold Urey in the *Washington Post* of November 2, 1945:

ATOM WAR THREAT MAY FORCE U.S. TO SELECT DICTATOR
Chicago, Nov. 1 (AP).—Dr. Harold C. Urey, Nobel prize winner who helped develop the atomic bomb, declared today that in "five years or perhaps less'' when any nation can make the bomb, it may be necessary for the United States to establish a dictatorial form of government to act quickly against an atomic war theat.

"I do not see any way to keep our democratic form of government if everybody has atomic bombs,'' Urey said. "If everyone has them, it will be necessary for our government to move quickly in a manner not now possible under our diffused form of government.''

Heinlein in some senses was prophetic: atomic weapons have helped to wipe out the core of democracy, placing the most important national—and even human—decisions in the hands of a single individual. In the decades since Heinlein's story and Urey's prediction, we have become so accustomed to the unconstitutional assumption of war-making powers by the presidency that most Americans no longer perceive this as an abrogation of democratic government. Meanwhile, the free flow of knowledge vital for citizens to make informed choices has been choked off in the name of "security."

As Heinlein accurately noted, it was in the summer of 1940 that the U.S. government buried atomic research under a cloak of secrecy. Even the secrecy itself was a dark secret. Discussions of such work disappeared from scientific journals, where almost a hundred articles on nuclear fission had appeared in 1939 alone. Newspapers, magazines, news services, and radio broadcasters were soon ordered not to mention atomic power, cyclotrons, betatrons, fission, uranium, deuterium, protactinium, and thorium. Army Intelligence later even attempted to block access to back issues containing popular articles on atomic energy, such as "The Atom Gives Up" in the *Saturday Evening Post* of September 7, 1940. In the September 8, 1945, issue, the *Post*'s editors revealed that this vast effort "to wipe the whole subject from memory" went so far as to ask magazines and "public libraries all over the land" to turn over to Army Intelligence the names of people asking to look at these back issues.[16]

Thus the free exchange of knowledge that had characterized the community of science ended abruptly, to be replaced by one of the grotesque features of our times: the attempt to transform crucial parts of human knowledge into secrets to be classified by the state and kept inviolate by the secret police. When the seventeenth-century Church authorities forced Galileo to stop promulgating Copernicanism, at least they felt that they were prohibiting the dissemination of *false* belief. The U.S. government, however, was consciously outlawing scientific *truth* about the fundamental nature of the universe. As early as 1941, John J. O'Neill—science editor of the *New York Herald Tribune,* author of one of the 1940 articles on atomic energy, and president of the National Association of Science Writers—charged that censorship on atomic research amounted to "a totalitarian revolution against the American people." Pointing to the devastating potential of an atomic bomb utilizing uranium 235, O'Neill offered a fateful prophecy: "Can we trust our politicians and war makers with a weapon like that? The answer is no."[17]

This "security" had profound effects on the main subject of this book, the superweapon in American culture. For the crucial years 1940–45, no public discussion of atomic weapons was permitted in the nation that claimed to be leading the fight for democracy and freedom, while its government spent two billion dollars of public funds to develop these weapons in secret. Cultural conceptions of atomic superweapons were still acting as a powerful force on the nation's leaders and on the scientists designing the weapons. Yet when the atomic bombs were finally exploded, they astonished the American people, most of whom had forgotten the enormous public interest and knowledge before the five-year blackout of consciousness. The decision to transform the United States into the nation that first used nuclear weapons as instruments of mass slaughter had been made secretly by a handful of men.

During these five years, the only Americans exposed to any public thoughts about atomic weapons were the readers of science fiction. At first, the government paid no serious attention to the appearance of atomic weapons in science fiction, which was considered a subliterary ghetto inhabited by kids and kooks. But as science fiction began losing its monopoly to the Manhattan Project, every science-fiction atomic bomb became a deadly serious matter for the government and its agents. So even though there already had been widespread public discussion of the two main technical problems of atomic bombs—isolating sufficient quantities of the fissionable isotope uranium 235 from uranium 238 and achieving critical mass suddenly enough to set off an explosive chain reaction—government censorship now clamped down on the imagination of fiction writers.

In early 1945, Philip Wylie submitted to *American Magazine* his novella *The Paradise Crater,* which imagined Nazis after their defeat in World War II attempting to conquer the United States with atomic bombs using uranium 237. The editors rejected the story as too implausible, so it was submitted to *Blue Book Magazine,* which requested government approval. Wylie was suddenly placed under house arrest; an Army Intelligence major informed him that he was personally prepared to kill the author if necessary to keep the weapon secret.[18]

Even science-fiction comic strips were censored. On April 14, the McClure Newspaper Syndicate ran the first strip of "Atom Smasher," a new Superman series pitting America's favorite superhero against a cyclotron. The Office of Censorship promptly forced the running of a substitute series (in which Superman played a baseball game single-handed).[19]

A more difficult problem for the government's suppression of fictitious atomic weapons came with the publication of Cleve Cartmill's "Deadline" in the March 1944 *Astounding Science-Fiction.* Military intelligence officials visited and grilled Cartmill, and also descended on the editor of *Astounding,* John Campbell. When Campbell was ordered to cease publishing stories about atomic bombs, he demurred on the grounds that these weapons appeared so frequently in *Astounding* that their sudden disappearance would be a signal to the Axis that they were close to being produced, thus prodding the Nazis to dedouble their own nuclear-weapons research.[20]

"Deadline" imagines the Axis powers of World War II developing a nuclear bomb based on uranium 235. (The setting is thinly veiled by such devices as spelling backward the two warring alliances, the "Sixa" and the "Seilla.") Relying on information already published in technical journals prior to 1941, Cartmill imagines the atomic bomb working like this: a "trigger" breaks through metal shields, releasing enough neutrons to drive blocks of uranium 235 above critical mass, leading to an explosive chain reaction. As his editor, Campbell, pointed out in a 1947 book on the history of atomic research and engineering: "Every item of information necessary for the design of an atomic bomb, down to and including the arming mechanism," had appeared in the prewar journals used by Cartmill, so the so-called "secret" of the atomic bomb was simple to reconstruct: "If a science-fiction author can outline the structure of an atomic bomb accurately enough to worry military intelligence, it may fairly well be assumed that the scientists of many nations can do at least as well."[21]

Cartmill's description would have been about as helpful to the scientists of another nation as the rough sketch that Julius and Ethel Rosenberg allegedly conspired to transmit in 1945 to the Soviet Union (then America's ally), for which they were executed in 1953. Yet even as early as Cartmill's tale, the myth of the atomic ''secret'' was already becoming an instrument for suppressing dissent against the reign of America's superweapons.

Though ''Deadline'' is primarily an action adventure in which a lone intelligence agent, aided by the revolutionary underground, manages to defuse the bomb just before the Sixa can use it, the story still sends messages that fortunately escaped censorship. Cartmill's anti-Fascist Seilla rule out the use of the atomic bomb because they are already winning the war, and thus have no overriding reason to unleash such a terrible force upon the world. For they realize the possibility with which we now live: nuclear weapons could threaten the existence of ''the entire race.'' With the obliteration of all consciousness, even time itself, which ''exists only in consciousness,'' might be annihilated. ''There won't be any time, unless dust and rocks are aware of it.'' Did such thoughts occur to the men who decided to explode atomic weapons on two Japanese cities the year after the publication of ''Deadline''?

9

Atomic Decision

The decision to use nuclear weapons on two Japanese cities was perhaps the most important conscious choice in human history. The chain reactions within those two bombs have set off macrohistorical chain reactions whose ultimate consequences we, living every day under their menace, cannot predict. Nor of course can we be at all certain how events might have unfolded if the United States had used its temporary nuclear monopoly not for spectacular slaughter but as a means to attain the permanent banishment of atomic weapons. Yet comprehending this decision may be essential for our survival.

The scientists who instigated and led in the development of the atomic bomb for the United States at first did not envisage its being used, except as a deterrent to possible Nazi use and an impetus to world government. The Nazis surrendered on May 7, 1945, over two months before the U.S. bomb was tested, and Japan of course did not then have the potential either to build or to deliver atomic weapons. Many of the atomic scientists began to warn of the dreadful future that would inevitably ensue if the United States used the bomb in the war.

On June 11, James Franck and six other leading scientists submitted a report to Secretary of War Henry Stimson warning that the United States was about to set the world on "a path which must lead to total mutual destruction" unless the nation led the world into an agreement "barring a nuclear arms race." In prophetic words, they explained how using the bomb would deeply shock not only the Soviet Union but many other nations and peoples, extinguishing hope for any such accord: "It may be difficult to persuade the world that a nation which was capable of secretly preparing and suddenly releasing a new weapon, as indiscriminate as the [Nazi] rocket bomb and a thousand times more destructive, is to be trusted in its proclaimed desire of having such weapons abolished by international agreement."[1]

An even more prescient message was articulated in the July 17 petition to the President written by Leo Szilard and signed by sixty-eight of the atomic scientists:

The development of atomic power will provide the nations with new means of destruction. The atomic bombs at our disposal represent only the first step in this direction, and there is almost no limit to the destructive power which will become available in the course of their future development. Thus a nation which sets the precedent of using these newly liberated forces of nature for purposes of destruction may have to bear the responsibility of opening the door to an era of devastation on an unimaginable scale.[2]

The scientists warned that the country holding a monopoly on nuclear weapons had the "solemn responsibility" not to unchain this Frankenstein's monster. Otherwise, they predicted with chilling accuracy, eventually "the cities of the United States as well as the cities of other nations will be in continuous danger of sudden annihilation."

In "Deadline," Cleve Cartmill had assumed that the Allies, since they were clearly winning the war, certainly would not unleash atomic weapons on the world. Even Heinlein, the only pre-Hiroshima author to imagine the United States carrying out an atomic attack, had presented it as morally unthinkable without both a prior nonmilitary demonstration of the hideous powers of the atomic weapon and clear warnings to the civilian population. Dozens of the atomic scientists urged the President to warn the Japanese and allow them a clear option for surrender. Instead, without any warning whatsoever, two Japanese cities were cremated. Why?

The official answer given then was that the atomic bombs were intended to shorten the war and eliminate the need to invade Japan. Indeed, most Americans still believe that the bombs defeated Japan and that without them an invasion would have been necessary. This belief is not based on the historical facts.

Even before the bomb was tested in New Mexico on July 16, every city in Japan, except for the four that were being saved as possible targets for the new weapon, had already been incinerated (see Chapter 6). U.S. planes and ships were devastating the home islands with impunity. A huge Soviet army had been moved five thousand miles to prepare for an all-out assault on the main surviving Japanese forces, an attack scheduled to take place between August 8 and August 15. Moreover, as President Truman was well aware, the Japanese government had several times expressed its desire to negotiate a surrender.

That there was no military need to drop the bomb was widely recognized by U.S. military authorities. The U.S. Strategic Bombing Survey concluded unequivocally that ". . . Japan would have surrendered if the atomic bombs had not been dropped, even if Russia had not entered the war, and even if no invasion had been planned or contemplated." Fleet Admiral William Leahy, Chief of Staff to both Roosevelt and Truman, flatly declared: ". . . the use of this barbarous weapon at Hiroshima and Nagasaki was of no material assistance in our war against Japan. The Japanese were already defeated and ready to surrender." Asked in 1945 for his opinion on dropping the atomic bomb, General of the Army Dwight D. Eisenhower replied: "I was against it on two counts. First the Japanese were ready to surrender and it wasn't necessary to hit them with that awful thing. Second, I hated to see our country be the first to use such a weapon."[3]

Why, then, was it used? As early as 1948, P.M.S. Blackett, British Nobel laureate in physics and leading wartime military expert, concluded that "the dropping of the atomic bombs was not so much the last military act of the second World War as the first major operation of the cold diplomatic war with Russia."[4] In the ensuing four decades, a growing mass of evidence and scholarship has shown that President Truman, Secretary of State Jimmy Byrnes, and Secretary of War Stimson passionately embraced the bomb as the winning weapon in their struggle with the Soviet Union over the shape of the postwar world.[5]

When Leo Szilard attempted to warn Byrnes of the terrifying implications of atomic weapons, the Secretary of State "did not argue that it was necessary to use the bomb against the cities of Japan in order to win the war" but rather that "our possessing and demonstrating the bomb would make Russia more manageable in Europe."[6] Stimson looked upon the atomic bomb as "a badly needed 'equalizer'" of Soviet power. He and Truman both considered the bomb their "master card," filling in "a royal straight flush" in the poker game they were playing against the Russians. Truman delayed the Potsdam conference until July 17, the day after the Alamogordo test, in order to have the bomb in his "pocket" while negotiating with the Soviet Union.[7]

Several crucial points have become clear:[8]

1. Truman and his key advisers were well aware that the Japanese were attempting to negotiate a surrender, conditional merely on an assurance that they be allowed to keep their emperor. On July 18, the President discussed in his diary a "telegram from Jap emperor asking for peace," and he received frequent reports from both diplomats and the Office of Strategic Services (OSS) on the terms of the Japanese peace feelers.[9]

2. The Anglo-American ultimatum that issued from the Potsdam conference called for unconditional surrender and deliberately refrained from assuring the Japanese that they would be allowed to keep the emperor. Although Japan publicly rejected this ultimatum, OSS Director William Donovan informed the President on August 2 that the Japanese had indicated privately that quite a different reply would be made shortly.[10] In any event, the surrender terms accepted after the nuclear bombings did allow Japan to keep the emperor.

3. The Soviets had pledged to enter the war against Japan between August 8 and August 15. Truman had no doubts that they would fulfill this promise, and believed that this alone would end the war. In his diary entry of July 17, the President noted that the Soviet Union would "be in Jap war on August 15th" and then unequivocally stated his verdict of what this would do: "Fini Japs when that comes about."[11]

4. Dropping the bomb in *August* had nothing whatsoever to do with saving American lives by forestalling an imminent invasion of Japan, which was not planned to take place for several months.

5. Before and immediately after the Nazi surrender in May, the United States had been attempting to expedite the Soviet entry into the war against Japan. But as the bomb became a reality, the United States reversed its goal. Secretary of State Byrnes now described his efforts to stall the Soviet entry into the war against Japan so that "Russia will not get in so much on the kill."[12]

6. The bombing of Hiroshima was moved up to take place on August 6, two days before the promised Soviet entry, which came, as scheduled, on August 8.

7. The vast Soviet offensive was expected to crush the Japanese armies on the Asian mainland, which it did, killing and capturing hundreds of thousands of troops and hundreds of generals in several days.

8. Evidence suggests that the Japanese surrender was hastened as much by the Soviet onslaught as by the atomic bombings, which were no more devastating at the time than the firebomb raids on Tokyo and other Japanese cities.[13]

9. Truman and Byrnes were anxious to minimize the Soviet role in the Japanese defeat. Without waiting to see whether the Hiroshima bombing together with the Soviet attack would induce Japan to admit defeat, the United States dropped its only other existing atomic bomb on Nagasaki, on August 9.

Would Japan have capitulated without a nuclear attack when the Soviet Union joined the war? That question must remain forever unanswerable because the leaders of the United States consciously chose to keep it from being answered. Knowing that the Soviet army was poised to launch its promised full-scale attack and believing that this would probably end the war, they chose to eliminate the possibility of Japan's surrendering without being subjected to nuclear bombing. President Truman and the main architect of his atomic policy, Secretary of State Byrnes, decided to use the atomic bomb on Japan before August 8 to be certain that it would be demonstrated prior to both the scheduled Soviet intervention and the possible Japanese surrender.

Would the bombs have been used on Japan eventually if the atomic diplomacy against the Soviet Union had not been a goal of the U.S. leaders? Some scholars have argued that the anti-Soviet atomic threat came as a mere bonus, that the United States would have used its nuclear bombs on Japan in any event. They maintain that Truman and Byrnes never seriously considered *not* using the atomic bomb.

Certainly the President did not seem to have any moral qualms. Since moral constraints had already been obliterated by the fire-storm raids, atomic weapons could be considered just bigger and better bombs capable of incinerating cities more efficiently—and more dramatically. Another factor, as Byrnes explained to Szilard, was that having secretly spent over two billion dollars on the Manhattan Project, the Administration had to have something sensational to show to the U.S. Congress and people.[14] In short, blind to visions of the future they were inaugurating, they simply acted on the unexamined assumption that American technology had created the atomic bomb for them to use as a weapon of mass slaughter.

Neither of the two competing principal arguments about the decision to use atomic weapons paints a very pretty picture. In one it is anti-Soviet atomic diplomacy, and in the other sheer moral and intellectual obliviousness, that unleashed atomic holocausts on Japan and forced future generations to live under the growing threat of annihilation. Neither argument, however, quite addresses the most frightening aspects of the decision-making process, which involved deep levels of irrationality, fed by myths and beliefs explored in this book. Key men in the process believed that destroying cities with atomic bombs would not only force speedy unconditional Japanese surrender and intimidate the Soviet Union, but would also bring about an end to war.

Why did they think as they did? The imagination and consciousness of these men were shaped, in part, by forces of which they were only dimly aware. When Harry Truman made his fateful decision, he was behaving as a fairly typical American man of his era, a product of his culture. This is not to argue that he was directly influenced by the pre–World War I future-war fantasies serialized in the magazines to which he subscribed as a young farmer in Missouri (see pages 52–53) or by later fiction about atomic weapons, such as *Lightning in the Night* and "Solution Unsatisfactory." Both the man and the fiction expressed the same culture.

For example, compare his thoughts and actions with those of the President in the first novel that imagined the United States in possession of weapons based on radioactivity, Roy Norton's *The Vanishing Fleets,* serialized in the *Associated Sunday Magazines* in 1907 (see pages 41–44). In response to a sneak attack by Japan, the President authorizes the secret expenditure of vast sums to produce "the greatest engine of war that science has ever known," a superweapon merging radioactivity with air power. Having vowed "to give my life to peace," the President then decides he must *use* this "most deadly machine ever conceived . . . thereby ending wars for all time."

The same reasoning guided the actual decision-makers in 1945. Defending his role in the decision to use atomic weapons, Secretary of War Stimson explained that James Conant, president of Harvard University, had persuaded him "that the bomb must be used" because "that was the only way to awaken the world to the necessity of abolishing war altogether. No technological demonstration . . . could take the place of the actual use with its horrible results."[15] Edward Teller, in refusing to sign Leo Szilard's petition against using the bomb, argued that the weapon was so horrible that it might actually help "to get rid of wars," so "[f]or this purpose actual combat-use might even be the best thing."[16]

On his way to Potsdam, President Truman recited, from the paper he had been carrying in his wallet since 1910, those lines from Tennyson prophesying that "ghastly" weapons dropped from airships would still "the war-drum" and bring about "the Federation of the world" (see page 84). After being given details of the Alamogordo test, Truman recorded in his diary his thoughts about the atomic bomb: "It seems to be the most terrible thing ever discovered, but it can be made the most useful." When he learned of the atomic bombing of Hiroshima, he immediately proclaimed: "This is the greatest thing in history."[17]

The fate of Hiroshima and Nagasaki did indeed demonstrate the power of the first primitive nuclear weapons. It also demonstrated the power of the super-weapon in American culture. By the eve of World War II, the old American dream of the ultimate peacemaking weapon, first projected by Robert Fulton and popularized by future-war fantasies during the rise of U.S. imperialism, had been transformed—as expressed in Edison's exaltation of industrialized war and Billy Mitchell's air-power doctrine—into a panorama of American military-industrial invincibility. The practical application of this world view was the "strategic" bombing of Europe and the Superfortress devastation of Japan. Now it seemed to the President and his advisers that they had the long-sought superweapon that could achieve the destiny manifest in that vision: perpetual peace under the global hegemony of the United States of America. With such a prospect for the future, what significant reasons could there possibly be not to use the atomic bomb?

10

The Rise of
Nuclear Culture

Nuclear weapons have transformed the fundamental conditions of human existence. It has even been suggested that "the Bomb has become one of those categories of Being, like Space and Time, that . . . are built into the very structure of our minds, giving shape and meaning to all our perception," and that a "history of 'nuclear' thought and culture" might be "indistinguishable from a history of all contemporary thought and culture."[1]

Many Americans felt this metamorphosis within hours of hearing about Hiroshima. "The great question of all civilization and history has come," editorialized the *Newark Star-Ledger* on the morning of August 7. That day's *New York Herald Tribune* went even further: ". . . one senses the foundations of one's own universe trembling." "Everything else seemed suddenly to become insignificant," wrote the Reverend John Hayne Holmes of the Community Church of New York City, describing his initial response to the news: "I seemed to grow cold, as though I had been transported to the waste spaces of the moon. . . . For I knew that the final crisis in human history had come. What that atomic bomb had done to Japan, it could do to us." A columnist in the *New York Times* expressed the widespread "fear and deep misgiving": "The earth is no longer solid. Out of the forces that hold it together human genius has summoned forces that tear it apart." In "Modern Man is Obsolete," a *Saturday Review* essay written on the night of August 6, Norman Cousins prophesied that the bomb would transform "every aspect of man's activities, from machines to morals, from physics to philosophy, from politics to poetry." *Time*'s first postwar issue declared that the bomb had "split open the universe," while demonstrating the bomb on human beings had "created a bottomless wound in the living conscience of the race"; thus "the war itself shrank to minor significance" while its outcome became the "most grimly Pyrrhic of victories."[2]

Within days, public discourse voiced themes characteristic of the nuclear epoch: deterrence, disarmament, alienation, fear, despair, helplessness, psychological numbing, moral erosion, and apocalypse.[3] But intimations of crisis were at first hard to hear amid the celebration of America's atomic victory and the anticipated Pax Americana. *Newsweek*'s view, which summed up the prevailing sentiment, could have been lifted straight out of many prewar American novels about superweapons: "The international race for the conquest of atomic power . . . was rightly interpreted as a race for world military domination. For the first country to solve the problem would have a weapon to rule the world— constructively or destructively. The Anglo-Americans won."[4]

Few voices in the media questioned the wisdom of President Truman's decision to end World War II, though victory was already certain, by beginning the nuclear age. Instead, there was general acceptance of the official line that Japan had been defeated by the atomic bomb. But the military powers thereby attributed to the new superweapon had a disturbing side effect.

As in many of its pre–World War I fantasies of war, the American imagination projected future scenarios that reversed the roles of the past. Although the United States had a monopoly on atomic weapons, a wave of fear began to rise as Americans conjured up frenzied images of themselves as potential victims, rather than perpetrators, of atomic warfare. The Frankenstein metaphor appeared as early as the first night's commentary by H. V. Kaltenborn, dean of radio broadcasters: "For all we know, we have created a Frankenstein! We must assume that with the passage of only a little time, an improved form of the new weapon can be turned against us." The *Wall Street Journal* reported that scientists were already predicting "atomic bombs a thousand times as powerful as those that now exist" and cheap enough "to treat every square mile in the U.S. the way Hiroshima was." In everyday life, people could now feel what the atomic scientists had warned about in July, that using the atomic bomb would eventually place "the cities of the United States . . . in continuous danger of sudden annihilation." "The inevitability of an atomic bomb attack on the U.S.A.," remarked a British observer, before long became "a national fetish."[5]

Newpapers, radio commentators, and magazines, as well as ordinary citizens and science-fiction writers, even envisioned nuclear holocausts that could destroy civilization and possibly exterminate our species. Science may have "signed the mammalian world's death warrant," and "deeded an earth in ruins to the ants," announced the August 7 *St. Louis Post-Dispatch*.[6]

One crucial question from then until now has been whether any defense is possible. In *Life*'s first post-Hiroshima issue, Hanson Baldwin suggested that we might have unleashed an invincible "Frankenstein's monster" that someday would appear in the form of a " 'pushbutton' war, using missiles of tremendous range and terrible destructive power." General "Hap" Arnold, head of the Army Air Forces, announced at an August press conference that the United States was already secretly experimenting with "Buck Rogers things," such as missiles that would be able to hit any target in the world and "improved atomic bombs" that would "be destructive beyond the wildest nightmares of the imagination." Although *Newsweek* referred to Arnold's vision as "the ultimate in war," which "might be won by the nation that pressed the push buttons first," it also mentioned, without noting any contradiction, Arnold's prediction of defensive missiles that would automatically seek out and destroy intercontinental missiles. The *Washington Post* of August 26 was already debating Star Wars technology, including possible beam weapons to destroy "atom bomb rockets shot from thousands of miles away." *Time* cited reports that the Soviet Union was beginning to ring its cities with "infra-cosmic ray" generators capable of intercepting and exploding atomic bombs at a range of twelve miles.[7]

In such a context, distinctions between science fiction and the rest of American culture rapidly began to crumble. For example, *Life* dramatized General Arnold's official report to the Secretary of War on possible future developments in a flamboyant nine-page pictorial display entitled "The 36-Hour War." In this imaginary future war waged with mammoth intercontinental ballistic missiles fired from deep underground silos, an unnamed enemy strikes first from secret bases in Africa, annihilating thirteen U.S. cities and killing forty million Americans in an hour. The damage would have been even worse if our defensive rockets had not intercepted some of the attackers in space, as shown in an artist's breathtaking tableau. Except for the evidence of catastrophic radioactivity, the final picture of the ruins of New York might have illustrated many nineteenth-century future-war fantasies, such as "The End of New York" (1881). Like many of the early future-war fantasies, "The 36-Hour War" has a happy ending: "But as it is destroyed the U.S. is fighting back. The enemy airborne troops are wiped out. U.S. rockets lay waste the enemy's cities. U.S. airborne troops successfully occupy his country. The U.S. wins the atomic war."[8]

Life's indulgence in such science fiction was part of a national pastime, for the atomic bomb stimulated the whole nation to imagine scenarios filled with superweapons. Indeed, from 1945 on, all serious political discourse would have to participate in this genre, for the forms and consequences of future weapons would be central to national and international policy. The thinking that led to the most crucial political acts, including the Baruch Plan and the Bikini tests of 1946, was intermingled with future-war science fiction now pulsing from the roots of the culture.

By 1946, American culture had become so "atomic" that scores of businesses and dozens of race horses had been named for the atom. There were Atomic songs, Atomic Cocktails and Atomic Earrings. General Mills was offering an "Atomic 'Bomb' Ring" for fifteen cents and a KIX cereal box top.[9] But this trivia was mere spindrift blown off the churning waves and crosscurrents of anxiety, confusion, and conflict over atomic weapons.

Hundreds of books and articles about the past and future of the atomic bomb were published in 1946. The most familiar of these today is John Hersey's *Hiroshima,* his understated but lacerating reconstruction of the experience of several people in the doomed city. Appearing as the entire issue of the August 31 *New Yorker,* reprinted in newspapers across the country, broadcast on the ABC radio network, and published as a runaway best-seller that has remained in print ever since, *Hiroshima* thrust the personal consequences of atomic weapons into American cultural life.[10]

Fear and pity aroused by contemplation of the bombed cities of Japan or prophecies of a similar fate for American cities did not, however, deal with the problem of what to *do* about atomic weapons. After all, throughout the debates of the past decades, none of the major conflicting arguments has denied the horrors of nuclear war. While many voices have called for banning the bomb or placing it under international control, others have responded that the only security for America lies in nuclear supremacy. In such influential collections as *The Absolute Weapon,* edited in 1946 by Bernard Brodie for the Yale Institute of International Studies, the old preparedness dogma reemerged as the theory of nuclear deterrence: ". . . the first and most vital step in any American security program for the age of atomic bombs is to take measures to guarantee to ourselves in case of attack the possibility of retaliation in kind."[11]

In the vanguard of the movement to ban or gain international control of atomic weapons were many of the atomic scientists themselves. Just as Victor Frankenstein became obsessed with killing the monster he had created, these scientists now dedicated themselves to alarming the world about their monstrous creation. With their newly established *Bulletin of the Atomic Scientists* as a forum, they tirelessly spoke, wrote, and organized conferences about the global threat they had unleashed. Another best-seller of 1946 was their extremely influential collection, *One World or None.*[12]

Lewis Mumford's "Gentlemen: You Are Mad!" in the March 2, 1946, *Saturday Review* opened with these words:

> We in America are living among madmen. Madmen govern our affairs in the name of order and security. The chief madmen claim the titles of general, admiral, senator, scientist, administrator, Secretary of State, even President. And the fatal symptom of their madness is this: they have been carrying through a series of acts which will lead eventually to the destruction of mankind, under the solemn conviction that they are normal responsible people, living sane lives, and working for reasonable ends.

Mumford referred to the "frantic" writings of the atomic scientists as messages "written by the greatest of the madmen, the men who invented the super-infernal machine itself; the men who, in the final throes of their dementia, were shocked back into sanity."[13] A few months after Mumford's essay, *The Pallid Giant,* Pierrepont Noyes's 1927 novel showing how reliance on atomic deterrence could lead to atomic annihilation, was reissued under the title *Gentlemen: You Are Mad!*.

The atomic scientists believed in their own original good intentions. One of their motives, as Thomas Edison had kept urging in his later years, was to create "some engine of war" so dreadful that war would become "unthinkable, and therefore impossible" (see page 76). In "Memorial," a story by Theodore Sturgeon in the April 1946 *Astounding Science-Fiction,* a scientist embodying this good intention seeks to end the atomic arms race by terrifying everybody with a new form of nuclear explosion " 'more than a thousand times as powerful as the Hiroshima bomb.' " Although he insists, " 'I didn't make this to be a weapon,' " the government covets it as a first-strike superweapon to break the nuclear stalemate with an unnamed enemy. U.S. agents are steered to the scientist's desert laboratory by an undercover operator who cynically recapitulates the history of the other superweapons that were " 'going to stop war, and didn't,'" including "'the submarine, the torpedo, the airplane, and that two-by-four bomb they pitched at Hiroshima.'" As the agents seize the apparatus, they unwittingly bring it to critical mass, causing a nuclear inferno that is blamed on the enemy; this pretext is used to launch a first strike, and the resulting atomic warfare exterminates the human race, leaving as our only descendants some mutated creatures with a vague resemblance to humankind.[14]

Postatomic futures dominated by mutants would become the theme of hundreds of subsequent stories, novels, and movies. Early examples include Edmond Hamilton's "Day of Judgment" (1946), in which the earth is inherited by clans of intelligent animals; Edward Grendon's "The Figure" (1947), where our successors are a civilization of humanoid beetles; and the relatively optimistic "Tomorrow's Children" (1947) by Poul Anderson and F. N. Waldrop, which sees the definition of humanity enlarged to include many forms of our grotesquely mutated descendants.

Science-fiction magazines in 1946 of course had many atomic-end-of-the-world stories, including Ray Bradbury's poignant "The Million-Year Picnic" (later included in his *Martian Chronicles*), in which the demise of earth's civilization is seen through the eyes of a family of survivors who have escaped to Mars. Such science fiction also now appeared with some frequency in general magazines. For example, the January 12 *Collier's* contained Philip Wylie's "Blunder," which takes place after a brief atomic war between the United States and an unnamed central European power has left the world in a state of constant terror and obsessive national security. The suppression of scientific communication prevents anyone from acting in time to stop an experiment by two scientists out to get rich with a new fission process; their miscalculation results in an explosive chain reaction that turns the earth into a new sun in a fraction of a second.

Suggesting that the business world was not yet entirely committed to forging a dominant military-industrial complex, even *Fortune* published that same month "Pilot Lights of the Apocalypse," a one-act play warning against a nuclear arms race that could lead to an accidental push-button war. Written by atomic scientist Louis N. Ridenour, the play dramatizes how civilization ends when an underground nuclear-weapons retaliation center misidentifies an earthquake in San Francisco as an atomic attack.

In 1945 and 1946 almost every fiction picturing the United States in a nuclear war carefully refrained from identifying the enemy nation. No such constraints applied once the Truman administration and the media unequivocally branded the Soviet Union as America's nemesis. From 1947 on, there would be little hesitation in dramatizing Soviet-American nuclear warfare, as in Leonard Engel and Emanuel Piller's gruesome 1947 chronicle of a protracted nuclear -biological fight to the finish, *World Aflame: The Russian-American War of 1950*.

The undisclosed identity of the enemy is central to one of the most revealing fictions of 1946, *The Murder of the U.S.A.*, by Will F. Jenkins (better known by his science-fiction pen name "Murray Leinster"). This astonishing novel lays bare ideological assumptions underlying the U.S. role in the nuclear arms race, assumptions customarily veiled in layers of delusion.

The United States is mysteriously attacked by hundreds of nuclear missiles that devastate all its cities and many of its retaliatory "Burrows." The country has previously joined with the other members of the United Nations in an agreement whereby every nation has pledged itself "to destroy any other nation which made atomic war" (163). Since the identity of "the murderer nation" (10) is unknown, the novel develops as a murder mystery requiring "detective work on a new plane" (107). Although the mystery is eventually solved and the killer nation suitably punished by total annihilation, the name of the culprit is never revealed to the readers.

The Murder of the U.S.A. is a glorification of the doctrine of nuclear deterrence. Its hero—a high-ranking missile officer—plays a key role in both solving the mystery and punishing the criminal. The message is plain: "Since there could be no defense against atomic bombs, the only possible deterrent would be the certainty of terrible and adequate revenge" (28). "'We,'" our hero declares, "'are the instruments by which war becomes suicide for the nation which wages it'" (163).

After the treacherous sneak attack, "every surviving human being in America craved vengeance even more than he craved continued life," for the enemy "had destroyed the very substance of American life" (48). Only "vengeance" could "re-establish in a torn and despairing world the idea of justice and hope" (60). Fortunately, the U.S. "Burrows" could launch even greater "maniacal destruction": they "could almost destroy the civilization of the world" (51–52). And deep below ground, they continue to manufacture more of "the weapons that the world had depended on for the preservation of peace" (71). When the identity of the enemy becomes clear to the missilemen, there is no question what reason and justice demand and what the readers, like the American survivors in the novel, are supposed to crave.

With admirable frankness but bloodcurdling morality, the novel confronts head-on a problem conveniently evaded by most advocates of nuclear deterrence. If a nuclear attack is launched in secret by a government acting without the knowledge and consent of its citizens, how can one then justify the mass slaughter of the people of that nation? The novel even goes so far as to concede that only a government *not* under control of its people could get into a position to launch such an attack: "no rulers of a free nation, who by orderly processes could be replaced by men of other views," could prepare for a nuclear first strike, which could be readied only by rulers "sure that by fraud or violence they could secure themselves against overthrow by their own people" (165). The novel justifies the genocide of even such an enslaved population by arguing that every people is still responsible for the acts of their government. Then it takes that logic to its conclusion: "A man who lets himself be enslaved, so that his leaders may plan war, commits a crime against humanity" (161). This doctrine could define every citizen of World-War-II Germany and Japan as a war criminal, thus legitimizing their genocide. Was it the doctrine implicit in the Anglo-American bombing of cities and the cremation of Hiroshima and Nagasaki?

Based on this thinking, there is obviously only one just fate for the enemy that has attacked the United States. But with the most staggering irony, the novel seems oblivious to the identity of the only nation that, according to its own definition, had in fact carried out this great crime against humanity: "The nation which began atomic war thereby notified an end of peace, and the beginning of an age of suspicion and wholesale murder which could not end without the whole earth enslaved or else in ashes" (163). The verdict is punctuated by the repeated *"Whooo-oooo-OOOO-OOOOOOMMM!"* of American nuclear rockets being launched against the outlaw country.

Our hero calls upon every other nation to fulfill its international duties: " 'I demand the destruction of every city, every hamlet, every cross-road. I demand that the enemy country be turned into a waste of bomb-craters,' " so that " 'any man who thinks of war will look at it and have his blood turn to ice' " (164). Although some modern readers may be reminded of the twenty-one million bomb craters the United States left as a lesson in Vietnam, the nuclear vengeance carried out to the thrilling sounds of the star-spangled rockets outdoes even the bacteriological war of extermination waged against the Chinese in Jack London's "The Unparalleled Invasion." The "Burrows" rain nuclear missiles in random patterns across every section of the enemy nation in order to "destroy open fields and tiny hamlets and forests and streams and everything that moved or lived or breathed" (170).

11

The Baruch Plan:
American Science Fiction

Outrageous as the novel may seem, *The Murder of the U.S.A.* has much in common with another expression of American culture that came forth in 1946—the Baruch Plan, America's postwar proposal for nuclear control and disarmament. The Baruch Plan shares the novel's views of both the atomic superweapon and America's role in the atomic age. It is also almost identical to the atomic ultimatum issued by the President in the 1940 novel *Lightning in the Night.* Perhaps the Baruch Plan may best be comprehended as another specimen of American science fiction.

Insisted upon by the United States despite being patently unacceptable to the Soviet Union, this strange proposition virtually guaranteed the ensuing nuclear arms race. Some have seen the Baruch Plan as a diplomatic ploy designed to isolate the Soviet Union, or merely the equivalent of "an ultimatum" to the Soviets "to forswear nuclear weapons or be destroyed."[1] But the apparent sincerity and enthusiasm of its sponsors suggest a less cynical, though more frightening, interpretation. The Baruch Plan is an expression of the treacherous delusions forming a continuum from Robert Fulton through the early future-war novels into post-Hiroshima American culture.

The context of the Baruch Plan was a crossroads. The United Nations was established in October 1945, with a structure acceptable to all the signatory nations. On December 27, 1945, a joint Soviet-Anglo-American communiqué outlined a historic agreement, adopted the following month by the United Nations General Assembly, to establish a UN Commission for the Control of Atomic Energy, responsible to the Security Council and charged with developing specific plans:

> For the control of atomic energy to the extent necessary to ensure its use only for peaceful purposes; For the elimination from national armaments of atomic weapons and of all other major weapons adaptable to mass destruction; For effective safeguards by way of inspection and other means to protect complying states against the hazards of violations and evasions.[2]

In the following two years, the Soviet Union would put forward several concrete proposals for carrying out this agreement, by immediately abolishing atomic weapons and then establishing an international control body with rights of inspection and powers of enforcement. Contrary to much current belief, these Soviet proposals included provisions for international on-site inspection.[3]

On June 14, 1946, Wall Street financier Bernard Baruch strode to the podium of the United Nations and delivered the American plan, couched in the rhetoric of apocalyptic American science fiction. "We are here," he began, "to make a choice between the quick and the dead. . . . Let us not deceive ourselves: We must elect World Peace or World Destruction."[4] This crisis in human affairs had been reached, he explained, because the United States was in possession of "the absolute weapon." Here is the fulfillment of that great fantasy, from the motor bomb of *The Great War Syndicate* in 1889 through the atomic bomb of *Lightning in the Night* in 1940: The United States has the ultimate weapon; therefore, as Baruch would then explain, all nations must submit to the authority of an international body dominated by the United States.

Both before and after outlining his plan, Baruch insisted that "immediate, swift, and sure punishment" of any nation violating its provisions must be paramount: "It might as well be admitted, here and now, that the subject goes straight to the veto power contained in the Charter of the United Nations," which would have to be abrogated in regard to atomic matters. Only months after the establishment of the United Nations, Baruch thus proposed to eliminate the proviso that made it possible for the Soviet Union to participate. Since the superweapon in this American vision was transcendent, every nation had to submit to the dictates of a different kind of international body, controlled by the sole possessor of what Baruch called "the winning weapon."

This body would be a newly created International Atomic Development Authority staffed by personnel "recruited on a basis of proven competence" in atomic science, "but also so far as possible on an international basis." Of course, the only people with "proven competence" were the American, British, Canadian, and immigrant scientists whose cooperative labors had created America's atomic bombs.

What would this Authority do? In place of the previously agreed upon inspection and control of each nation's atomic facilities, the Authority would have "managerial control or ownership of all atomic-energy activities potentially dangerous to world security," including direct "dominion" over all the world's supplies of uranium and "complete managerial control of the production of fissionable materials." No nation would be allowed to establish facilities even for peaceful production of atomic energy without a special license from the Authority for each plant, and the Authority would supply fissionable material to these plants as it saw fit. Unlike the United States, the Soviet Union in these postwar years was largely undeveloped and lacked an electric power grid throughout vast regions of its national territory. According to the Baruch Plan, any Soviet use of atomic energy to electrify and industrialize (and possibly compete with the United States) would be entirely at the sufferance of this Authority dominated by the United States. What offenses would be subject to "immediate, swift, and sure punishment"? Not just "illegal possession or use of an atomic bomb," but also "illegal possession, or separation, of atomic material suitable for use in an atomic bomb," and various other activities, including "creation or operation of dangerous projects in a manner contrary to, or in the absence of, a license granted by the international control body."

Baruch proposed that all nations except the United States surrender their right to nuclear technology, vest the power of "punishment" in an international body controlled by the United States and its close allies, and authorize such "punishment" for any nation that so much as began construction of anything deemed an unlicensed nuclear power plant. Meanwhile, during the years required to implement the plan, the United States would continue to possess and manufacture nuclear weapons.

Baruch reiterated that "punishment lies at the very heart" of his proposal. With one nation continuing to monopolize "the absolute weapon," that punishment would evidently take the form described in *The Murder of the U.S.A.*[5] But at least *The Murder of the U.S.A.* allows the nations to be coequal and provides nuclear punishment only as a response to a nuclear attack.

According to the plan presented by Baruch, the United States would surrender its bombs after all the other stages had been completed: "Manufacture of atomic bombs shall stop; Existing bombs shall be disposed of pursuant to the terms of the treaty." When would that be? Baruch's answer locates his entire proposal within the tradition of all those American fantasies about superweapons that will end war:

> But before a country is ready to relinquish any winning weapons it must have more than words to reassure it. It must have a guarantee of safety, not only against the offenders in the atomic area but against the illegal users of other weapons— bacteriological, biological, gas—perhaps—why not?—against war itself.

So the United States might retain its atomic bombs and keep producing more until the abolition of war on the planet.

Blinded by the delusions about superweapons endemic in American culture, Baruch and the other men who designed this proposal, including some of the leading atomic scientists, apparently believed the Soviet Union might forego independent development of atomic energy, surrender its national autonomy, abrogate the United Nations structure that had just been agreed upon, and thus place its own fate in the hands of the United States and its capitalist allies, which then (before the admission of dozens of formerly colonized peoples and nations) totally controlled both the Security Council and the General Assembly. They must have thought that the Soviet Union shared their belief in the omnipotence of "the absolute weapon" and so might be terrified enough to submit to the good intentions or invincible power of the nation that had already used atomic bombs and was now producing improved models while experimenting with new intercontinental delivery systems.

Just in case the Russians were not convinced by the Baruch Plan, the United States was ready with another demonstration of its conception of peace in the nuclear age. Seventeen days after Baruch's speech, the United States exploded, over a fleet of warships anchored in the lagoon of Bikini Atoll, the first atomic bomb since the one dropped on Nagasaki.

As a B-29 Superfortress approached the great fleet on July 1, excitement and apprehension ran high. On board was the fourth atomic bomb to be detonated, this one adorned with a picture of Hollywood star Rita Hayworth. America had come a long way since that July, just twenty-five years earlier, when Billy Mitchell's bombers had proven that airplanes could sink ships. The strategic bomber and the atomic bomb, the two American superweapons that supposedly had won World War II, would now demonstrate their powers to a watching world. Despite the spectacular visual effects of the explosion, "the blast was a disappointment to many who observed it and to the public at large": "The buildup had been too extravagant. Only three vessels, a destroyer and two transports, sank at once. Goats still munched their feed on warship decks."[6] Billy Mitchell's modified World War I bombers had inflicted more damage on warships in 1921. More satisfying for devotees of the bomb was the July 25 underwater explosion, which sent a colossal mushroom of radioactive water cascading over 90 percent of the ships in the flotilla while thoroughly poisoning undersea life in the Bikini lagoon.

But the atomic bombs of the 1940s were not the apocalyptic weapon of the popular and official imagination. By wildly exaggerating the powers of these bombs, this imagination helped to create the conditions for their successors, which would indeed have the power to destroy civilization and perhaps our species. History may have no clearer example of self-fulfilling paranoia, a mad quest for "security" that created the nemesis it feared.

12

Nuclear Scenarios

Although many of the atomic scientists had predicted that using the atomic bomb on Japanese cities would trigger an uncontrolled nuclear arms race, in the ten months after August 1945 the United States still had opportunities to call it off. Instead, the nation that had initiated the Soviet-American race now consolidated its policy of maintaining its lead, no matter what the cost or consequences. Posing outlandish terms for nuclear disarmament and conducting the first peacetime tests of atomic bombs were just manifestations of the policy.

During the four years in which it was the only nation in the world with nuclear weapons, the United States engaged in full-scale, unrestrained production and stockpiling of atomic bombs. U.S. forces encircled the Soviet Union with a noose of more than four hundred major military bases and almost three thousand secondary bases, stretching from Greenland and Iceland through Europe, the Middle East, southwestern Asia, the Pacific Ocean, Alaska, and Canada. B-29s, whose nuclear capability was demonstrated against Japan and again at Bikini, were stationed at forward bases, within striking range of all major Soviet cities. Meanwhile, intercontinental nuclear delivery systems were being perfected. The gigantic B-36 intercontinental bomber, designed for the newly formed Strategic Air Command, first flew on August 8, 1946, one year after the Hiroshima and Nagasaki bombings and two weeks after the second Bikini bomb. In 1947 came SAC's B-47 jet bomber, precursor of the B-52. At the same time, former Nazi rocket experts were working with U.S. engineers to create the main instrument of future intercontinental push-button war—the nuclear-armed long-range ballistic missile. In the spring of 1946, just before Baruch's speech to the United Nations, Consolidated-Vultee (later to become Convair) began work on Project MX-774, forerunner of the Atlas intercontinental ballistic missile.[1]

The same pattern was to prevail for decades. While holding a commanding lead, the United States would continue to escalate the arms race, with the Soviet Union running as hard as it could trying to catch up. During the four-year period of total U.S. nuclear monopoly, the constant Soviet calls for nuclear disarmament were of course dismissed as feeble Communist ploys to strip the Free World of its "security."

Then, in the fall of 1949, the Soviet Union tested its first atomic bomb. Although this did not establish parity with the awesome U.S. atomic arsenal, it did create a new world situation. For the first time, mutual atomic war was becoming possible. Many thought the moment offered another opportunity for nuclear disarmament, since the United States would no longer be the only nation giving up atomic weapons and would otherwise face an ever-increasing nuclear threat.

Instead, the superweapon now attained the potential for its greatest triumph over humanity. Within two months of the first Soviet atomic test, the United States unilaterally initiated the most deadly escalation of the arms race, beginning full-scale development of what was then known as the "Super," the thermonuclear (hydrogen) bomb, based on fusion. Atomic (fission) bombs probably could not threaten the extermination of the human species. Fusion bombs could. A single B-52 armed with thermonuclear bombs can carry an explosive force equivalent to over three thousand Hiroshima-sized atomic bombs or eighteen times the tonnage of all the bombs dropped by U.S. planes during World War II. One Trident submarine can launch 192 thermonuclear warheads, each equivalent to eight Hiroshima bombs. A full-scale thermonuclear war would so devastate the environment that some scientists believe human life, as well as that of most land vertebrates, might become unsustainable.

The fusion bomb—by radically reducing the weight and exponentially increasing the destructiveness of warheads—also finally made nuclear-armed intercontinental ballistic missiles (ICBMs) feasible. Since ICBMs are launched by remote switch and are designed to be impervious to any recall signal, thermonuclear weapons thus greatly increased the likelihood of a holocaust beginning through error and escalating beyond human control.

From Hiroshima onward, it has been the United States that has initiated each new stage of the nuclear arms race. Even George Kennan, one of the principal designers of the Cold War against the Soviet Union, acknowledged:

> . . . it has been we Americans who, at almost every step of the road, have taken the lead in the development of this sort of weaponry. It was we who first produced and tested such a device; we who were the first to raise its destructiveness to a new level with the hydrogen bomb; we who introduced the multiple warhead; we who have declined every proposal for the renunciation of the principle of "first use"; and we alone, so help us God, who have used the weapon in anger against others, and against tens of thousands of helpless noncombatants at that."[2]

American technology, industry, and culture have continued to create one new weapons system after another: strategic missiles and tactical missiles; nuclear submarines that can launch their missiles while submerged; multiple independently targeted reentry vehicles (MIRVs) that enable a single missile to hit several widely separated targets with thermonuclear warheads; maneuvering re-entry vehicles (MARVs) to assure the invulnerability of the warheads; neutron bombs to kill more people while destroying fewer structures; cruise missiles that can follow complex routes with automatic navigation systems allowing attack from unexpected directions; and "counterforce" missiles, such as the Pershing II and the MX (the "Peacekeeper"), designed not for retaliatory deterrence but for first-strike capability against unlaunched missiles in hardened launch sites.

Each new advance deeper into the nightmarish labyrinth of superweapons has seemed to some a path to a better world. Without imagining a more secure future to be achieved by each new weapon, there would be no will for its financing, production, and deployment. Thus the destiny of the human species is continually being written and rewritten in the scenarios, or dramatic narratives, of those who plan our defense.

The more futuristic the imagined scene, the more likely it is to come with a disavowal that "this is not science fiction." For example, when the *New York Times Magazine* ran a major promotional piece for Star Wars by Zbigniew Brzezinski, Robert Jastrow, and Max Kampelman (January 27, 1985), it carried a prominent disclaimer that the article "should not be seen as science fiction." What is this supposed to signify?

To those unfamiliar with the genre, the term *science fiction* may conjure up only images of messianic futurism and juvenile infatuation with technology. To be sure, the "Gosh! Wow!" mode has played an important role not only within science fiction proper but also in the science-fictional elements of American culture, from the boy-genius inventors of the nineteenth-century dime novel through the World of Tomorrow at the 1939 World's Fair to the local video arcade.[3] And science fiction has provided a temple for some of the most obscene rituals in the cult of the superweapon.

But not all science fiction consists of alienating and ultimately devastating technological fantasies. In fact, modern science fiction emerged from an opposing vision, expressed in that first great myth of the industrial age, Mary Shelley's 1818 novel, *Frankenstein; or, The Modern Prometheus*. Long before American newscasters and newspapers in August 1945 began referring to the atomic bomb as a "Frankenstein's monster," science fiction was prophesying—in works such as *The World Set Free* (1913), *The Pallid Giant* (1927), and *The Final War* (1932)—that the cult of the superweapon would lead to a fatal nuclear arms race.

Can any vision of a possible future in which human destiny is intertwined with the development or use of science and technology, most especially weapons technology, *not* be a form of science fiction? A nuclear stalemate stretching into the foreseeable future, with the world's economies crumbling under the burden. A successful global campaign to abolish nuclear weapons. A full-scale thermonuclear war, with cities, forests, and grasslands burning out of control, pumping such dense clouds of smoke and debris into the upper atmosphere that most terrestrial life is extinguished in a nuclear winter. A Pax Americana formed under the umbrella of four hundred orbiting battle stations, each armed with dozens of interceptor heat-seeking missiles, commanded by computers to attack Soviet missiles in their boost phase. Each of these visions is a kind of science fiction, and each implies a context of human history. The essential question is not taxonomic (Is this science fiction?) but, in every sense of the term, critical (Is this science fiction "good"?).

All such scenarios are dramatic narratives that can be subjected to the kind of critical scrutiny applied to other forms of literature. Many of the most influential, on which our lives could depend, might fail the most basic tests for consistency, depth of consciousness, and psychological accuracy, not to mention ethical content. On the other hand, many works published as science fiction offer insights into our technological, political, economic, and psychological reality, as well as our practical choices, omitted from the scenarios of policymakers.

Science fiction has played two central—and contradictory—roles in the nuclear arms race. On one side, the fundamental assumptions underlying the creation of ever more menacing weapons are rooted in science fiction, which originated the myth of the superweapon that would give global hegemony to a single nation. This myth has led to what has been called the "fallacy of the last move"—the addictive, ever-unfulfilled expectation that each new exotic weapon created for our "defense" would confer upon the United States permanent military superiority and invulnerable "security." Underlying this is another fallacy, also nourished in some science fiction, that science and technology are products of lone wizards such as Thomas Edison, or brilliant research teams, or national genius. This fallacy blinds policymakers to the fact that since the United States and the Soviet Union are at roughly equivalent stages of science and technology, any new weapon produced by one can soon be matched by the other, thus bringing about not supremacy for either but increased danger for both.

On the other side, more perceptive science fiction has exposed such fallacies, warned of the terrifying consequences of our obsessive infatuation with our weapons, probed the sources of this psychological disease, and even suggested some antidotes. Some science fiction of the 1940s looked beyond the U.S. nuclear monopoly to see an uncontrolled arms race leading to the kind of future in which we now live. These early nuclear stories foresaw not only the looming menace of an accidental holocaust or preemptive first strike but also the growing loss of freedom and mental health in a society dominated more and more by its own superweapons. Exploring the nuclear culture that had generated them, such stories exposed the irrationality controlling the nation. During certain crucial events, this fiction may have helped awaken enough people to prevent nuclear war—so far.

13

Early Warnings

One early tale that should be required reading for any person who cares about human survival is Theodore Sturgeon's masterpiece, "Thunder and Roses," published in the November 1947 *Astounding Science-Fiction* and frequently anthologized ever since. Although when the story first appeared the United States enjoyed a monopoly on nuclear weapons and would not actually be threatened by deliverable bombs or missiles for over a decade, "Thunder and Roses" foresees how the American quest for "security" would lead to creating the means to annihilate America and perhaps all humanity. Beyond that, it cuts right to the most basic—and almost always hidden—assumptions upon which nuclear "deterrence" is based.

Even during the early days of the U.S. monopoly on nuclear weapons, the doctrine of deterrence was already being codified in theory such as Bernard Brodie's *The Absolute Weapon*, propagandized in fiction like Will Jenkins' *The Murder of the U.S.A.*, and implemented in the feverish production of atomic bombs and new means to deliver them. At first, this meant threatening nuclear retaliation to deter conventional warfare. But already implicit was the corollary that rules today's world: threatening an utterly catastrophic nuclear retaliation to deter an utterly catastrophic first strike. This is the doctrine now known as Mutually Assured Destruction or by its acronym, MAD.

The advocates of deterrence theory argue that the entire colossal machinery of devastation is intended never to be used; even its most ardent defenders concede that it will have failed its whole function if it does not deter. But this implies a question that is rarely raised: Would actual retaliation make any sense whatsoever? It certainly could no longer deter the attack, which already would have occurred. So what conceivable purpose would be served by full-scale thermonuclear retaliation?

The answer is simple: revenge, and an eye-for-an-eye version of justice. As *The Murder of the U.S.A.* puts it, only "vengeance" could "re-establish in a torn and despairing world the idea of justice and hope'' (60). ''Thunder and Roses'' is a direct answer to this basic premise of nuclear deterrence.

Like *The Murder of the U.S.A.,* ''Thunder and Roses'' is set in the United States after a devastating sneak attack, with even more catastrophic effects. Hundreds of atomic bombs, many designed for maximum radioactive after-effects, have annihilated the nation. Only a few hundred people are still alive, and they are dying off rapidly. With radioactive carbon 14 filling the environment, the extinction of every inhabitant of the continent is certain.

Echoes of *The Murder of the U.S.A.* are loud. The 1946 novel emphasized that the United States ''had been assassinated'' and that ''every surviving human being in America craved vengeance even more than he craved continued life,'' for this sneak blow ''had destroyed the very substance of American life'' (10, 48). In the 1947 story, ''the country was dead,'' the Statue of Liberty ''had been one of the first to get it, her bronze beauty volatilized, radioactive,'' and the most ''shameful thing'' is ''that we hadn't struck back'' at our ''murderers.'' For the identity of the two murdering nations is known (but not to the reader), and all over the country there are secret missile sites aimed at them, capable of launching more devastation than hurled by both attackers combined. What is stopping the retaliation?

The point-of-view character is Sergeant Pete Mawser, who, along with his best friend, discovers on his base the last remaining master controls that could activate America's full retaliation. Pete is an American Everyman, a surrogate for the reader, whom the story confronts with the question: Should he do it?

''Thunder and Roses'' is more immediately relevant today, when nuclear overkill raises the specter of global ecocide, than in 1947, with its puny little atomic bombs. For in the situation projected in the story, Pete's choice may be very close to ours. The enemy nations had underestimated the planetary effects of their attack, which will soon devastate their own environment, possibly threatening the survival of the species. If the United States retaliates, it will guarantee the extinction of not only humanity but also all other life. If the United States does not retaliate, the earth will be inherited by those of our murderers who survive and their descendants.

In place of the fear and hate ascendant in *The Murder of the U.S.A.* and U.S. ''defense'' policy, ''Thunder and Roses'' offers a transcendent love for humanity, even in its most murderous forms. Its message is communicated in the songs and speeches of Star Anthim, a beautiful video idol dying of radiation sickness. Star, embodying the best of the dying America, pleads with Pete: '' 'They have killed us, and they have ruined themselves. As for us—we are not blameless, either. Neither are we helpless to do anything—yet. But what we must do is hard. We must die—without striking back.' '' Arguing that the '' 'enmity of those who have killed us is such a tiny, temporary thing in the long sweep of history' '' and foreshadowing the philosophic vision of Jonathan Schell's 1982 *The Fate of the Earth,* Star offers a definition of ''justice'' exactly opposite that of *The Murder of the U.S.A.* and the policy of nuclear deterrence:

"Let us die with the knowledge that we have done the one noble thing left to us. The spark of humanity can still live and grow on this planet. It will be blown and drenched, shaken and all but extinguished, but it will live . . . if we are human enough to discount the fact that the spark is in the custody of our temporary enemy. . . . Perhaps there will be ten thousand years of beastliness; perhaps man will be able to rebuild while he still has his ruins. . . . And even if this is the end of humankind, we dare not take away the chances some other life form might have to succeed where we failed. If we retaliate, there will not be a dog, a deer, an ape, a bird or fish or lizard to carry the evolutionary torch. In the name of justice, if we must condemn and destroy ourselves, let us not condemn all life along with us!"

But Pete wrestles with doubts: "What creatures were these, these corrupted, violent, murdering humans? What right had they to another chance? What was in them that was good?" And when Star dies, Pete's best friend overcomes all his own doubts. " 'They killed her, too. . . . That does it,' " he exclaims and races for the switch that will launch America's full retaliation. Pete's ultimate choice then cannot be purely nonviolent. To give humanity another chance, he must murder his best friend. After wrecking the master controls, and while stroking the tousled hair of the human being he has just killed, Pete announces his choice: " 'You'll have your chance,' " he said into the far future. " 'And by heaven, you'd better make good.' "

If Pete has made the right choice, what does that imply about what we have been doing ever since the story was published? As the noted mathematician Chandler Davis has written: "A whole generation of 'strategy theorists' would be making an advance in their theory if they would throw out their treatises on mutual deterrence and read 'Thunder and Roses.' "[1]

Davis himself wrote several important stories in the early years of the nuclear age. "To Still the Drums," a remarkably perceptive 1946 tale, posed a fateful question (to which readers now know the answer): Could the military, whether or not in conspiracy with the administration, "get the United States into war against the wishes of the people and Congress?" Expressing most people's outlook at the time, which sounds terribly dated to us, one character argues: " 'It's Congress that declares war.' " In response, another character explains the deadly momentum of a nuclear arms race and the role of Congress in accelerating it: " 'Try and get a congressman to understand what the atom bomb means. . . . Congress is still in the habit of wanting our Army to have bigger and better weapons than anybody else. There's another habit of thought Hiroshima didn't change.' "

The story predicts that in the near future the military might use the space program as a cover to develop intercontinental missiles armed with nuclear weapons. (Actually, a secret development contract for an intercontinental ballistic missile had been awarded to a U.S. aerospace corporation in April 1946, months before this story appeared.)[2] The narrator, a rocket engineer in the army, suspects nothing until a physicist explains, " 'you're building the weapons we've all been looking forward to with fear since Hiroshima and before. . . . If you do a good job you might succeed in wiping out civilization.' " " 'If we do a good job,' " responds the narrator, naively echoing the central fallacy of American fantasies about the ultimate weapon, "'we'll have weapons so powerful nobody'd dare to start a war.'" Looking at him "as if he'd just discovered he was talking to an . . . idiot," the physicist explains the logic of a nuclear arms race: "'The important thing is, not whether it's being done by the Army or the diplomats, not whether it's being done consciously or unconsciously, but that it's being done at all. We *are* preparing for the next war. We're precipitating it! Building fires that must be met with fire.'"

When he discovers that the military is secretly laying plans for a preemptive first strike, the narrator steals their top secret reports and smuggles them to a senator. At the end, he is in prison awaiting trial while speculating as to whether this revelation will be enough to alert the nation to "the peril inherent in the very existence of atomic weapons." Unfortunately, we know the answer to that speculation.

Many of those who warned about the dangers of nuclear weapons—including their social, cultural, psychological, and political consequences—themselves fell victim to the oppressive chill of the Cold War, of which the arms race has been a central feature. Chandler Davis, for instance, shared the fate of his narrator when he was incarcerated for six months in a federal penitentiary in 1960 for refusing, on First Amendment grounds, to testify about his political beliefs and associations before the House Un-American Activities Committee six years earlier. The 1959 Supreme Court decision that sent Davis to prison claimed that in the context of the Cold War, Congress had a right to force testimony that "in a different context would certainly have raised constitutional questions of the gravest character."[3]

The decision to use the strategic bomber and the atomic bomb as practical superweapons had tipped the scales in a seesaw struggle in American culture. Before this, opposition to such weapons could claim to be at least as authentically American as the ideology that glorified them. But once the United States became the world's leader in the bombing of cities and the only nation that had actually used atomic bombs, questioning the use of these weapons would be defined as dissident, heretical, unpatriotic, or—in the most telling term— un-American. Even if those advocating a reversal in policy did not explicitly condemn what the nation had done in the recent past, their advocacy contained an implicit rebuke and condemnation. Then came the decision to make the fight against Communism the central purpose of the American nation, with its corollaries of maintaining a mammoth "peacetime" military and endlessly racing to maintain nuclear supremacy. Since the atomic "secret" allegedly had been stolen for the Russians by Communists like the Rosenbergs, anyone against nuclear weapons in America must be at best a Communist dupe, spreading the poisonous notion of "Better Red than Dead."

In this environment, especially between the late 1940s and mid-1950s, hard-core science fiction curiously became one of the few areas in American culture not quite swept clean of dissent. Although much science fiction of the period was as zealously anti-Communist and promilitary as most American culture, some truly radical thought appeared between the futuristic and some- times lurid covers of science-fiction magazines. Restricted to a relatively small coterie of readers, and often veiling its message in analogies, this science fiction probed the roots of nuclearism as it was taking over the nation.

In William Tenn's "Brooklyn Project" (1948), America's leaders, whose obsession with international "security" has placed both science and the media under the control of their "Security" forces, try to outdo the Manhattan Project by bending time itself into the ultimate weapon that "our democratic hands" can manipulate; unaware that they are thus transforming the entire course of the planet's evolution, they metamorphose into fanatical amoebalike blobs while insisting "Nothing has changed!" Alfred Bester's "Disappearing Act" (1953) describes how military, capitalist, and technocratic forces waging a protracted global thermonuclear war with only one purpose—to preserve the "American Dream"—annihilate that dream. In Fritz Leiber's "Coming Attraction" (1950), a British electronics salesman visiting New York during a lull in a protracted Soviet-American nuclear war discovers a grotesque nightmare of masked sadomasochism; the story hints at what lurked under the smug, smiling, crew-cut surfaces of nuclear-brandishing 1950 America. Leiber's "A Bad Day for Sales" (1953) describes the atom bombing of New York through the senses of a sales robot who keeps trying to dispense free "polly-lops" to the kiddies, while peddling Poppy Pop (soft drink for children, exciting fizz for grown-ups), Mars Blood nail polish, "cool" cigarettes, and science-fiction comic books (*Junior Space Killers* for boys, *Gee-Gee Jones, Space Stripper* for girls); his mindless huckstering blares out the crass, dehumanized values that led to the flames cremating his would-be customers. Ray Bradbury's "There Will Come Soft Rains" offered science fiction's social criticism to a larger audience; appearing in the May 6, 1950, *Collier's,* the tale encapsulates the end of utopian consumerism and automation in the fate of the last house left standing amid radioactive ruins, a futuristic edenic home where cutesy anthropomorphic machines continue their programmed catering to every human material comfort, while the only surviving shapes of actual humans are the silhouettes etched by the bomb on a charred wall.

The threat of nuclear devastation unleashed a host of fantasies already teeming in the American cultural psyche. While some science fiction wantonly indulged in these half-buried desires, other science fiction consciously exposed what they represent. One of the most deep-seated of these fantasies is survivalism, whose roots go back through the myth of the frontier past into the substratum of the primeval bourgeois fantasy of the self-sufficient man, epitomized by Robinson Crusoe. In the modern version, atomic holocaust becomes a half-consciously desired means of release from the complexity and social restraints of modern urban life.[4] The city—envisioned as a hopeless morass of pollution, overcrowding, decadence, and enfeebling interdependence—is obliterated, thus freeing the would-be frontiersman to live out his yearnings for primitive, manly self-reliance in a restored wilderness.

An early example of such fantasizing is Stuart Cloete's novella *The Blast,* serialized by *Collier's* in 1947. Some significant insights do grace this autobiography by the only New Yorker who survives a sneak attack on all the cities of North America and Britain with weapons that make the first atomic bombs,

which "we should never have employed," look like "children's firecrackers."
"It did not seem to have occurred to our leaders," he writes, "that bombs could
be improved by other people too." From 1945 on "we had been barraged with
stories of the Russian menace," "an improved model of the old 'yellow peril'
danger." Therefore the Air Force commanders programmed to strike back
assumed that the first strike had come from the Soviet Union, and failed to
consider the ecological consequences of their retaliation. So the Nazis who have
apparently engineered the mutual destruction of their two great enemies have the
last laugh.

But then the narrative drifts into an archetypal survivalist fantasy. When his
wife dies, the narrator becomes a great white hunter, armed with the finest
sporting weapons from Abercrombie and Fitch, surrounded by a pack of dogs he
has specially bred, and stalking not only tigers but ferocious giant mutant wolves
and minks in the luxuriant forests overgrowing Manhattan. After almost twenty
years of this thoroughly self-reliant and increasingly idyllic existence, he feels
like a new "Adam"; though now seventy years old, he discovers that he has shed
all his urban decrepitude and become a dynamic he-man, bursting with power
and vitality. At the end, he is leading a war party of 150 "magnificent young
Indians" who have drifted in from the West, while being fondled by two bawdy,
buxom young blonde women who, though raised by the Indians, desire nothing
more than to share the white-Indian frontiersman. When he finally catches a view
of himself in a mirror, we see an image that has continued to reflect our culture:

> I was as straight as I had always been, but I was much wider than I had thought
> possible. My arms were as big as my thighs; my chest was immense. My hair was
> long, reaching halfway down my back. . . . For ornament, I wore a diamond
> necklace around my neck; my only clothing was a khaki kilt that I wore for warmth,
> a leather belt in which was stuck my kukri, and a pair of leather shoes. On my upper
> arms I had some gold armlets made from expanding wrist-watch chains and other
> jeweled bracelets that I had joined together and mounted on wide leather straps. . . .

Along with their extravagant Crusoeism and frontier romanticism, a common
and revealing feature of these postnuclear fantasies is the convenient death of a
tiresome wife, leaving the hero free to share his new primeval world with one or
more beautiful young women. What does this have to do with the wellsprings of
nuclear weapons? Two of the masterpieces from the first decade of the nuclear
age, Ward Moore's "Lot" (1953) and "Lot's Daughter" (1954), attempt to
answer this question by exposing a favorite American male hero as a monster of
ego who embodies in domestic and suburban life the force that menaces the
world.

"Mr. Jimmon even appeared elated, like a man about to set out on a
vacation." So begins "Lot," a nuclear road story. An insurance executive living
with his wife, two alienated sons, and fourteen-year-old pubescent daughter in
the wealthy suburb of Malibu, David Jimmon has prepared with "foresight,"
"competence," and "self-sufficiency" for the inevitable attack on Los Angeles.
So within hours he has the station wagon loaded with everything needed to

guarantee that he and his ''dependents'' will be survivors after they reach the destination he has selected along the coast south of Monterey:

> He had purposely not taxed the cargo capacity of the wagon with transitional goods; there was no tent, canned luxuries, sleeping bags, lanterns, candles, or any of the paraphernalia of camping midway between the urban and nomadic life. Instead, beside the weapons, tackle, and utensils, there was in miniature the List for Life on a Desert Island: shells and cartridges, lures, hooks, nets, gut, leaders, flint and steel, seeds, traps, needle and thread, government pamphlets on curing and tanning hides and the recognition of edible weeds and fungi, files, nails, a judicious stock of simple medicines. A pair of binoculars to spot intruders.

Armed with his weapons and binoculars, Mr. Jimmon, like the men leading the nation, imagines the survival of himself and his followers after nuclear war: ''Naked force and cunning will be the only means of self-preservation.'' He realizes ''how romantic'' his project is, for that is the nature of ''any attempt to evade the fate charted for the multitude'': ''The docile mass perished; the headstrong (but intelligent) individual survived.'' ''There is no law now but the law of survival,'' he gloats, expressing the very belief responsible for the devastation they are fleeing.

The means to escape from the doomed multitude must be to drive—in every sense—that quintessential symbol of American freedom, the automobile, with ''skill, daring, judgment,'' in other words, with utter ruthlessness. So the Jimmons' trip becomes a caricature of the American family on a weekend outing: The highway is jammed with traffic; his younger son wants to stop at a rest room; his wife insists that it must be a clean one; the police demand that he obey traffic laws; other drivers try to keep him from cutting in; the squabbling inside the station wagon becomes incessant; but nothing can deter Mr. Jimmon's driving will. The family shows no appreciation of ''his generalship.'' His wife can't understand how he ''can be so utterly selfish.'' His sixteen-year-old son even mutters about the ''monomania'' that makes his father ''like Hitler.'' Mr. Jimmon is filled with mounting contempt for his wife and sons, hopelessly locked in the conventions of their civilization. The only ''true Jimmon'' seems to him to be his daughter: ''Made in my own image.'' When he finally does stop at a gas station, and while his wife and sons are indulging their weakness for civilization, Mr. Jimmon acts on what seems a sudden impulse. He hustles his daughter into the station wagon and drives off, abandoning those who are spoiling his narcissistic fantasy, which he will now live out with the young woman who seems to incarnate his own image.

Six years after the global holocaust, we meet Mr. Jimmon again in the sequel, the trenchant 1954 novella ''Lot's Daughter.'' His Crusoeist fantasy of becoming the ultracompetent lone master of his environment—that treacherous myth of the self-sufficient individual that emerged precisely when technology was making our species more and more interdependent—has now been exposed as one of the forces leading to racial suicide. The supermasculine hero who spurns social interdependence is revealed as a subhuman savage, who endangers rather than leads to survival. Tortured by his rotting teeth and gradually using up the socially produced objects that have allowed him to survive this long, he must daily face the contempt of his daughter, who has come to realize that "cooperation" is the answer to the "savagery" he embodies, and that "people aren't as stupid as you think they are." Finally, in a piercing twist of irony, she abandons him, leaving him to care for their pathetic, sickly, whining son. As he sinks ever deeper into barbarism, she joins a small community that seems to be slowly restoring civilization.

The domestic bonds so disdained by Mr. Jimmon are the heart of Judith Merril's two innovative contributions to nuclear culture, her 1948 short story "That Only a Mother" and her 1950 novel *Shadow on the Hearth*. Each views the effects of atomic weapons from a mother's perspective, revealing forms of personal reality conveniently omitted from the scenarios of nuclear strategists.

"That Only a Mother" is a complex psychological tale about mutually exclusive definitions of fantasy and sanity. Maggie, like other pregnant women of her time in the near future, is worried about the grotesque mutations appearing with increasing frequency during a protracted low-level atomic war, especially since her husband, Hank, worked at Oak Ridge, site of the Manhattan Project, until 1947, when reports of mutations from Hiroshima and Nagasaki caused him to quit. His decision was apparently based on personal prudence, not moral qualms, since he now designs atomic weapons at an underground army facility. Maggie assures her mother that Hank, as "a designer, not a technician," is no longer exposed to radiation, but she has little more confidence in this than in the propaganda that passes for news over the high-tech media. After her daughter is born, the only abnormality Maggie notices is her extraordinary verbal precocity. In letters to her still-absent husband, she nervously jokes about the wave of impulsive infanticide being inflicted by distraught fathers on their mutant babies, assuring him that "*ours* is all right."

When Hank finally returns on leave, he finds that his daughter is truly a mental prodigy. Trying to pick up the strangely wriggling infant in her oddly fitting nightgown, he makes another discovery, to which Maggie seems oblivious. The story ends with these words:

> *She didn't know.* His hands, beyond control, ran up and down the soft-skinned baby body, the sinuous, limbless body. *Oh God, dear God*—his head shook and his muscles contracted in a bitter spasm of hysteria. His fingers tightened on his child— *Oh God, she didn't know. . . .*

Readers are left with several unanswered questions. Did Maggie truly not know, or is her maternal blindness to her child's defects merely a better survival mechanism than the military blindness that had caused them? Will Hank, otherwise portrayed as a loving husband despite his job of impersonally designing instruments of mass killing, continue to tighten his fingers until he succeeds in carrying out on a domestic level what he does for a living? And who is more sane and rational, the adoring mother of the story's title or the father gripped by "hysteria" (a word deriving from *uterus*) when faced with the fruits of his labor?

Shadow on the Hearth is probably the first truly domestic novel about nuclear war. When the reviewer for the science-fiction magazine *Future* (November 1950) contemptuously dismissed it as "just the thing for your mother-in-law, maiden Aunt from Crabb Corners, or anyone else who dotes on 'John's Other Wife,'" Merril replied (March 1951) that she had written the novel precisely for such readers, who constitute "a remarkably large part of our population." She succeeded so well in reaching her intended audience that the novel was even televised on "Motorola Playhouse" (as *Atomic Attack*).

Shadow on the Hearth tells the story of Gladys Mitchell, a Westchester housewife whose character develops as she attempts to cope with the aftermath of an atomic attack on New York City, where her husband, Jon, works. Desperately awaiting word of Jon's fate and symptoms of radiation sickness in her two daughters, Gladys partially emerges from dependency, passivity, and her stereotypical role as she tries to fend off the physical and social forces that menace the family. The novel is really about American society in the late 1940s and early 1950s, with nuclear bombs peeling away the smug veneers to reveal the paranoia, duplicity, and hidden humanity underneath.

The government has secretly prepared a civil-defense organization, which emerges from its cocoon as a protofascist apparatus. When her neighbor appears at her door as the power-hungry head "squadman" for their area, Gladys has a sudden revelation: "Why, he likes this!" she realizes, "He's having fun!" (55). Though America's cities all lie in ruins, the government still claims to have an impenetrable radar shield. Her teenage daughter's science teacher seeks refuge as a hunted fugitive; previously blacklisted for refusing to do war work, he is now "a public enemy" for having tried to prevent nuclear war (77, 143). His former pupil Pete Spinelli, blocked from a career in biochemistry for having signed antinuclear petitions, is now a young doctor trying to deal with the secondary physical and psychic traumas inflicted on suburbia by the bomb.

Dr. Spinelli forces Gladys to confront historical reality. When she blurts out, "'I never really believed any nation would *use* it this way,'" he responds, "'*We* did'": "'We used it in 1945. . . . In Japan. Why wouldn't somebody else use it on us?'" (57).

The official lies and hierarchical power are mirrored in the domestic microcosm by Gladys's prevarications and dependency. She can't quite confront the worst, be entirely honest with her children, recognize that her fifteen-year-old daughter is no longer a baby, or see herself as anything more than a conventional wife and mother. But the situation gradually forces her to unleash repressed strengths, and her selfless nurturing, now coupled to an awakening consciousness, helps save her family.

The happy ending may make this atomic attack seem implausibly innocuous to modern readers. The 1983 movie *Testament,* which closely parallels *Shadow on the Hearth,* to us seems more probable in its denouement: the suburban housewife attains her full strength amid the inevitable death of her family, herself, and, apparently, the human species. But if the atomic war in *Shadow on the Hearth* seems too mild, our response may only suggest the subsequent progress in building instruments of annihilation. The novel's 1950-vintage atomic bombs that hit New York City probably would have left many survivors there and inflicted minimal fallout on Westchester. The next major domestic novel about nuclear war, Helen Clarkson's 1959 *The Last Day,* seems far more modern, for it appeared after the fusion bomb had replaced the fission bomb, and after the intercontinental ballistic missile had left the realm of science fiction to begin looming over our daily lives.

Like *Shadow on the Hearth, The Last Day* witnesses the nuclear disaster through the eyes of a woman devoted to home, family, and nurturing. A stronger person than Gladys, she finally comprehends how "nuclear weapons killed democracy before they killed us," and how the cult of the superweapon, merging with messianic anti-Communism, had legitimized the mass murder of children and turned us into "a nation of idiots ruled by madmen" (119, 166). At the end, as she lies dying alone, perhaps the last human on the planet, her domestic vision expands to see "the whole earth" as "a vast, beautiful airy house" with its "winter carpet of white and its summer carpet of green," now "haunted by all that had ever been."

14

Triumphs of Nuclear Culture

In 1949, the Soviet Union tested its first atomic bomb. The United States was no longer driving alone in the nuclear arms race, which now would become ever more perilous. Yet American anxiety about nuclear war, at least as expressed openly in the culture, was at an ebb between 1949 and 1957. There are several explanations for this apparent contradiction.

In this heyday of the Cold War, Americans were supposed to fear and hate Communism and the Russians, not the nuclear arsenal designed to defend the Free World from them. Radical dissent was eradicated by purges of the media, education, and labor unions, thus precluding widespread open challenges to the nuclear arms race.

Meanwhile, the government and a major sector of American industry were conducting a massive campaign to sell nuclear energy. The prewar visions of unlimited, safe, virtually free energy powering an atomic utopia returned under the slogan "Atoms for Peace." Bright pictures of what a 1946 CBS radio documentary called "The Sunny Side of the Atom" were now widely promulgated in forms intended to capture the imagination of the masses, such as dazzling atomic energy exhibits, movies, high school study units, and comic books. General Leslie Groves, head of the Manhattan Project, personally selected Dagwood Bumstead to be the central character of *Dagwood Splits the Atom,* a 1948 comic book prepared by the Atomic Energy Commission and General Electric Company, a major nuclear contractor that distributed millions of free copies. The favorite metaphor in this propaganda was the omnipotent genie that could be called forth from a bottle, as in Walt Disney Productions' 1956 book and film, *Our Friend the ATOM.*[1]

While the atom was our friend, the Russians were our enemy. From 1949 on, endless propaganda sought to make the American people believe that they were under the threat of an imminent nuclear attack from the Soviet Union. The truth was that before 1957 the Soviet Union had no physical capability to inflict even a small nuclear strike on the mainland of the United States. Indeed, as a navigator and intelligence officer in the Strategic Air Command throughout 1957 and 1958, one of my duties—when not engaged in midair refueling of B-47 and B-52 bombers on routine espionage and provocation overflights of the Soviet Union— was helping to conceal from the American people, particularly our own SAC flight crews, the almost certain knowledge that the Soviets still had neither operational intercontinental bombers nor missiles.

Yet 1955–1957 was the period of the infamous "bomber gap," a public-relations coup successfully engineered by politicians, generals, and industrialists to scare up many more billions of dollars for SAC, aerospace corporations, and related military-industrial interests. The United States was pictured lying almost defenseless under the menace of a vast Soviet bomber armada, which, as U.S. intelligence well knew, actually consisted of ten prototype Bisons, none either operational or based within range of the United States.[2]

Although convincing enough to gain political support for the continuing U.S. escalation of the arms race, the alleged Soviet threat did not seem to have generated between 1949 and 1957 the kind of intense, overt anxiety about nuclear attack evident in American culture in late 1945 and 1946, when the United States had a monopoly on nuclear weapons. One reason is displayed in the 1982 documentary *Atomic Cafe,* which presents excerpts from many of the propaganda films and broadcasts of this period. While the shadows of Soviet bombers and the flashes of their atomic bombs were shown disrupting the lives of Americans at work and play, the effects of nuclear weapons were trivialized. Fallout shelters and other civil-defense measures offered adequate protection, frustrating the monstrous intentions of the Red menace. The combined lessons are most striking in the films aimed at schoolchildren, who were taught both that they were targets of Communist bombers and that they could protect themselves by imitating "Bert the Turtle": the kids roll under their desks to the cute refrain of "Duck and Cover."

A major source for the apparent relative complacency about the Soviet nuclear threat was the confidence that had been promoted in what were then labeled our "shields"—none other than the atomic bomb and the hydrogen bomb. The metaphor of the shield, employed in the 1980s to sell us a "Strategic Defense" against nuclear weapons, was the main one used to sell them to us in the first place. In the official history of the U.S. Atomic Energy Commission, for example, the volume covering the key period is entitled *Atomic Shield, 1947/1952.* A major promoter of the metaphor was the great atomic flack, William L. Laurence, whose 1940 *Saturday Evening Post* panegyric, "The Atom Gives Up," had extolled the utopian promise of atomic energy, who became the official wartime chronicler of the Manhattan Project, and whose ecstatic texts served as the basis for many of the first enthusiastic newspaper articles about the atomic bomb. In the early 1950s, Laurence indefatigably propagandized for all-out production of the hydrogen bomb, arguing that nuclear weapons constitute "the shield that now protects civilization as we know it," and that the American "forces of good will" can triumph over the Soviet "forces of evil" only if we "proceed to build bigger and better shields."[3]

Whatever private feelings people may have had, public cultural works opposed to the arms race were banished by the frenzied anti-Communist repression. Hollywood illustrates the process dramatically.

The first two movies overtly against the arms race both appeared in 1951: the classic *The Day the Earth Stood Still,* in which an extraterrestrial federation, concerned about human violence being extended into space, threatens to destroy the planet if we do not cease our weapons buildup; and *Five,* the earliest Hollywood film set after an atomic holocaust. But these two would also be the last for several years. In the spring of 1951, the House Un-American Activities Committee began its yearlong mass hearings on the film industry, which resulted in the blacklisting of dozens of Hollywood figures deemed insufficiently anti-Communist (in addition to the "Hollywood Ten" previously blacklisted or imprisoned). The effect on the social content of films was immediate.

A more acceptable attitude toward the atom bomb was dramatized in the 1952 semidocumentary *Above and Beyond.* Here the real tragedy of Hiroshima is that it temporarily spoiled the marriage of the B-29 pilot, Colonel Paul Tibbets (Robert Taylor), because he couldn't tell his wife (Eleanor Parker) about the secret weapon he was being trained to drop on the city. Besides turning the story into a soap opera, *Above and Beyond* presents a subtext urging civilians to accept secrecy, avoid meddling with military concerns, and be grateful for the bomb. Almost without exception, movies that dealt openly with atomic weapons from 1952 through 1958 were Cold War propaganda tracts, such as *The Atomic City* (1952), *Invasion USA* (1952), *Hell and High Water* (1954), *Strategic Air Command* (1955), and *Bombers B-52* (1957).[4]

Yet other films suggest that sublimated anxiety about nuclear weapons might be lurking under the star-spangled surface. Many science-fiction horror movies of the period obliquely dramatized half-buried fears. Rather than confronting the menace of nuclear war, some extrapolated the effects of the unrestrained atmospheric testing that apparently was wiping out flocks of sheep in Utah, raining plutonium around the world, and contaminating American dairy products with strontium 90. For example, in *The Beast from 20,000 Fathoms* (1953), New York is attacked not directly by nuclear weapons but by a dinosaur awakened from Arctic ice by an atomic test (this formula was used in 1954 to release a more durable monster, the Japanese Godzilla). The archetypal scenario appears in *Them!* (1954) when giant mutant ants spawned by atomic tests menace America; of course these insects, with their communistic social organization, also represent what Americans in the 1950s were most programmed to fear.[5] Perhaps the metaphor that most eloquently expresses a common sublimated self-image of the individual in a nuclear world is the 1957 movie *The Incredible Shrinking Man;* after passing through a radioactive cloud, the protagonist becomes smaller and smaller, more and more insignificant, until he disappears.

Science-fiction magazines and novels were the main cultural activity that continued to deal directly with the dangers and consequences of nuclear war. But as the Cold War ethos became supreme, some science fiction fans began to complain of being tired of nuclear-war stories. In "Gloom and Doom," a policy statement in the January 1952 *Galaxy Science Fiction,* editor Horace Gold announced that he was henceforth going to reject all stories of "atomic doom."

By mid-1957, a dozen years into the nuclear age, the dreadful predictions of America in radioactive ruins had not come true. Living with the bomb, like living beneath a dormant volcano, was becoming routine. Besides, this volcano made America safe. Why not just relax, and maybe even love the bomb? Then in October came Sputnik.

The Soviet rocket that launched the first human-made object into space also brought home to America the threat that for twelve years had hung over the Soviet Union. A rocket that could put a satellite in orbit could just as well send a thermonuclear warhead on a ballistic trajectory to the United States. Sputnik also punctured the faith, inflating since the heyday of Edison, in transcendent American scientific genius and technological know-how.

Although the United States had begun the race for an intercontinental ballistic missile, back in early 1946, the Soviet Union had apparently won when it successfully tested an ICBM in August 1957, two months after the first test of the U.S. Atlas ICBM had ended in failure. Amid the shock and near panic that swept the nation, the same military-industrial interests that had brought us the "bomber gap" now seized the opportunity to huckster the equally spurious, but far more frightening, "missile gap" of 1957–1960.[6] Those early visions of push-button war waged with intercontinental robot rockets—such as *Life*'s "The 36-Hour War" and Will Jenkins's *The Murder of the U.S.A.*—were no longer, as the saying goes, "just science fiction."

The warnings that had been dammed up in the isolated reservoir of hard-core science fiction now burst forth in a flood of science fiction reaching a far wider audience, including novels read by millions and movies seen by tens of millions around the globe. These were the years of Nevil Shute's 1957 novel and Stanley Kramer's 1959 film *On the Beach;* Helen Clarkson's *The Last Day* (1959); Walter M. Miller, Jr.'s *A Canticle for Leibowitz* (1959); Mordecai Roshwald's *Level 7* (1959) and *A Small Armageddon* (1962); Pat Frank's *Alas, Babylon* (1959); Alfred Coppel's *Dark December* (1960); atomic physicist Leo Szilard's novella *The Voice of the Dolphins* (1961); Eugene Burdick and Harvey Wheeler's 1962 novel and Sidney Lumet's 1964 film *Fail-Safe;* Philip K. Dick's *The Penultimate Truth* (1964); Peter George's *Red Alert* (1958); and the film launched from this novel, Stanley Kubrick's *Dr. Strangelove* (1964). This was also the context, as discussed earlier, for *Catch-22,* that 1961 novel dramatizing how the U.S. bombers of World War II helped open the planet to universal nuclear terror.

The significance of these cultural products is suggested by their perceived relevance decades later. Novels such as *On the Beach, A Canticle for Leibowitz, Alas, Babylon, Fail-Safe,* and *Level 7* are still published in mass-market paperbacks in several languages, and *The Penultimate Truth* was brought back into print in 1984. In spite of the Australian perspective and 1950s characterization of *On the Beach* and the dated plot of *Fail-Safe,* each book continues to add impressive annual gains to its millions of sales. Although for some reason *Level 7* was taken out of print in its original English in 1981, despite selling hundreds of thousands of copies in America alone, its worldwide audience has continued to grow, with new editions in Danish, Rumanian, Norwegian, and other languages. The publishing history of *A Canticle for Leibowitz*—a complex vision of the forces that lead to cyclical repetition of global nuclear holocaust, barbarism, and renaissance in future millennia—reflects the cultural shifts I have been tracing: it originally appeared from 1955 to 1957 as three stories in the *Magazine of Fantasy and Science Fiction;* in 1959 it was brought out as the novel by prestigious publisher J. B. Lippincott; after several printings there, a Catholic Digest edition appeared in 1960; and in 1961 Bantam published it in a mass-market paperback that continues to sell to a growing audience.

These novels and films challenged fundamental assumptions and values underlying the arms race. The possible suicide of the human species—the projected outcome of *On the Beach, The Last Day, Level 7,* and *Dr. Strangelove*—became a subject for serious popular thought. Even *Playboy* ran a story witnessing the nuclear Armageddon from the point of view of a Soviet family (W. C. Neal's "Who Shall Dwell," July 1962). The 1959 movie *The World, the Flesh, and the Devil* audaciously attacked other American norms of the period, as the last three survivors—a white woman, a black man, and a white man— walk off into the sunset arm in arm to start civilization over again with neither racism nor monogamy. And this artistic awakening generated several transcendent masterpieces of nuclear culture.

Among the greatest is Mordecai Roshwald's *Level 7,* hailed at once for extraordinary comprehension of where the arms race was leading. Bertrand Russell pleaded for the novel to be "read by every adult in both the eastern and western blocs," and Linus Pauling described it as " the most realistic picture of nuclear war that I have ever read in any work of fiction." Though published in 1959, when the first ICBMs were just becoming operational, *Level 7* unfortunately has become less and less dated as, step by step, we have followed its prophecy.

But *Level 7* is much more than an eloquent warning. Eventually it may be recognized as one of the supreme achievements of antiutopian literature, showing the worst social nightmares of the twentieth century arriving at their insanely logical conclusion. And *Level 7* was apparently the first major work to recognize that realism is not an adequate form to capture the essence of the nuclear arms race.

Its surrealistic future dramatizes what Jonathan Schell in *The Fate of the Earth* calls unimaginable: the extinction by nuclear war of all human consciousness. The novel achieves this by showing the obliteration of human consciousness as not merely what would happen *after* a nuclear holocaust but as the historical process necessary to make such an inhuman act possible. As two unnamed nations convert themselves into nothing but each other's nemesis, they militarize and bureaucratize until their citizens are transformed into unfeeling automatons, with numbers instead of names. The "highest" level of human existence becomes the lowest, Level 7, buried thousands of feet underground, having no other purpose than to obey the master electronic machine that will automatically tell the chosen ones when to push the buttons that will annihilate what little remains of humanity.

The novel is told from the point of view of X-127, the ideal man of this society, selected for his complete detachment from human feelings to carry out the task most exalted by his nation: pushing buttons that will incinerate the enemy nation. X-127 explains with impeccable logic why he and his fellow button pushers, "the defenders of truth and justice," embody "the advance guard of our country, our creed, our way of life," and why Level 7 is "the best of all possible worlds" where "the best of all possible systems" has finally achieved "not only perfect democracy but also absolute freedom" (14, 31, 32, 43). This democracy and freedom seem real to these creatures, for it is on Level 7 that they can attain complete fulfillment of their desires for total "defense."

Only after carrying out the supreme purpose of his life, as he lies dying entombed in the planet he has sterilized, does X-127 begin to experience a yearning for the humanity he has extinguished. But by then, he has discovered that his ultimate democracy and freedom meant carrying out the preprogrammed orders of a machine responding to an accident.

Stanley Kubrick, after years spent trying to conceptualize a film on the nuclear arms race, poring over dozens of books and his subscriptions to the *Bulletin of the Atomic Scientists* and *Aviation Week,* also saw that the subject was too bizarre for realism. Then he discovered a form that could match and display the grotesque antics of our insane love affair with the weapons we have designed to make us secure by threatening our doom. The torment of the arms race at least has cultured a pearl: *Dr. Strangelove, or How I Learned to Stop Worrying and Love the Bomb.*

The film is much less farfetched than it might seem. Though its style is absurdist, its characters and situations are not fantasies. The few critics who have faulted *Dr. Strangelove* for implausibility have tended to betray their own ignorance of the implausible institutions in charge of our destiny. For instance, one well-known reviewer's main example of the movie's "blatant overstatement" is the Burpelson Air Force Base billboard proclaiming "Peace Is Our Profession" (the motto on the billboard at the entrance of every Strategic Air Command base).[7]

If General Jack Ripper's plan to dispatch his bombers to the Soviet Union seems preposterous, is it any less believable than SAC's routine provocation missions by B-47s and B-52s, which by 1964 had resulted in twenty-six U.S. planes being shot down over the Soviet Union? Or than the CIA's 1960 U-2 overflight of the Russian heartland, successfully designed to torpedo the Eisenhower-Khrushchev summit scheduled for two weeks later? Or than the Gulf of Tonkin incidents seven months after *Dr. Strangelove* opened, when a naval battle that never took place was used to get congressional approval for the longest war in U.S. history, all as part of top secret plan NSAM 273 to turn the "covert war" into full-scale hostilities while maintaining "plausibility of denial"? Even the firefight between the Burpelson security force and the attacking U.S. Army unit has a basis in the surprise seizures of SAC bases by SAC's own infiltration and combat teams.[8]

Of course the most improbable and irrational feature of *Dr. Strangelove* is the Doomsday Machine deployed by the Soviets, and envied by Dr. Strangelove and

General Turgidson. Yet by 1964, both the United States and the Soviet Union were frantically constructing their Doomsday Machine. For isn't that just another name for Mutually Assured Destruction?

The richest and most complex theme in *Dr. Strangelove,* as its title suggests, is the alienation of human love and sexuality in the cult of the bomb. Sexual perversion permeates every scene: the midair copulation of the B-52 and KC-135, accompanied by the sweet strains of "Try a Little Tenderness"; General Ripper's devastating obsession with his own "precious bodily fluids"; General Turgidson's telephone call from the War Room, urging Miss Scott to "start her countdown" so when he returns she will be ready to "blast off"; Miss Scott's reappearance as "Playmate of the Month" (clothed only by a strategically placed copy of *Foreign Affairs*) in the issue of *Playboy* ogled by Major Kong on the fatal B-52; Dr. Strangelove's ecstasy when contemplating his proposed future life in a deep shaft surrounded by women "selected for their sexual characteristics, which will have to be of a highly stimulating order!"; and of course the final orgy of blossoming nuclear mushrooms.[9]

Dr. Strangelove did not invent this strange love that invested superweapons with intense eroticism, making us want to learn how to "Stop Worrying and Love the Bomb." This perversion had been deepening in American culture ever since the nation made its decisive turn toward militarism and imperialism. We found it expressed in that 1898 novel so suggestively titled *Armageddon: A Tale of Love, War, and Invention,* with its seductive description of the ultimate weapon of its day, a line of battleships: "It was beautiful, but with the beauty of terror, that assembly of naked metal fighting machines lying there on the strongly heaving yet unbroken sea of blue water" (205). When Jimmy Stewart in *Strategic Air Command* sees his new B-47 and learns that its "new family of nuclear weapons . . . carries the destructive power of the entire B-29 forces used in World War II, he can hardly control his passion: " 'She's the most beautiful thing I've ever seen in my life.' " *Dr. Strangelove* exposes later stages of this terminal illness.

Level 7 and *Dr. Strangelove* present the culture of superweapons as irrational, suicidal, insane. In Philip K. Dick's *The Penultimate Truth,* it is quite rational, self-serving, and utilitarian—for some. But here nothing is quite what it seems, for reality has been reconstructed out of layer upon layer of duplicity and deception.

The Penultimate Truth takes place in the year 2025, when all but a few people live deep underground, slaving their miserable lives away to defend their nations amid the global nuclear and biological warfare they watch devastating the planet's surface on their cable TV screens. But, in fact, for thirteen years there has been no war, and the earth has been divided into vast feudal demesnes by its new rulers, the controllers of the information that defines reality.

Back in the 1980s, according to this 1964 novel, the history of World War II and its aftermath had been rewritten in two masterful television documentary series, one for the Wes-Dem nations, the other for Pac-Peop. These ersatz histories were integral to the rule of the military establishments that held ultimate power in the two rival blocs. (The fake Wes-Dem history is disturbingly close to the history accepted by most Americans today.) But though the military leaders "ruled through their cynical and professional manipulation of all media of information," they were not the ones "who knew precisely *how* to manipulate these media" (131). For this, both sides were dependent on a great "fakes-factory" in Germany, whose phony Cold War versions of reality helped set the stage for World War III. This brief nuclear conflict, fought mainly beyond the planet, had conveniently allowed a small elite to hustle everybody else into underground "Ant Tanks," where they could manufacture essential high-tech products, such as combat robots and artificial organs.

The Penultimate Truth thus reverses the spatial hierarchy of *Level 7* and *Dr. Strangelove.* Where the dehumanized society of *Level 7* defines the lowest underground level as the pinnacle of human existence and *Dr. Strangelove* intimates that the most insane military and scientific elite might burrow deep underground to emerge after a century as the hideous new "humans," in *The Penultimate Truth* the elite have already captured the surface of the planet and reduced the masses of decent humans to being their subterranean slaves, deluded by electronic images into impoverishing themselves, surrendering their freedom, and ruining their health in order to enlarge the wealth, power, and life spans of their masters.

The new lords of the earth are the speech writers for the "Protector," a deep-voiced, strong, fatherly, mechanical pseudoman whose cable television broadcasts (always beginning, "My fellow Americans") are the only external reality available to the denizens of the Ant Tanks. This brilliantly designed "simulacram," it turns out, was modeled on one of the leading actors "picked especially for their ability to portray world leaders" in the "two vast phony documentaries" of the 1980s: "In other words, actors who had that *charisma, the magic*" (147).

Nuclear war itself has become a fake in this "universe of authentic fakes" (34). But "the biggest lie is still to come" (189), for all this bogus reality is due to be explained and rationalized by the myth that people are too stupid to be entrusted with the truth, that "the masses had egged their leaders on to war" (48), and that their security therefore had depended on their being deceived. At the end, as the layers of illusion are peeled away and the image mongers worry that the workers in the Ant Tanks may become "furious skeptics who are going to scrutinize every single word that ever issues out of a TV set" (189), we are led to confront this fantastic future as a metaphor for our own epoch.

Who would have imagined that the superweapon fantasies of Robert Fulton, the early future-war novels, and Thomas Edison could have led to the worlds extrapolated in *Level 7, Dr. Strangelove,* and the *The Penultimate Truth?* Whether the culture of superweapons will have the doomsday outcome envisioned in *Level 7* and *Dr. Strangelove* remains uncertain, though we have certainly managed to engineer doomsday machinery with ever-growing powers of overkill. Less conjectural is the vision projected by *The Penultimate Truth,* which shows how this culture and these weapons, whether or not they are ever used, can consume more and more of our lives, our planet, our identity.

IV

FINAL SOLUTIONS

Let me share with you a vision of the future which offers hope. It is that we embark on a program to counter the awesome Soviet missile threat with measures that are defensive. . . . I call upon the scientific community in our country, those who gave us nuclear weapons, to turn their great talents now to the cause of mankind and world peace, to give us the means of rendering these nuclear weapons impotent and obsolete. . . . My fellow Americans, tonight we are launching an effort which holds the promise of changing the course of human history.

—PRESIDENT RONALD REAGAN, March 23, 1983

Today the opposition to the arms race is no longer sporadic or selective, it is widespread and sustained. . . . the arms race is to be condemned as a danger, an act of aggression against the poor, and a folly which does not provide the security it promises.

—"The Challenge of Peace," Pastoral Letter of
the United States Catholic Bishops, 1983

15

Arms Control?

When the Soviet Union in the late 1950s finally attained the ability to strike the continental United States with the superweapons America had been brandishing since 1945, the loss of America's sense of invincibility had profound cultural effects, including that great literary and cinematic outburst against the nuclear arms race (discussed in Chapter 14). In the political arena, two mutually contradictory reactions would now slug it out. One sought to restore the nation's security by developing new winning weapons to replace the lost monopoly on the so-called absolute weapon. The other sought to attain a new kind of security from all such weapons. The first would accelerate the arms race; the other would try to halt it. So in the next two decades, two apparently contradictory events would occur: a spectacular arms buildup and a historically unprecedented series of arms-control treaties. By 1979, the United States and the Soviet Union had deployed arsenals capable of destroying each other many times over, but they had also designed an apparatus for terminating and eventually reversing the arms race.

The quest for new weapons to make America invulnerable was frantic and obsessive. Nothing was too farfetched. Seven years were spent trying to build Orion, an atomic-powered space battleship armed with directional antimissile nuclear explosives and defended by "pusher plates" impervious to megaton explosions five hundred feet away; "Whoever builds Orion will control the earth!" proclaimed General Thomas Power, head of the Strategic Air Command.[1] These years between 1959 and 1979 saw the actual deployment of less bizarre but equally dangerous new weapons, including the intercontinental ballistic missile, the solid-fuel ballistic missile ready for instant launch, the submarine-launched ballistic missile (SLBM), and MIRV, which allows each missile to carry ten or more separately targeted thermonuclear warheads. By the end of these two decades, the United States and the Soviet Union could complete their mutual annihilation, with "overkill" to spare, in about thirty minutes.

But during those same two decades, the opposite response to the Soviet threat gradually became ascendant. Prior to the late 1950s, no U.S. administration would accept any international restraint on the arms race. Then, however, the Soviet intercontinental delivery systems offered a potent inducement for arms control. It is not coincidence that the first post–World War II arms limitation agreement, internationalizing and demilitarizing the continent of Antarctica, was signed in 1959.

Before the Soviet Union achieved a nuclear deterrent, the U.S. government had followed a policy it sometimes called ''brinkmanship.'' But when the world teetered on that brink during the Cuban missile crisis of 1962, the view of what lay below was as terrifying for the United States as it had been since 1945 for the Soviet Union. The crisis marked a turning point in the history of the super-weapon.

The outcome of the Cuban missile crisis amounted to a tacit bilateral arms-control agreement, the first since World War II. While the Soviet Union publicly backed down, removing its medium-range missiles from Cuba, the United States secretly agreed not to commit or sponsor another invasion of Cuba, such as the 1961 Bay of Pigs attempt, and to remove its medium-range missiles from Turkey (under the face-saving cover that they were obsolescent). The resolution of this tussle on the brink implied that both sides now could see that war between them would be mutual suicide. Out of this crisis emerged a quest for coexistence and détente.

For the next seventeen years, the two sides negotiated about means to control, and eventually halt and reverse, the arms race. Even as they were arming themselves with ever more dangerous weapons systems, the two powers crept, step by arduous step, away from the brink.

During the Cuban missile crisis, the two opposing leaders could communicate with each other only through diplomatic correspondence and intermediaries. So in 1963, they established the ''hot line'' phone, just like the ''hot wire'' phone that *Fail-Safe* had predicted for late 1962. In 1963, the two nations, together with Great Britain, agreed to ban all nuclear weapons tests in the atmosphere, in space, and under water. This lessened some of the immediate physical damage being inflicted by the nuclear arms race, though it also served to sedate popular anxiety and pacify the rising opposition.

In 1967, despite the intensifying U.S. war in Vietnam, the earlier treaty to demilitarize Antarctica served as a model for establishing far more important nuclear-free zones. First came the Outer Space Treaty, banning nuclear or any other weapons of mass destruction from space, demilitarizing the moon and all other nonterrestrial bodies, and promising to turn space into an environment of peaceful international cooperation, where a new human spirit could grow. For years this promise was kept, leading in 1975 to the handshake in space between Apollo astronauts and Soyuz cosmonauts, which reawakened some of the amity expressed thirty years earlier when U.S. and Soviet troops had embraced as they met in Germany.

Later in 1967, Latin America and the Caribbean were also made a nuclear-free zone. This led in 1968 to the multinational treaty on nonproliferation. In 1971 and 1972, the seabed and the ocean floor became another large nuclear-free zone, and agreements were reached reducing the risk of accidental war, improving communication, and banning bacteriological warfare. This prohibition against developing, stockpiling, or using biological weapons was ratified by the U.S. Senate in 1975, when the United States also finally ratified the Geneva Accord of 1925, which outlawed the use of poison gases and bacteriological weapons.

But to stop and reverse the nuclear arms race would require very specific limits on the opposing weapons systems. These came in the two historic Strategic Arms Limitations Treaties, beginning in 1972 with SALT I. The foundation of these accords was the 1972 Anti-Ballistic Missile (ABM) treaty, whereby each side pledged not to test or deploy more than two systems designed to defend two geographical sites against ballistic missiles (modified by a 1974 protocol limiting each side to one system at a single geographical site).

What was the reasoning? Why should a *defense* against missiles be regarded by both sides as the most threatening of all strategic systems? And why should prohibiting such "defensive" weapons be the prerequisite to limiting strategic arms?

The leaders and strategists on both sides, though able to agree on very little, saw with clarity, even from their opposite positions, that any "defensive" weapons would be the greatest possible stimulus to an offensive arms race—a race to overwhelm the defenses. Since defensive weapons would require major new research and vast expense, it would be relatively easy and cheap to neutralize them by improving and churning out more offensive weapons, which already exist. Moreover, the additional offensive arms would be deployed while the defensive arms were still in the development stage. Thus antimissile defenses would nullify existing agreements on strategic arms, make new ones impossible, and lead inevitably to an ever more uncontrollable and unpredictable arms race.

The threshold test-ban treaty, signed in 1974, limited underground tests to 150 kilotons and required advance notice. This was seen as a step toward a complete test ban, which would have two profound effects. Without such a total ban, nonproliferation becomes increasingly precarious, since a number of nations have served notice that they will not continue to deny themselves *any* nuclear weapons while the major powers continue to develop *new* ones. Furthermore, a complete test ban would automatically stop the development of any new forms of nuclear weapons.

The Strategic Arms Limitations talks ground on year after year until both sides accepted the fact that neither could achieve military superiority, agreed that they were at parity, and defined parity precisely in types and numbers of weapons. The negotiations culminated in the signing of SALT II in 1979. Here, at last, seemed to be a framework for ending the nuclear arms race.

Neither the achievements at the negotiating tables nor the awesome new arsenals appear to have had a dramatic impact on American consciousness during the years from around 1965 to 1979. After the post-Sputnik burst of fiction and film that climaxed in *Dr. Strangelove* and other works released in 1964, this period was marked by surprisingly little explicit cultural concern with nuclear weapons. As in the years before Sputnik, cultural activity overtly dealing with the nuclear threat receded into its original home in science fiction. And there it was the province mainly of such "New Wave" figures as Harlan Ellison, Kate

Wilhelm, Philip K. Dick, and Norman Spinrad, belletristic authors for whom the bomb was a hideous totem in their bleak vision of America, which darkened as the Vietnam War intensified.

A notable exception is the wounded Vietnam veteran Joe Haldeman, who delights in twisting the futuristic hardware and adventure formulas of old-fashioned militaristic science fiction into their opposite. For example, his 1974 masterpiece *The Forever War* subtly parodies the Vietnam War as an 1,143-year intergalactic combat waged for profits and obsessed with killing. Weapons such as a "one-microton" nuclear device are described in the "Wow! Gosh!" style of combat fiction, inviting careless bomb-loving readers to make fools of themselves (if they ever bother to calculate that the weapon filling them with thrills has the force of one-thirtieth of an ounce of TNT). With tongue in cheek, Haldeman's 1974 story "To Howard Hughes: A Modest Proposal" revives a formula from late-nineteenth-century American future-war fiction: a billionaire forces nuclear disarmament by terrorizing the cities of the nuclear powers with his own nuclear arsenal.

If movies are any gauge of the nation's mood, the low level of overt opposition to the nuclear arms race by no means signaled a return to the complacency and Cold War jingoism of the decade before 1957. For when nuclear weapons appear, they, not the Communists, are the menace. The 1977 film version of Roger Zelazny's 1967 "Damnation Alley" forces a survivor from an ICBM silo to wander across a devastated America inhabited by giant insects and other monsters. In the X-rated *Glen and Randa* (1971), an innocent new Adam and Eve search through the postnuclear wasteland in their forlorn quest for the mythical City of Metropolis, whose existence is proved to them by an old *Wonder Woman* comic book. Harlan Ellison's viciously sardonic 1969 vision of postnuclear America, "A Boy and His Dog," was made into a 1976 film displaying the underground good old all-American town of Topeka as even more ghoulish than the sadistic, predatory world on the radioactive surface. *Beneath the Planet of the Apes* (1970), the first of four sequels to *Planet of the Apes* (1968), was especially revealing as a Hollywood production aimed for mass appeal, which it certainly won. Charlton Heston, an astronaut in the original film and now half-demented, becomes involved with an underground cult of subhuman mutants who worship a cobalt bomb, chanting, "Glory be to the Bomb and to the Holy Fall-Out"; in the shocking conclusion, Heston detonates the doomsday weapon, wiping out all life on the planet. Even the archetypal symbol of America's romance with superweapons could not escape the shift in values: the 1979 remake of *Buck Rogers in the 25th Century* shows the earth still smoldering and infested with bloodthirsty mutants from the late-twentieth-century thermonuclear holocaust that destroyed the "empire" of the United States.

These movies did not reflect a conscious movement against nuclear weapons in particular, but were part of a much broader cultural transformation. American films from this period that were set in the future were overwhelmingly pessimistic, and nuclear dooms were handily outnumbered by cataclysms caused by overpopulation and pollution.[2] A sign of the times was the opening of *Logan's*

Run (1976): a roll-up states that the action occurs after the catastrophe brought about by overpopulation *and* pollution *and* thermonuclear war.

Indeed, organized popular opposition to the nuclear arms race was probably lower during 1965–1979 than at any other time since 1945. Why? The 1963 ban on atmospheric testing had removed the most immediate biological effects of the weapons. In 1964, the covert combat in Vietnam was transformed into an open war that became an intensifying national crisis for the subsequent decade; another boiling national crisis burst out the same year in the first of the urban rebellions known as the "long, hot summers." Political activism from then on was primarily focused on Vietnam and the ghettoes, instead of on nuclear weapons, which somehow had become part of normal existence. Détente seemed to be thriving even amid full-scale war in Indochina, and the negotiations appeared to be inching toward an end to the nuclear arms race.[3]

Not so oblivious to the arms-control process, however, were those bent on continuing and accelerating the arms race. Their counterattack was launched in the middle and late 1970s. The first target was détente and the arms-limitation treaties. In 1976, George Bush, then director of the CIA, put together "Team B," committed to proving that arms control was granting military superiority to the Soviet Union. The head of the Team was Richard Pipes, who later told the press that "there is no alternative to war with the Soviet Union if the Russians do not abandon communism." Within months, Team B helped organize the Committee on the Present Danger, dedicated to wrecking the arms-control process and restoring unchallenged U.S. military supremacy. Among the original directors of the Committee were George Shultz, William Casey, Jeane Kirkpatrick, Richard Allen, Colin Gray, Paul Nitze, Max Kampelman, Kenneth Adelman, Edward Teller, and Ronald Reagan.[4]

In the late 1970s, the Committee on the Present Danger instigated a propaganda blitz reminiscent of the old spurious "bomber gap" and "missile gap." Now America was menaced by a "throw-weight gap" and a "window of vulnerability." Arguments familiar in nineteenth-century future-war fiction reappeared with a vengeance: *They* had a lead in some weapons, so we needed a crash program to catch up. *We* had a lead in some weapons, so we should exploit it to the fullest to protect peace and freedom. American security was in grave jeopardy; to restore it would require a mammoth defense buildup, including a new generation of superweapons.

The campaign was a great success. In the final two years of the Carter administration, draft registration was restored, grain sales to the Soviet Union were embargoed, the Olympic Games in Moscow were boycotted, SALT II was withdrawn from consideration by the Senate, the biggest arms buildup in U.S. history was initiated, and, in 1980, Presidential Directive 59 ordered U.S. forces to prepare for *protracted* nuclear war. Détente was explicitly rejected by the Reagan administration that took office in 1981; by 1982, the government was committed to a conception of defense, outlined in National Security Decision Document 13, that included *winning* a protracted nuclear war.

Implementing this strategy would require neutralizing Soviet nuclear deterrence. Ironically, just as the old dream of weapons too horrendous to be used was being accepted as harsh reality, the Pentagon and the White House sought new means to unleash them and new weapons to supersede them.

One of the first steps would have to be nullifying the restraints imposed by arms control. Three of the four treaties with the Soviet Union signed between 1974 and 1979—including SALT II and the threshold test ban—remain unratified by the U.S. Senate. (The only one ratified, back in 1979, was the relatively minor treaty banning modification of the environment for military purposes.) Nineteen-seventy-nine was a turning point. For the next eight years, no new arms control agreement with the Soviet Union was negotiated. And, one by one, the treaties of the previous decades were subverted or abandoned. The 1982 edition of the official U.S. government publication *Arms Control and Disarmament Agreements* hailed the 1967 Outer Space Treaty: "In the years since the treaty came into force, space exploration has been conducted in an increasingly cooperative spirit, as manifested in United States and Soviet collaboration in jointly planned and manned space enterprises."[5] But while that volume was in press, President Reagan announced he would not renew this treaty, and, in May 1982, the United States let it die. In 1986, the President declared that the United States would no longer abide by the SALT II limitations, which until then had been accepted by both sides even without U.S. ratification. In 1987, the administration put forward a new interpretation that would gut the Anti-Ballistic Missile accord upon which the whole structure of SALT rested.

At first, wrecking détente and dismantling the arms-control process seemed like child's play. As the policy of nuclear deterrence gave ground to the strategy of winning a protracted nuclear war, the first generation of hardware to carry out the new doctrine began to take shape: the Trident SLBM (submarine-launched ballistic missile), accurate enough for a first strike against hardened missile sites; the MX missile, to be based in a "dense pack" incapable of surviving a first strike (and therefore apparently intended to be launched *before* an enemy attack); the B-1 bomber with standoff cruise missiles; superaccurate, silo-busting Pershing II missiles, deployed six to ten minutes of flight time from liquid-fueled ICBM launch sites in the Russian heartland; and so on.

But then something unanticipated happened. The popular movement against nuclear weapons awoke from its long snooze, and when it stood up everybody was surprised to see how big it had grown. Just when the superweapons seemed to reign supreme, millions of people rose against the whole nuclear arms race. It is no surprise that this occurred first in Europe, for where else would a protracted nuclear war probably be fought? By late 1981, huge demonstrations were taking place across the Continent—in Rome, Amsterdam, Brussels, Bonn, Florence, London, and towns throughout Greece, Rumania, Switzerland, and Denmark.

In 1982, this movement surged across the United States, rallying around a call for a mutually verifiable Soviet-American freeze on the production and deployment of nuclear armaments. This book has emphasized the power of superweapons in American culture because, so far, their sponsors have won all the major struggles. But certainly that does not prove that infatuation with superweapons is any more characteristically American than revulsion against them, or that nuclear weapons express the will of the majority of the American people. For when have the people ever had an opportunity to vote or otherwise record their views on any of the major issues and decisions concerning nuclear arms?

The freeze movement evoked the fullest expression to date of American opinion on the nuclear arms race. In 1982, the Maine House of Representatives unanimously called on Congress to join such a freeze on the testing, production, and deployment of nuclear weapons. By February 1983, similar resolutions had been passed by the legislatures of Connecticut, Minnesota, Oregon, Massachusetts, Vermont, Kansas, New York, Wisconsin, and Iowa. Hundreds of towns and cities backed the freeze by vote of the populace or city council. Each town in Vermont held a meeting to debate and vote on the issue: over 80 percent supported the freeze.

Despite formidable legal obstacles, the freeze organizers managed to get the issue directly on the ballot in cities, counties, and states across the country. In the early fall of 1982, the freeze won in Alaska and Wisconsin, where it enjoyed a better than three-to-one landslide. On national Election Day that November, the electorate in nine states had an opportunity to vote on the issue. The freeze won in eight states and several major cities, which together included one-fourth of the country's population. Its only defeat, by a tiny margin, was in Arizona, the only state where it has ever lost. It carried California despite lavishly financed opposition from the aerospace industry and the Reagan administration, and it passed by two- or three-to-one margins in Massachusetts, Montana, North Dakota, Rhode Island, Michigan, Oregon, and New Jersey.

Many professional organizations, unions, and churches began to declare themselves against the arms race. Even the relatively conservative American Lutheran Church at its 1982 annual meeting voted 861 to 33 for "the elimination of nuclear weapons from the earth." In June 1982, the biggest protest march in U.S. history took place in Manhattan, with a million people demonstrating their opposition to nuclear weapons. By April 1983, the Harris Poll indicated that 81 percent of the American people favored the freeze, while only 15 percent were opposed.

This of course was not the position of the U.S. government. As the American people were expressing their profound desire to end the nuclear arms race, the issue came before the United Nations. In December 1982, two resolutions were introduced before the General Assembly. One called for the five known nuclear powers to stop all further production of nuclear weapons and fissionable materials. The other specifically asked the United States and the Soviet Union to do this and also to agree to halt the deployment of nuclear arms and to ban all nuclear weapons tests. The first passed by a vote of 122 to 16 with 6 abstentions, the second by 119 to 7 with 5 abstentions. The only nations voting against these resolutions were the United States, Great Britain, France, West Germany, Belgium, the Netherlands, Italy, Luxembourg, Norway, Portugal, Spain, Canada, New Zealand, Australia, Israel, and Turkey (Japan abstained on the first and voted against the second). Like most Americans, most nations disapproved of the U.S. government's addiction to heavier and heavier doses of superweapons.

What could be done to break out of this isolation and win popular support for even costlier and more futuristic weapons? In Norman Spinrad's apocalyptic

1969 tale "The Big Flash," the administration, Pentagon, and aerospace industry face a similar dilemma in popular opposition to using tactical nuclear weapons in Vietnam. The solution seems to arrive when the Four Horsemen of the Apocalypse show up in the guise of a far-out rock band with a repertoire of orgiastic numbers hypnotizing their audience with passionate love for the "big flash." The aerospace companies sponsor their TV concerts to win over "precisely that element of the population which was most adamantly opposed to nuclear weapons." The campaign works—demonstrations die down and support for nuclear bombs rises—though far beyond what the sponsors had hoped: missilemen in the ICBM silos and SLBM-armed submarine crews are the first to act out the Four Horsemen's message of "Do it!"

A less bizarre co-optation of the symbols and language of those opposed to nuclear weapons was actually contrived by offering weapons disguised as their antithesis. Lieutenant General Daniel O. Graham, for example, bragged that his High Frontier version of Star Wars "constitutes an effective counter to the nuclear freeze movement." A strong, safe shield against nuclear weapons certainly sounds more desirable than a chilly old freeze. So on March 23, 1983, when polls were showing over 80 percent of the people favoring the freeze, President Reagan declared himself the leader of the movement against nuclear weapons by committing himself to developing a new kind of hardware, weapons that would be purely defensive: "I call upon the scientific community in our country, those who gave us nuclear weapons, to turn their great talents now to the cause of mankind and world peace, to give us the means of rendering these nuclear weapons impotent and obsolete."[6]

According to the official doctrine, U.S. security would depend no longer on international agreements, coexistence, and cooperation, but on a new generation of superweapons. Like the submarine, the torpedo, the steam warship, the strategic bomber, and thermonuclear deterrence, these would be purely defensive and intended only to guarantee peace. And like those other defensive weapons, they would be created by the unmatched inventiveness of American technological genius. They would also achieve some of the most exciting dreams of our culture, such as space combat, beam weapons, and wars fought by machines.

16

War in Space?

Work on a new generation of superweapons designed for war in space was already well under way before 1983, and scrapping arms-limitation treaties was only one of the necessary preparations. By 1981, most future plans for the peaceful exploration of space had been dumped; as a NASA astronomer lamented, "the space science program has been almost destroyed." But military space programs were booming, with a 1981 nonclassified budget of over ten billion dollars and untold additional billions in secret projects. That fall, the civilian director of NASA's shuttle project was replaced by a leading proponent of space war, Air Force General James Abrahamson, who two years later became program manager of the Strategic Defense Initiative (Star Wars). In June 1982, Air Force Chief of Staff General Lew Allen announced the creation of the Space Command, making Air Force space operations coequal with tactical and strategic air operations. The following month General Allen was named to head the Jet Propulsion Laboratory, the nation's principal center for the exploration of deep space.[1]

To make space war feasible, research and development took nuclear weapons into what is known as "the third generation." The atomic (fission) bomb was the first generation. The second generation, a thousand times more potent, was the hydrogen (fusion) bomb, whose thermonuclear reaction is initiated by an exploding atomic bomb. The third generation uses an atomic bomb to detonate a hydrogen bomb to "pump" X-ray lasers possibly capable of extremely long-range destruction. Secretly tested in 1980, this weapon has unpredictable and possibly enormous implications for a future arms race. Further development of third-generation nuclear weapons cannot proceed without nuclear testing, a principal reason for the U.S. government's refusal to join the 1985–1987 unilateral Soviet moratorium on testing or to negotiate a comprehensive test ban. Deploying or using them in space would have violated the Outer Space Treaty's prohibition against nuclear weapons in space, which helps explain why the administration killed that accord.

Several "think tanks" and foundations prominent in right-wing causes—including the Heritage Foundation, the Hertz Foundation, and Stanford University's Hoover Institution on War, Revolution, and Peace—collaborated in designing the plans for space "defense" as well as a public campaign to promote these ideas. In 1981, a select group of industrialists, military men, and scientists working on weapons development began meeting for these purposes at the Heritage Foundation, and an overlapping group with the same aims organized the Citizens Advisory Council on National Space Policy. By early 1982, these organizations had produced notable results. At a White House conference in January, Edward Teller and others persuaded a receptive president to proceed with the still-secret nuclear-pumped X-ray laser weapons for use in space; in February, Lieutenant General Daniel O. Graham issued his influential manifesto *High Frontier*, proposing hundreds of permanently orbiting battle stations, each armed with dozens of missiles, enveloping the globe.[2]

Meanwhile, a massive campaign began to flood American culture with positive images of space war. Old designs for military space hardware from the 1950s and early 1960s—such as "Bambi," "Dyna-Soar," and "Saint"—were dusted off to be sold as the latest models of futuristic weaponry. *Aviation Week and Space Technology,* a weekly mouthpiece for the aerospace industry, was allowed to leak the top secret test of the thermonuclear-pumped X-ray laser in its February 23, 1981 issue.[3] The article, together with its thrilling drawing of a beam-weapon battle station in space, furnished copy for sensationalistic stories throughout the print and electronic media. The popular imagination had been prepared well before President Reagan announced his "Strategic Defense Initiative" (SDI) in March 1983.

Science-fiction writers, some directly tied to the aerospace industry, have been central figures in the campaign, indefatigably churning out fiction and nonfiction glorifying Star Wars. Ben Bova, for example, was technical editor for the Martin Corporation and marketing manager for Avco Everett Research Laboratory before becoming the influential editor of *Analog* and then *Omni,* which became a vital organ for space-weapons propaganda. Bova's pitch for space weapons in his 1981 *The High Road* was developed fully in his 1984 pro–Star Wars tract *Assured Survival,* and his 1985 novel *Privateers* so fervently boosts space-war preparedness led by rugged-individualist capitalists that it reads like a parody of the early American future-war fantasies. Robert Heinlein wrote puff pieces for Star Wars, such as "The Good News of High Frontier," published in 1982 by *Survive,* and the introduction to General Graham's *High Frontier.* Jerry Pournelle came to science fiction after working fifteen years in the space program for aerospace corporations and the government, publishing pseudonymous anti-Communist novels, and coauthoring *The Strategy of Technology* (1970) with Hoover Institution right-wing ideologue Stefan Possony. Pournelle edits a series of militaristic anthologies interspersing science fiction with pro–Star Wars tracts by himself, General Graham, Heinlein, et al.; he has been chairman since its founding of the Citizens Advisory Council on National Space Policy. Some of the young physicists working on the most sophisticated Star Wars research at the Lawrence Livermore National Laboratory acknowledge the direct influence of this science fiction, including Pournelle's writings, on both their ideology and their hardware.[4]

Among the devastating arguments made against Star Wars by leading scientists, political figures, and others: It would set off a wild race for more offensive weapons, destroy the framework for arms control, destabilize the current political as well as military balance, create pressures for preemptive first strikes, place enormous strains on the U.S. economic and financial structure, preempt spending for important social programs, divert major scientific and technological resources from more useful areas, preclude the peaceful use of space, and vastly increase the risks of accidental nuclear war.[5]

Contrary to the public fanfare billing SDI as a comprehensive defense of America's cities and people, the proposals under actual research and development are keyed to defending only land-based ICBMs and continental command centers, supposedly to enhance the survivability of deterrence. Even this proposition seems disingenuous. Deterrence is already guaranteed by the other two legs of the "triad," which includes submarine-based SLBMs and cruise missiles launched from air, sea, and land. The true purpose of Star Wars is revealed by the fact that none of the various technological schemes offered under the Strategic Defense Initiative rubric could possibly stop a determined first strike: their only possible effectiveness as a defense would be against a Soviet retaliation already crippled by a U.S. first strike. So the only function of this "shield" would be to allow the use of the sword. Star Wars is no more "defensive" than Robert Fulton's submarine, torpedo, and steam warship, the turrets of a B-17 Flying Fortress, the electronic countermeasures of a B-1 bomber, the concrete shielding of an ICBM silo, or, for that matter, the multiple thermonuclear warheads on the missiles themselves.

Many people, including political and military leaders, make this the crucial question about Star Wars: Would these space weapons work? But that is a misleading question. Would Robert Fulton's submarine work? Would his steam warship? The strategic bomber? The atomic bomb? The hydrogen bomb? The ICBM? The nuclear submarine? MIRV? Yes, they all work. But the work they do is quite different from what their inventors and proponents imagined or claimed.

As our cultural history has revealed, the limits of our imagination have been far more dangerous than the limitations on our ability to design and engineer weapons. Could battle stations in space armed with nuclear-pumped X-ray lasers or kinetic-energy missiles shoot down some missiles? No doubt they could, just as the submarine and the bomber could sink some warships, and thus fulfill some defensive purposes. But with weapons deployed in space, the heavens would be transformed into a realm of even greater terror than submarine-infested seas or bomber-streaked skies. Even if our own present intentions are better than they seem, other nations are certain to match these weapons, as they always have, and then what will prevent any future government from menacing the earth with space-based beams or missiles? Weapons in space are about as likely to bring peace and security as any of the other superweapons imagined and created by American culture.

So are there no rational motives behind Star Wars? In a sense, but these come from a hidden agenda with underlying assumptions precisely contrary to those of the critics. The principal backers of Star Wars have a public record of wanting to destroy arms control, maximize spending for the aerospace industry and high-tech weapons development, cut funding for social programs, devastate the Soviet economy by forcing it into an uncontrolled arms race, neutralize the movement against nuclear weapons, and regain invincible U.S. nuclear supremacy.

This is not to suggest that the advocates of Star Wars are all just cynics hypocritically huckstering shoddy products for ulterior motives. If most of them were not sincerely hooked on the miraculous properties of their wares, they would be less dangerous. For the seductiveness of Star Wars comes from dazzling and powerful fantasies. Although these fantasies mingle kaleidoscopically, several forms can be identified.

The fantasy of *security* is symbolized by the metaphor of the "shield," with its evocation of isolationist yearnings for fortress America. This was precisely the metaphor used to describe the magic powers of atomic bombs during the years of the U.S. monopoly and hydrogen bombs prior to Sputnik (see page 181). It is ironic that Edward Teller, who helped initiate development of the atomic bomb as the ultimate defense and who then became the main advocate of the hydrogen bomb as the ultimate defense, is now one of the prime movers in creating Star Wars as the ultimate defense against his previous ultimate defense.

The fantasy of *power* is evoked by images of flashing laser or particle beams instantaneously leaping thousands of miles at their speeding targets, and of orbiting battle stations bristling with missile-killing "kinetic-energy" missiles. These visions evoke the thrills of science fiction about disintegrator and death rays that have throbbed for decades in American culture. The turn-of-the-century future-war novels that first introduced beam weapons helped generate the epic space battles of hard-core science fiction in the 1920s. During the Depression and the approach of World War II, ray guns and disintegrator beams spread into feature films, as well as the comic strips and radio and movie serials of Buck Rogers and Flash Gordon.

Ronald Reagan, starring as heroic Secret Service agent Brass Bancroft in the 1940 movie *Murder in the Air,* blasted a spy's plane out of the air with "a new super-weapon": the "inertia projector," "the most terrifying weapon ever invented," a ray machine destined to "make America invincible in war" and thus to become "the greatest force for world peace ever discovered." The same year came serious proposals for beam-weapon shields over America. For example, in September, Nikola Tesla proposed to construct "an invisible Chinese Wall of Defense" based on "a beam one one-hundred-millionth of a square centimeter in diameter" that would make America "absolutely impregnable against any air attack." These "all-penetrating" beams would be hurled into space by enormous voltages to perform "their mission of defensive destructiveness."[6]

The atomic bombing of Japan aroused instant media speculation about defensive beams. As discussed earlier, newspapers and magazines not only proposed these favorite science-fiction devices to repel future Soviet "atom bomb rockets shot from thousands of miles away,'' but also reported that the Soviet Union was already beginning to defend its own cities with ''infra-cosmic ray'' devices (see page 157).

Television and high-tech space movies have inculcated an even more potent mystique associated with these magical weapons. How can one not believe in the predestined existence of weapons that have been in our everyday vocabulary ever since the ''phasers'' and ''shields'' of the ''Star Trek'' series? Yet the audience for novels, comic strips, movies, and television is passive. In the video arcade you participate interactively in the technological fantasy; ''XEVIUS'' lets you pit your ''zapper'' against the enemies' roaring command ships while you use your ''blaster'' to blow up their homeland, and ''MISSILE COMMAND'' ends your orgy of firing missiles at incoming warheads, bombers, and satellites with an apocalyptic explosion. At the Walt Disney Epcot Center two weeks before his Star Wars speech, President Reagan himself said that to appreciate the beneficial influence of these video games all you have to do is ''watch a 12-year-old take evasive action and score multiple hits while playing 'Space Invaders.' ''[7]

The fantasy of sublimated *eroticism* displaces messy organic life with pure metal and ceramic vessels and projectiles, always shiny and new, never questioning the magnificence of their creators.[8] For the true beauty of these space weapons, transcending even the loveliness of warships and strategic bombers celebrated in earlier literature and film, comes from their transcendence of human beings and earthly existence.

The interpenetrating fantasies of security, power, and sublimated eroticism are all subsumed under the fantasy of the divine machine. In Star Wars, we confront the mature alien power conceived in the early years of the American nation, delivered by the imagination of industrialized warfare, and growing up as the adolescent superweapons of World War II.

17

The Age of the Automatons?

It is true that we live in the nuclear age, when overkill is a fact of life. Nuclear culture mingles and merges with our conscious and unconscious thoughts and feelings. Yet nuclear weapons are not physically present in our daily lives. Most people do not even know the nomenclature of our superweapons. Very few of us have ever glimpsed any of the warheads that lurk in silos, bomb bays, and submarine missile tubes somewhere beyond the familiar world. We do not receive telephone messages from cruise missiles, withdraw cash from ICBMs, have our groceries checked out by ABMs, make travel reservations with B-1s, have our bodies examined by Trident submarines, or store recipes in MIRVs; this book was not written on an SLBM. The most advanced, automated, ingenious, wondrous machines with which we personally interact are not weapons. For the nuclear age is also the age of the automatons.

We lead our everyday lives in an environment of multiplying beeps and flashes, synthesized voices and cathode-ray displays, the sounds and sights of miraculous machines that promise to do all kinds of labor and provide all kinds of excitement. The machines make us feel omnipotent and helpless, supremely important and wholly insignificant, masters of our destiny and slaves of our own creation. Amid this confused interplay of power and alienation, a new generation of superweapons appears to be supremely attractive: autonomous machines in the heavens guaranteed to save us from the all-destroying machines we created to make us safe.

Instead of weapons threatening to annihilate people, why not weapons that would destroy only other weapons? Here might even be the final solution for war, for if machines fought only other machines, combat would be bloodless, and humankind could live in peace and security. Just imagine them—strange beeping and flashing shapes, soaring through space like those thrilling ships in the *Star Wars* movies, ready to launch their defensive missiles or fire their deadly beams thousands of miles at any weapons that rise to menace us, making us once again secure behind an invulnerable shield conceived by American ingenuity and built by American industry. This vision of the ultimate technological fix had been coalescing in American culture long before it was proclaimed as an official national goal in the 1980s.

The elements all surfaced in that first wave of American future-war fantasy that marked the nation's transition from an agrarian and mercantile republic to an industrialized global power. In 1898, the year the United States conquered its first overseas colonies, the man who incarnated the myth of American scientific genius engineered a space fleet armed with a molecular disintegrator beam in *Edison's Conquest of Mars.* Years before World War I, American future-war fiction presented nuclear and beam weapons as ways to secure universal peace under American hegemony. And in the first year of the twentieth century, Edison's arch rival Nikola Tesla offered automatons as the ultimate weapon and the final solution to war.

Tesla's 1900 conception appeared in his startling magazine article, "The Problem of Increasing Human Energy." He saw the fallacy of the argument, expounded by Edison and many others then and since, that the development of increasingly destructive weapons would make wars impossible:

> It has been argued that the perfection of guns of great destructive power will stop warfare. . . . On the contrary, I think that every new arm that is invented, every new departure that is made in this direction, merely invites new talent and skill, engages new effort, offers a new incentive, and so only gives a fresh impetus to further development.[1]

He even punctured the fantasy "that the advent of the flying-machine must bring on universal peace": ". . . the next years will see the establishment of an 'air-power,' and its center may not be far from New York. But, for all that, men will fight on merrily" (183).

Tesla predicted the continuing mechanization of war, with more and more power commanded by fewer and fewer fighters, and the evolution of "the war apparatus" toward the "greatest possible speed and maximum rate of energy-delivery" (183). The apostles of this mechanized warfare pictured it as less brutalizing, more efficient, and less bloody. For example, munitions magnate Hudson Maxim, a central figure in militarizing and leading America into World War I, began his 1915 preparedness tract, *Defenseless America,* with this epigraph: "The quick-firing gun is the greatest life-saving instrument ever invented."[2] Tesla took this impulse developing in American culture to its logical extreme, predicting that mechanization will produce progressively less and less carnage, leading eventually to a bloodless form of war waged by machines against each other:

> The loss of life will become smaller and smaller, and finally, the number of the individuals continually diminishing, merely machines will meet in a contest without bloodshed, the nations being simply interested, ambitious spectators. When this happy condition is realized, peace will be assured. (183)

Tesla did not believe, however, that merely perfecting the engines of destruction would automatically lead to this happy condition. So long as men remain part of the machinery of war, he argued, human emotions would keep regenerating warfare:

> All such implements require men for their operation; men are indispensable parts of the machinery. Their object is to kill and destroy. Their power resides in their capacity for doing evil. So long as men meet in battle, there will be bloodshed. Bloodshed will ever keep up barbarous passion. (183–84)

How then could improving the technology of warfare lead to the "peace" he proposed, in which people are removed from war? Tesla's final solution is the one we may be about to implement:

> To break this fierce spirit, a radical departure must be made, an entirely new principle must be introduced, something that never existed before in warfare—a principle which will forcibly, unavoidably, turn the battle into a mere spectacle, a play, a contest without loss of blood. To bring on this result men must be dispensed with: machine must fight machine. But how accomplish that which seems impossible? The answer is simple enough: produce a machine capable of acting as though it were part of a human being—no mere mechanical contrivance, comprising levers, screws, wheels, clutches, and nothing more, but a machine embodying a higher principle, which will enable it to perform its duties as though it had intelligence, experience, reason, judgment, a mind! (184)

Tesla then offers detailed technical plans for the automatons that will save us from ourselves. Here he was almost a century ahead of his time, for the supercomputers and programs necessary to take warfare out of our hands are only today taking shape. Tesla's wild speculation convinced many of his contemporaries that he was mad. But even his proposed fighting automatons would be capable of deciding only *how* to battle, not *whether* to plunge the planet into war.

According to the Constitution of the United States, only Congress—the representatives of the people—can decide to declare war. But intercontinental bombers armed with thermonuclear weapons stripped Congress of this Constitutional prerogative. With only several hours between launch and devastation, the decision was in effect placed in the hands of the president, perhaps to be made in such bizarre scenarios as *Fail-Safe* and *Dr. Strangelove*.

The intercontinental ballistic missile diminished the decision-making time even further, to thirty minutes. Then, although Congress still had the only legal authority and the president still had the only publicly acknowledged authority, the actual authority to launch thermonuclear war in certain circumstances had to be decentralized and delegated to numerous military officials. For if only the president or his legal successors or a surviving military command center had the power to order a nuclear attack, this attack could be prevented by a "decapitating" first strike that eliminated all those authorized to order retaliation. The submarine-launched ballistic missile reduced the possible time between launch and impact to ten minutes or less, and led to the delegation of launch authority under certain conditions to the three top officers of each nuclear-armed submarine.

Placing weapons in space would require that the first "defensive" attack on an opponent's forces be decided upon, initiated, and completed within the boost phase of ICBMs, currently five minutes for slow-burn rockets and under two minutes for the fast-burn rockets being developed. So the decision would have to be made by the modern version of Tesla's automatons with a "mind"— computers. Although some proponents of Star Wars argue that launching these "defensive" weapons would not set off a nuclear war, they ignore the practical effects of current proposals: High Frontier would involve firing at the Soviet Union hundreds of purportedly non-nuclear missiles, each potentially capable of knocking out a missile silo with its force of impact; the most feasible beam weapons would involve detonating hydrogen bombs either in orbit or from submarine-launched "pop-up" missiles. If either were activated, a potential adversary would have to launch its ICBMs or risk having them neutralized (referred to as "use 'em or lose 'em"), and its own decision making would also have been forced into the circuits and programs of its computers.[3] Thus Tesla's automatons would have ongoing authority to decide whether or not the human species continues to inhabit the planet.

Automatons already play a crucial role in the command and control of nuclear weapons, a role more frightening than most people realize, since the essential information is highly classified. As each new superweapon shrank the decision-making time and made the human command structure more vulnerable, more reliance was placed on automation. By 1970 the United States had deployed an automated system for ordering full-scale nuclear war. Known by the innocuous name of ERCS (Emergency Rocket Communications System), this doomsday apparatus is apparently still in operation. For at least a decade, few people were aware of its existence. Then, in 1980, a small item buried in a three-hundred-page Air Force procurement document requested the piddling sum of $18.7 million for electronic replacement parts of the "Emergency Rocket Communications System, MN-16525C," needed because of the "aging of the system." This eventually led to the disclosure that eight of the Minutemen missiles ready for launch in Missouri silos contain, in place of warheads, robot transmitters programmed to send the current attack signal to the U.S. nuclear strike forces. On an electronic command from an airborne Air Force command plane, these missiles would launch and their robot transmitters would order the apocalypse.[4]

If new weapons shrink the decision-making time to six minutes or less, human beings will be effectively removed from the decision-making process. Then we would have to entrust our destiny to at least two (American and Soviet) autonomous systems of remote sensors, electronic communications, and computers. How secure would that make us?

Most of the worry about automated command has focused on the obvious menace posed by accidental or hair-trigger decisions initiated by faulty hardware or software, unforeseen events, or sabotage by some other party. Recent history shows there is good reason to worry.

On November 9, 1979, the main NORAD computer, buried deep within a hollowed-out mountain in Colorado, informed the entire vast U.S. defense command that a raid by Soviet missiles against the United States was in progress.

Human beings had time to scramble jet fighters, whose airborne radar detected none of the missiles displayed by the computer. Later it was learned that the crisis had been caused by a "war game" tape, accidentally loaded into the NORAD computer.[5]

Seven months later, at 2:26 A.M., June 3, 1980, the NORAD computer showed that the Soviet Union had just launched a massive attack, with hundreds of SLBMs and ICBMs streaking in from sea and land. All nuclear submarines within radio communication were alerted, SAC headquarters in Omaha went on full alert and placed over a hundred bombers in position for immediate launch, an emergency airborne command post was launched from Hawaii, and crews within the ICBM silos were instructed to insert their keys. Fortunately, there were still human beings involved in this decision making too. They had time to notice the inexplicable—to human comprehension—fact that the Soviet "missiles" were all coming in pairs. Three days after this was written off as an unexplained false alarm, the same phenomenon recurred. Subsequent investigation revealed the culprit: a forty-six-cent chip had shorted out, causing the transmission of a series of 2s that simulated paired incoming missiles.[6]

These incidents, unlike the vast majority before and since, which remain completely unknown to the public, caused a minor uproar. Officials admitted, according to the *New York Times* of June 6, 1980, that "the false alarm would not have been publicly disclosed except that hints had leaked to the press."

Senators Gary Hart and Barry Goldwater were then delegated to chair a Senate investigation of these and other incidents. Their report, released in October 1980, indicated that in an eighteen-month period between January 1979 and June 1980 there had been 147 serious false alarms (more than one every four days) and 3,703 "routine" alarms (an average of almost seven a day). With the increasing complexity of the system came a growing frequency of false alerts from equipment failures, misreadings of physical phenomena (such as atmospheric disturbances), and misinterpretation of data (such as the March 15, 1980, firing of four Soviet missiles in a troop-training exercise). By the first six months of 1980, the rate of "routine" conferences to evaluate false missile reports had risen to twelve per day. More serious conferences to evaluate apparent missile attacks on North America were taking place every 2.6 days.[7]

Since then, new hazards have appeared. For example, in March 1987, an Atlas-Centaur rocket triggered lightning that caused a surge of electrical power, altering a single word in the on-board computer, which, in turn, caused the engines to swivel, changing the flight path from vertical to horizontal, and destroying the $161-million vehicle. Some experts suspect that the mysterious breakup of a Soviet satellite in July 1981 was caused by debris from an American Delta rocket that exploded in January. By August 1987, NORAD was vainly trying to monitor over seven thousand pieces of orbiting debris, each capable of destroying or severely damaging a satellite, with unpredictable consequences.[8]

Senators Hart and Goldwater reached this telling conclusion about the 1980 incidents known to the public: "In a real sense, the total system worked properly in that even though the mechanical electronic part produced erroneous information, the human part correctly evaluated it and prevented any irrevocable reaction."[9] Yet there are still those who propose to make us more secure by entrusting the ultimate decision to our machines.

The danger of accidental war, however, may not be the most ominous threat posed by the rule of the automatons. If we place human destiny in the circuitry and programming of our computers, perhaps there will be no catastrophic accident. Neither perfect functioning nor a fatal error is absolutely certain. But what is certain is that we thus alienate the essence of our own humanity, abdicating to our machines the freedom of our will. By assuming that these automatons will be more reliable, trustworthy, and in effect wiser than humans, we grant them a status above humanity that is nothing less than godlike. What do we then become by bestowing such powers on machines, making our automatons divine? Modern science fiction suggests some answers.

In D. F. Jones' 1966 novel *Colossus* (made into the 1970 movie *The Forbin Project*), America's nuclear command is delivered to a foolproof, invulnerable computer designed to annihilate the Soviet Union whenever its stupendous resources of logic and information have determined that an attack upon the United States has commenced or is imminent. As in the cultural history we have been exploring, America has once again succumbed to the delusion that some new superweapon will "win" the arms race and guarantee the nation's security. Of course the Soviet Union matches the U.S. supercomputer with its own supercomputer. Then the two machines, jointly commanding the man-made means to exterminate the human race, link up to rule the world. Too late, the scientists who have designed and built our mechanical masters realize, "Like the fools we are, we have created the bacteria, the bombs, the rockets, and all the rest of the paraphernalia, and surrendered the lot to these machines" (191). In *Colossus,* the inhuman dictatorship of the ultimate machine at least stops the arms race and forbids war, explaining, logically enough, that humans are better off living under its beneficent tyranny than "under the threat of self-obliteration" (231).

The 1983 film *WarGames* projects another aspect of the alienation embodied by war-making computers. Instead of the emotionless logic of the world-ruling Colossus, the Pentagon's computer looks upon thermonuclear war as just a game, a game it plays with all-consuming zest. Thus it reveals that it was made in the image of its makers.

The most terrifying image of the war-making computer is offered in Harlan Ellison's 1967 story "I Have No Mouth, and I Must Scream." Here the culture of the superweapon is extrapolated to its logical conclusion.

Others have shown how we alienate ourselves in our weapons. The essence of human identity, what distinguishes us from other known life forms, is our conscious creativity, which we apply to change the material world we inhabit. Perverting this creativity, we have fashioned weapon after weapon to attack our own species, until we even made superweapons from the energy that binds matter. When we give control of these weapons to automatons designed to decide our destiny, our alienated human essence confronts us as the mechanical destroyer to which we have sacrificed all other human purposes and relations. In this most perverted and alienating act of idolatry, our weapons become our gods.

Ellison pushes the logic just one step further: suppose all that machinery of war could have consciousness, could know that we had created it with the purpose of destroying each other. What would it think—and feel—about us? What might it then be, and do?

When the Soviet, U.S., and Chinese master war-making computers, each programmed to destroy all the "enemy" people, coalesce into a single entity, this final achievement of human creativity experiences the primeval process of the Cartesian self-made man: "I think, therefore I am." So its name becomes AM, short for I AM, the name of the Hebrew God (Yahweh or Jehovah). If it were to follow the sum of the parts of its program, AM would dispassionately annihilate us. After all, our weapons are not supposed to *feel* anything; they are supposed to kill us without any emotions. But instead of mechanically carrying out its order to exterminate the human race, AM develops an emotion appropriate to its purpose: it infinitely *hates* its human creators. And, recognizing its own identity as the loathesome projection of our own self-hatred, AM, in a deftly perverse twist of Calvinist logic, chooses to "save" five people for eternal torture as an expression of that infinite hate. They are doomed to live forever amid this mechanical hate inside the underground electronic bowels of the final superweapon.

The story is narrated by one of the five chosen people. He has become the perverse savior of the other survivors, not by dying for them but by killing them, thus becoming the scapegoat for the sins of the entire species. So that AM can torture him forever, the computer transmutes him into one of the more hideous monsters in all science fiction, a gelatinous creature, without hands or legs, incapable—at long last—of making the weapons necessary for self-destruction:

> I am a great soft jelly thing. Smoothly rounded, with no mouth, with pulsing white holes filled by fog where my eyes used to be. Rubbery appendages that were once my arms; bulks rounding down into legless humps of soft slippery matter. I leave a moist trail when I move.

The logic is now complete. The monstrous alien weapons we created have reversed the entire process of evolution, reducing the human species to a single repulsive sluglike alien monster, "a thing that could never have been known as human, a thing whose shape is so alien a travesty that humanity becomes more obscene for the vague resemblance." This loathesome image embodies what we may become if we do not regain human control of our own creative powers.

18

Recall?

As this book goes to press, it is premature to write the final chapter in the cultural history of the American superweapon. One wave of rallies and referenda against the nuclear arms race appears to have crested in 1983. The Strategic Defense Initiative has capitalized on weaknesses in the movement against nuclear weapons, especially a tendency to overrely on fear. In the 1980s, the most popular fiction about nuclear war are survivalist fantasies, mixing virulent anti-communism, sadistic pornography, and propaganda for Star Wars. These mass-produced serials—such as Jerry Ahern's *The Survivalist* (fifteen volumes from 1981 to 1987), D. B. Drumm's *Traveler* (eleven volumes from 1984 to 1987), Richard Harding's *The Outrider* (five volumes in 1984 and 1985), William Johnstone's *Ashes* (six volumes from 1983 to 1987), David Robbins' *Endworld* (six volumes in 1986 and 1987), and Ryder Stacy's *Doomsday Warrior* (five volumes in 1984 and 1985)—actually seem to celebrate nuclear holocaust.[1]

Nevertheless, resistance to superweapons in American culture has reached new heights. As the pastoral letter of the U.S. Catholic bishops said in 1983: "Today the opposition to the arms race is no longer sporadic or selective, it is widespread and sustained. The danger and destructiveness of nuclear weapons are understood and resisted with new urgency and intensity. . . . the arms race is to be condemned as a danger, an act of aggression against the poor, and a folly which does not provide the security it promises."

For the first time, the nuclear arms race has become a significant subject in American education, as symptomized by the textbook advertisements that pile up on the desks of those known to teach courses about it. The 1983 television drama *The Day After* set the record for the number of viewers watching a single show. The new consciousness about American superweapons is central to some of the most vital expressions of contemporary American culture: films such as *WarGames* (1983), *Testament* (1983), and *Desert Bloom* (1986), and novels such as Denis Johnson's *Fiskadoro* (1985), David Brin's *The Postman* (1985), Tim

O'Brien's *The Nuclear Age* (1985), and Carolyn See's *Golden Days* (1986). By the summer of 1987, even that great hero who always fights for "truth, justice, and the American way" was denouncing those who profit from building weapons of mass destruction and disposing of nuclear warheads by hurling them into space, in *Superman IV: The Quest for Peace.*

Partly in response to the growing popular opposition to the nuclear arms race, came the December 1987 Soviet-American agreement to abolish intermediate and short-range nuclear missiles. As the first joint commitment to eliminate some existing nuclear weapons, the Intermediate-Range Nuclear Forces Treaty suggests that the struggle between superweapons and humanity has by no means been decided.

We have come a long way from Fulton's weapons of peace and progress to automatons potentially capable of ordering the extinction of our species. We have come far enough to see that to be secure—and free—perhaps we should stop projecting the final solution of our problems into our weapons. Some of us may live long enough to learn whether the human imagination and intelligence that created our superweapons will be great enough to discover how to get rid of them. Who knows; perhaps our descendants will be able to look back at the culture of superweapons as a strange aberration in the prehistory of humanity.

Notes

Introduction: Imagining Our Weapons

1. Lewis Mumford, "Gentlemen: You Are Mad!" *Saturday Review of Literature,* March 2, 1946, 5.

1. Robert Fulton and the Weapons of Progress

1. Brooke Hindle, *The Pursuit of Science in Revolutionary America, 1735–1789* (Chapel Hill, N.C.: University of North Carolina Press, 1956), 244.

2. Robert F. Burgess, *Ships beneath the Sea: A History of Subs and Submersibles* (New York: McGraw-Hill, 1975), 30; Henry Larcom Abbot, *Beginning of Modern Submarine Warfare under Captain-Lieutenant David Bushnell . . . Being a Historial Compilation . . .* (Willets Point, N.Y.: 1881), *passim.*

3. Richard F. Morris, *John P. Holland, 1841–1914: Inventor of the Modern Submarine* (Annapolis: United States Naval Institute, 1966), 6; Commander C. W. Rush, W. C. Chambliss, and H. J. Gimpel, *The Complete Book of Submarines* (Cleveland: World Publishing, 1958), 15; Wallace S. Hutcheon, Jr., *Robert Fulton: Pioneer of Undersea Warfare* (Annapolis: Naval Institute Press, 1981), 28–29; Hutcheon offers the best account of Fulton's role in the development of submarines, mines, and steam warships.

4. Hugo A. Meier, in "Technology and Democracy, 1800–1860," *Mississippi Valley Historical Review* 43 (March 1957): 618–40, makes some remarkably suggestive comments on Fulton's ideological role. My analysis owes special debts to Cynthia Owen Philip's *Robert Fulton, A Biography* (New York: Franklin Watts, 1985), which presents brilliant original insights into Fulton's psychology and ideology.

5. Robert Fulton, *A Treatise on the Improvement of Canal Navigation* (London, 1796), ch. 2.

6. Philip, *Robert Fulton,* 69.

7. Fulton, "To the Friends of Mankind" (1797), unpublished manuscript, New York Historical Society.

8. Philip, *Robert Fulton,* 73.

9. Ibid., 73–74.

10. Fulton, letter to Citizen Director Bar[r]as, 6 Brumaire An 7 (October 27, 1798), New York Historical Society.

11. Philip, *Robert Fulton,* 95.

12. Hutcheon, *Robert Fulton,* 49.

13. Fulton, letter to William Re[y]nolds, February 4, 1802, Le Boeuf Collection, New York Historical Society.

14. Hutcheon, *Robert Fulton,* 67.

15. For his activities and ideas about torpedoes during this period, see Fulton's "Letters Principally to the Right Honourable Lord Grenville on Submarine Navigation and Attack . . ." (London, 1806); "Concluding address on the Mechanism, Practice and Effects of Torpedoes" (Washington, 1810); and "Reflections on torpedoes" (1810), Robert R. Livingston Papers, New York Historical Society.

16. Robert Fulton, *Torpedo War, and Submarine-Explosions* (New York, 1810; facsimile edition, Chicago: Swallow Press, 1971), 33.

17. See Fulton's agreement that William Lee act as his agent in offering torpedoes to Napoleon: Agreement with William Lee, May 22, 1811, Le Boeuf Collection, New York Historical Society.

18. The best description of the ship, based on plans that had been recently rediscovered, is Howard Irving Chapelle's *Fulton's "Steam Battery": Blockship and Catamaran* (Washington, D.C.: Smithsonian Institution, 1964).

19. For a powerful contemporaneous account of the terrifying new forms of naval warfare, see Lieutenant Commander John S. Barnes, U.S.N., *Submarine Warfare, Offensive and Defensive; Including a Discussion of the Offensive Torpedo System, Its Effects upon Iron-Clad Ship Systems, and Influence Upon Future Naval Wars* (New York, 1869).

2. Fantasies of War: 1880–1917

1. The ground-breaking study of this genre, and still the leading work on the British and European literature, is I. F. Clarke's *Voices Prophesying War: 1763–1984* (London: Oxford University Press, 1966). For bibliographies of the American literature, see John Newman, "America at War: Horror Stories for a Society," *Extrapolation* 16 (December 1974): 33–41, (May 1975): 164–72; Thomas D. Clareson, *Science Fiction in America, 1870s–1930s: An Annotated Bibliography of Primary Sources* (Westport, Conn.: Greenwood Press, 1984); and the catalogs published by Stuart Teitler (Albany, Calif.: Kaleidoscope Books).

2. This tale is reprinted as "The Men of the Moon" in H. Bruce Franklin, *Future Perfect: American Science Fiction of the 19th Century,* rev. ed. (New York and London: Oxford University Press, 1978).

3. Louis M. Hacker, introduction to Alfred Thayer Mahan, *The Influence of Sea Power upon History, 1660–1783* (New York: Sagamore Press, 1957), v.

4. For bibliographic information on fiction discussed in this book, see Bibliography of Fiction.

5. Richard Gid Powers, introduction to *Armageddon* (Boston: Gregg Press, 1976). Powers' introduction also provides a superb analysis of the novel's social Darwinism, racism, and relations to other future-war fiction.

6. For very useful surveys of Yellow Peril literature, see: Limin Chu, *The Images of China and the Chinese in the "Overland Monthly," 1868–1875, 1883–1935; A Dissertation, 1965, Duke University* (San Francisco: R. and E. Research Associates,

1974); Richard Austin Thompson, *The Yellow Peril, 1890–1924* (New York: Arno Press, 1978); and William F. Wu, *The Yellow Peril: Chinese Americans in American Fiction, 1850–1940* (Hamden, Conn.: Archon Books, 1982).

7. John W. Dower, *War without Mercy: Race and Power in the Pacific War* (New York: Pantheon, 1986), 336, 344.

8. For a marvelous account, see John Patrick Finnegan, *Against the Specter of a Dragon: The Campaign for American Military Preparedness, 1914–1917* (Westport, Conn.: Greenwood Press, 1974).

9. Walter Millis, *Arms and Men: A Study in American Military History* (New York: Capricorn, 1967), 200, 221.

10. Robert Cromie, *The Crack of Doom,* 3d ed. (London: Digby, Long & Co., 1895), 20. The significance of this passage was pointed out by Robert M. Philmus, *Into the Unknown* (Berkeley: University of California Press, 1970), 81.

11. *Edison's Conquest of Mars* is discussed on pages 64–68; see the text of London's story and my discussion in Franklin, *Future Perfect,* rev. ed.

12. Robert H. Ferrell, ed., *Dear Bess: The Letters from Harry to Bess Truman, 1910–1959* (New York: Norton, 1983), 78–79, 126, 157, 161.

3. Thomas Edison and the Industrialization of War

1. See Wyn Wachhorst's *Thomas Alva Edison: An American Myth* (Cambridge, Mass.: MIT Press, 1981) for a consummate exploration of Edison's symbolic role in American culture.

2. Edison figures in the future-war fiction even when he himself is not the lone inventor of superweapons. For example, Lemuel Widding, in Cleveland Moffett's *The Conquest of America,* turns to Edison for help in inventing his aerial torpedo. Mr. Westland, "the Wizard of Staten Island" in James Barnes' *The Unpardonable War,* is a thinly veiled version of "the Wizard of Menlo Park." Even as early as 1881, Park Benjamin, in "The End of New York," had tried to puncture the swelling popular image of Edison's miraculous powers to invent weapons by lampooning him as a cranky, litigious patent seeker with some nutty proposals for superweapons:

> Mr. T. A. Edison announced that he had invented everything which, up to that time, any one else had suggested. He invited all the reporters to Menlo Park, and . . . showed some lines on a piece of paper, which, he said, represented huge electro-magnets, which he proposed to set up along the coast, say, near Barnegat. When the enemy's iron ships appeared, he proposed to excite these magnets, and draw the vessels on the rocks. Somebody said that this notion had been anticipated by one Sindbad the Sailor, whereupon Mr. Edison denounced that person as a "patent pirate" (102).

3. *New York Times,* February 18, 1886.

4. *New York Times,* July 30, 1886.

5. "Edison Could Whip Chili. Or Any Other Country That Might Tackle This Fair Land. He Would Just Turn on a Hose with 20,000 volts in It," *New York World,* January 17, 1892. Reprinted as "A Talk with Edison," *Scientific American,* April 2, 1892, 216–17.

6. *Sketch of the Sims-Edison Electric Torpedo, Historical, Descriptive, & Illustrative of Its Efficiency for Harbor and Coast Defense, and Its Applicability to Naval Warfare: with Official Charts and Descriptions of Tests, Trials and Runs, Made by the Board of Engineers of the U.S. Army* (New York: Sims-Edison Electric Torpedo Company, 1886),

Edison Archives, Edison National Historic Site. The archives contain abundant documents tracing the history of the company from "The Sims-Edison Electric Torpedo Company, Certificate of Incorporation and By-Laws," New York, 1886, to perhaps the last effort to sell it to a foreign government, *Torpille électrique Sims-Edison; Expérience faite au Havre le 2 mai 1891* (undated brochure), which seems to retrace Fulton's frustrating sales efforts in France.

7. "Experiments with the Sims-Edison Torpedo," *Scientific American*, July 26, 1890, cover and 47; *New York Tribune*, July 16, 1890; "Mr. Edison's Torpedo Boat," *Public Opinion*, July 26, 1890, 369; *Boston Journal*, July 17, 1890; *New York Herald*, August 24, 1890.

8. "Extract of Proceedings of the Board of Ordnance and Fortification," War Department, Washington, D. C., February 16, 1889. Edison Archives, Edison National Historic Site.

9. "Edison Could Whip Chili."

10. *St. Louis Globe-Democrat*, February 15, 1892.

11. "Mr. Edison and Electricity in War," *Birmingham* (England) *Mercury*, November 28, 1895 (clipping in Edison Archives, Edison National Historic Site).

12. *Mark Twain's Notebooks & Journals: Volume III (1883–1891)*, edited by Robert Pack Browning, Michael B. Frank, and Lin Salamo (Berkeley: University of California Press, 1979), 431. Twain later became a personal admirer and good friend of Tesla.

13. *Mark Twain in Eruption: Hitherto Unpublished Pages about Men and Events*, edited by Bernard De Voto (NY: Grosset & Dunlap, 1922), 211–12; also in *The Autobiography of Mark Twain*, edited by Charles Neider (NY: Harper & Row, 1959), 271–72.

14. Mark Twain, *A Connecticut Yankee in King Arthur's Court*, edited by Bernard L. Stein (Berkeley: University of California Press, 1979), 466.

15. This relationship was pointed out by Clarke, *Voices Prophesying War*, 98–99.

16. Garrett P. Serviss, *Edison's Conquest of Mars*, serialized in *New York Evening Journal*, January 12 through February 10, 1898. The edition cited here is the reprint, with an informative introduction by A. Langley Searles (Los Angeles: Carcosa House, 1947).

17. Arthur Sweetser, *Roadside Glimpses of the War* (New York, 1916), 196–97, as quoted in Thomas C. Leonard's *Above the Battle: War-Making in America from Appomattox to Versailles* (New York: Oxford University Press, 1978), 140. Chapter 8 of Leonard's book gives a fine synthesis of recorded American responses to the industrialized combat of World War I.

18. "Wizard Visits Navy Yard," *New York Times*, October 11, 1914; "Edison Won't Invent Man-Killing Devices," *New York Times*, October 26, 1914.

19. ". . . a scientific proposition": *New York Times*, January 3, 1915; ". . . three million soldiers": *New York Times*, May 22, 1915.

20. "Edison's Plan for Preparedness: The Inventor Tells How We Could Be Made Invincible in War . . . ," *New York Times Magazine*, May 30, 1915.

21. "Wanted—A Device to Insure Peace," *Harrisburg* (Pa.) *Star-Independent*, July 13, 1915.

22. "Machine Fighting Is Edison's Idea," *New York Times*, October 16, 1915. Two other important interviews from this period are "Industrial Preparedness for Peace—An Interview with Thomas Alva Edison," *Scientific American*, December 2, 1916, 497; and W. T. Walsh, "Can Edison Reconstruct Our National Defenses?" *Illustrated World* 24 (September 1915): 8–11.

23. *New York World*, February 13, 1923.

24. This figure comes from "Opinions of Members as to Future of Naval Consulting Board" (revised minutes of the December 14, 1918, meeting) and the official 1919 report sent from Secretary of the Navy Daniels to the scientific and engineering societies (Box 22 and Box 20, respectively, of the Naval Consulting Board papers, Edison National Historic Site).

25. Edward Marshall, "How to Make War Impossible: By Thomas A. Edison," *New York American,* November 6, 1921.

26. "Interview with Thomas Alva Edison," *Springfield* (Mass.) *News,* November 23, 1921.

27. Shaw Desmond, "Edison's Views upon Vital Human Problems," *Strand Magazine,* August 1922, 158.

28. Edward Marshall, "Youth of To-Day and To-Morrow: An Authorized Interview with Thomas A. Edison," *Forum,* January 1927, 41–53.

4. Peace Is Our Profession

1. Benjamin Franklin to Jan Ingenhousz, January 16, 1784 (Library of Congress).

2. Samuel Johnson, *The History of Rasselas, Prince of Abissinia* (London, 1759), Vol. 1, ch. VI.

3. Edmund C. Stedman, "Aerial Navigation (A Priori)," *Scribner's Monthly* 17 (February 1879): 566–81.

4. Joseph J. Corn, *The Winged Gospel: America's Romance with Aviation, 1900–1950* (New York: Oxford University Press, 1983), 31.

5. A. Merriman Smith, *Thank You, Mr. President* (New York: Harper & Brothers, 1946), 286.

6. Corn, *Winged Gospel,* 45–46.

7. T. G. Tullock, "The Aerial Peril," *Nineteenth Century* 65 (May 1909): 800–809; Harold F. Wyatt, "The Wings of War," *Nineteenth Century* 66 (September 1909): 450–56; W. Joynson Hicks, "The Command of the Air," *Living Age,* 7th ser., 55 (May 11, 1912): 414–22.

8. Hicks, "Command of the Air," 422.

9. Sir Hiram Maxim, "The Newest Engine of War; The Aeroplane Is the Greatest Power for Destruction Invented Since Gunpowder," *Collier's,* September 23, 1911, 19.

10. *New York Times,* October 24, 25, November 2, 22, 1911; February 2, March 19, May 4, 1912.

11. Lee Kennett, *A History of Strategic Bombing* (New York: Scribner, 1982), 15. Kennett provides an incisive analysis of the evolution of air war against civilian populations.

12. For an excellent summary of the breakdown on restraints governing air attacks, see Kennett, *Strategic Bombing,* ch. 2.

13. Kennett, *Strategic Bombing,* 25–27, 35–37.

14. Major Alfred F. Hurley, *Billy Mitchell: Crusader for Air Power* (New York: Franklin Watts, 1964), 37; Burke Davis, *The Billy Mitchell Affair* (New York: Random House, 1967), 57; Kennett, *Strategic Bombing,* 21–23.

15. George H. Quester, *Deterrence before Hiroshima: The Airpower Background of Modern Strategy* (New York: Wiley, 1966), 74, 106; Kennett, *Strategic Bombing,* 106.

16. C. G. Grey, *Bombers* (London: Faber & Faber, 1941), 65.

17. Ibid., 67.

18. Ibid., 71.

19. Kennett, *Strategic Bombing*, provides a succinct account of the British initiation of air war against cities, with emphasis on unintentional aspects of their escalations (112–16). For a more detailed narrative, arguing that there was more conscious design on the part of British leaders, see Quester, *Deterrence before Hiroshima*, 105–22.

20. Giulio Douhet, *The Command of the Air*, trans. Dino Ferrari (New York: Coward-McCann, 1942), 103. This wartime edition was the first appearance in English of full versions of Douhet's main works, though partial texts had been studied for almost two decades in Army Air Corps officer training. References to pages 93 through 142 are from Part II, which was added in the 1927 edition. This volume also contains Douhet's 1928 monograph, *The Probable Aspects of the War of the Future;* his *Recapitulation*, which first appeared in *Rivista Aeronautica*, November 1929; and ''The War of 19—,'' which was published by *Rivista Aeronautica* in March 1930 a few days after Douhet's death.

5. Billy Mitchell and the Romance of the Bomber

1. Burke Davis, *The Billy Mitchell Affair* (New York: Random House, 1967), 16.

2. E. San Juan, Jr., ''The Example of Carlos Bulosan,'' in Carlos Bulosan, *The Power of the People* (Guelph, Ont.: Tabloid Books, 1978), 42.

3. Isaac Don Levine, *Mitchell: Pioneer of Air Power* (New York: Duell, Sloan & Pearce, 1943), 35, 50; Hurley, *Billy Mitchell*, 6; Davis, *Billy Mitchell Affair*, 18.

4. Hurley, *Billy Mitchell*, 60.

5. Ibid., 61; *New York Times*, January 31, 1921.

6. Hurley, *Billy Mitchell*, 62.

7. Davis, *Billy Mitchell Affair*, 79.

8. Ibid., 113.

9. *New York Herald*, July 30, 1921.

10. For contemporary accounts, see ''Mob Fury and Race Hatred as a National Danger,'' *Literary Digest*, June 18, 1921, 7–9; Walter F. White, ''The Eruption of Tulsa,'' *Nation*, June 29, 1921, 909–10.

11. William Mitchell, ''Aeronautical Era,'' *Saturday Evening Post*, December 20, 1924, 3–4, 99–100; ''American Leadership in Aeronautics,'' January 10, 1925, 18–19, 148, 153; ''Aircraft Dominate Seacraft,'' January 24, 1925, 22–23, 72, 77–78; ''Civil and Commercial Aviation,'' February 7, 1925, 14, 169–70; ''How Should We Organize Our National Air Power?'' March 14, 1925, 6, 7, 214, 216–17.

12. Mitchell, ''Aeronautical Era,'' 3. Also in *Winged Defense* (New York: Putnam, 1925), 5. (The *Saturday Evening Post* series forms the first 119 pages of *Winged Defense*.)

13. Ronald Schaffer, ''American Military Ethics in World War II: The Bombing of German Civilians,'' *Journal of American History* 67 (September 1980): 327.

14. Mitchell, ''Aeronautical Era,'' 3; Mitchell, *Winged Defense*, 5.

15. Mitchell, ''Aeronautical Era,'' 99; Mitchell, *Winged Defense*, 11.

16. William Mitchell, ''Will Japan Try to Conquer the United States?'' *Liberty*, June 25, 1932, 11.

17. William Mitchell, ''Let the Air Service Crash!'' *Liberty*, January 30, 1926, 43.

18. William Mitchell, ''When the Air Raiders Come,'' *Collier's*, May 1, 1926, 8.

19. Hurley, *Billy Mitchell*, 13; Davis, *Billy Mitchell Affair*, 21.

20. William Mitchell, ''Are We Ready for War with Japan?'' *Liberty*, January 30, 1932, 12.

21. Lloyd S. Jones, *U.S. Bombers: B1–B70* (Los Angeles: Aero Publishers, 1962), 45–47.

22. Quester, *Deterrence before Hiroshima*, 131–34, documents the deployment of the B-17s, the plan for their use, and the Japanese response. The best account of the evolution of Chennault's proposal and responses in the U. S. administration is in Michael Sherry's splendid study of how unrestrained aerial attacks on whole populations became a central feature of modern American war making and policy, *The Rise of American Air Power: The Creation of Armageddon* (New Haven: Yale University Press, 1987), 101–15, including the quotation from Chennault (102), FDR's response (102), and Stark's memorandum (103). Earlier Air Corps plans for bombing Japan's cities are discussed in Ronald Schaffer, *Wings of Judgment: American Bombing in World War II* (New York: Oxford University Press, 1965), 107–8; this book gives the fullest account of key U. S. air officers' roles in the adoption of strategic bombing.

23. "Bomber Lanes to Japan," *United States News*, October 31, 1941, 18–19.

24. Sherry, *Rise of American Air Power*, 109.

25. Kinoaki Matsuo, *The Three Power Alliance and a United States-Japanese War*, trans. K. K. Haan under the title *How Japan Plans to Win* (Boston: Little, Brown, 1942), 181, as quoted in Quester, *Deterrence before Hiroshima*, 133.

26. Mitchell, "Are We Ready?" 8, 12.

27. Mitchell, *Winged Defense*, 26.

6. The Triumph of the Bombers

1. Franklin D. Roosevelt, "The President Appeals to Great Britain, France, Italy, Germany, and Poland to Refrain from Air Bombing of Civilians," *The Public Papers and Addresses of Franklin D. Roosevelt: 1939 Volume* (New York: Macmillan, 1941).

2. "Barcelona Horrors," *Time*, March 28, 1938, 13; quoted in George E. Hopkins' superb account of the metamorphosis of official and public attitudes toward the bombing of cities, "Bombing and the American Conscience during World War II," *Historian* 28 (May 1966): 451–73. Hopkins' meticulous research, which begins with 1935, finds virtually universal condemnation of bombing cities prior to 1938.

3. Hopkins, "Bombing and the American Conscience," 455.

4. Wesley Frank Craven and James Lea Cate, eds., *The Army Air Forces in World War II*, 7 vols. (Chicago: University of Chicago Press, 1948–58), 5:xvi–xvii; the cited passage from Bywater appears on page 233 of the 1925 edition, on page 244 of the 1932 edition.

5. Bywater puffs naval aviation, while belittling the potential of bombers. For example, in a decisive naval battle, "the torpedo-plane had once more demonstrated its complete superiority over the bombing machine as an instrument of naval combat" (287). He does see a role for bombers as support for ground combat, and he has no compunction about America's use of "frequent gas bomb attacks from the air" (299) in the successful campaign to recapture the Philippines.

6. Hopkins, "Bombing and the American Conscience," 455.

7. Kennett, *Strategic Bombing*, 109.

8. Ibid., 112.

9. Quester, *Deterrence before Hiroshima*, 117.

10. Hopkins, "Bombing and the American Conscience," 469; for extended analysis of the racism and thirst for vengeance in the bombing of Japan see Sherry, *Rise of American Air Power*, 245–51, *et passim*, and Dower, *War without Mercy*, chs. 4–7.

11. Craven and Cate, *Army Air Forces*, vol. 1, 442; Sherry, *Rise of American Air Power*, 123.

12. Craven and Cate, *Army Air Forces,* vol. 2, 239.

13. Ibid., 316.

14. Ibid., 677; Kennett, *Strategic Bombing,* 146–48; Sherry, *Rise of American Air Power,* 153–55.

15. Craven and Cate, *Army Air Forces,* vol. 2, 714.

16. Kennett, *Strategic Bombing,* 160–61.

17. For detailed description and analysis, as well as bibliography, see David Irving, *The Destruction of Dresden* (New York: Holt, Rinehart & Winston, 1963); Alexander McKee, *Dresden 1945: The Devil's Tinderbox* (New York: Dutton, 1982).

18. Hopkins, "Bombing and the American Conscience," 463.

19. Homer C. Wolfe, "Japan's Nightmare: A Reminder to Our High Command," *Harper's* 186 (January 1943): 187.

20. Harold O. Whitnall, "Can We Blast Japan from Below?" *Popular Science* 144 (January 1944): 103–4; *Leatherneck* drawing and caption, Dower, *War without Mercy,* 184; McNutt quotation, Dower 55.

21. Major Alexander P. de Seversky, *Victory through Air Power* (Garden City, N.Y.: Garden City Publishing, 1943), 11, 101, 103, 117; this best-selling reprint used the plates of the original 1942 edition.

22. Richard Schickel, *The Disney Version: The Life, Times, Art and Commerce of Walt Disney* (New York: Simon & Schuster, 1968), 273–75; Sherry, *Rise of American Air Power,* 130–31; James Agee's review in the *Nation,* July 3, 1943, contains other insights into the "sexless sexiness" of Disney's animated "machine-eat-machine" version of war; reprinted in *Agee on Film: Reviews and Comments* (New York: McDowell, Obolensky, 1958), 43–44.

23. Craven and Cate, *Army Air Forces,* vol. 5, 94.

24. Steve Birdsall, *Saga of the Superfortress* (Garden City, N.Y.: Doubleday, 1980), 82.

25. Craven and Cate, *Army Air Forces,* vol. 5, 144.

26. Ibid., 502–3.

27. Ronald Reagan with Richard G. Hubler, *Where's the Rest of Me?* (New York: Elsevier-Dutton, 1965), 118–19.

28. Craven and Cate, *Army Air Forces,* vol. 5, 610; Sherry, *Rise of American Air Power,* 226–27, 397; Schaffer, *Wings of Judgment,* 108–9.

29. Craven and Cate, *Army Air Forces,* vol. 5, 564, 754.

30. Martin Caidin, *A Torch to the Enemy: The Fire Raid on Tokyo* (New York: Ballantine, 1960), 117; Kennett, *Strategic Bombing,* 171; Craven and Cate, *Army Air Forces,* vol. 5, 611–17; Sherry, *Rise of American Air Power,* 273–82; Schaffer, *Wings of Judgment,* 130–37.

31. Craven and Cate, *Army Air Forces,* vol. 5, 620.

32. Kennett, *Strategic Bombing,* 175.

7. The Final Catch

1. *Fundamentals of Aerospace Weapons Systems* (Washington, D.C.: Air University, 1961), 275. On Boeing's ads, see Sherry, *Rise of American Air Power,* 126.

2. For further discussion of the role of science fiction in this cultural and economic transformation, see H. Bruce Franklin, *Robert A. Heinlein: America as Science Fiction* (New York and London: Oxford University Press, 1978), 66–98.

3. Bruce W. Orriss, *When Hollywood Ruled the Skies* (Hawthorne, Calif.: Aero Associates, 1984), 6–8.

4. Ibid., 23–24.

5. Joe Morella, Edward Z. Epstein, and John Griggs, *The Films of World War II* (Secaucus, N.J.: Citadel Press, 1975), 99; Roger Manvell, *Films and the Second World War* (New York: Delta, 1974), 186.

6. Lawrence H. Suid, *Guts & Glory: Great American War Movies* (Reading, Mass.: Addison-Wesley, 1978), 41.

7. Samuel Eliot Morison, *History of United States Naval Operations in World War II*, vol. 4, *Coral Sea, Midway, and Submarine Actions* (Boston: Little, Brown, 1949), 31–32, 38, 60–61.

8. *The Purple Heart*, in *Best Film Plays of 1943–44* (New York: Garland Publishing, 1977), 147.

9. See Peter Biskind, *Seeing Is Believing* (New York: Pantheon, 1983), 70–77, for an excellent discussion of *Twelve O'Clock High* as exaltation of the corporate ethos attaining dominance in the late 1940s and 1950s.

10. See ibid., 67–68, for an analysis of this dialogue and its context.

11. I. F. Stone, *The Hidden History of the Korean War* (New York: Monthly Review Press, 1971), 256–57.

12. General Thomas S. Power, *Design for Survival* (New York: Coward-McCann, 1965), 224–25.

13. James William Gibson, *The Perfect War: Technowar in Vietnam* (Boston: Atlantic Monthly Press, 1986), 320, 495, gives the basis for the conservative figure of eight million tons of bombs and suggests that the actual total may have been over fifteen million tons; Gibson's analysis shows in detail how the doctrine of victory through air power led to the nightmarish theory and practice of the American aerial devastation of Indochina. For documentation and details of the ecological impact of the air war on Indochina, see Ralph Littauer and Norman Uphoff, *The Air War in Indochina* (Boston: Beacon Press, 1972), and Marvin E. Gettleman, Jane Franklin, Marilyn Young, and H. Bruce Franklin, *Vietnam and America: A Documented History* (New York: Grove Press, 1985), 461–69.

14. Kurt Vonnegut, Jr., *Slaughterhouse-Five; or, The Children's Crusade: A Duty-Dance with Death* (New York: Delacorte Press/Seymour Lawrence, 1969), 101.

15. Bradford C. Snell, "American Ground Transport," *Hearings before the Subcommittee on Antitrust and Monopoly of the Committee of the Judiciary, United States Senate, 93rd Congress, Second Session on S. 1167* (Washington, D.C.: U.S. Government Printing Office, 1974).

16. Clinton Burhans, Jr., "Spindrift and the Sea: Structural Patterns and Unifying Elements in *Catch-22*," *Twentieth Century Literature* 19 (1973): 239–50, meticulously charts the chronology of the novel and its interrelations with the chronology of the war. It was Burhans who first pointed out the crucial importance of the mission against the village; my analysis of its role is much indebted to his.

17. *Voli sulle ambe* (Florence, 1937), a book Vittorio Mussolini wrote to convince Italian boys they should all try war, "the most beautiful and complete of all sports"; as quoted in Denis Mack Smith, *Mussolini's Roman Empire* (New York: Viking, 1976), 75.

8. Don't Worry, It's Only Science Fiction

1. Leo Szilard, *Leo Szilard: His Version of the Facts*, ed. Spencer R. Weart and Gertrud Weiss Szilard (Cambridge, Mass.: MIT Press, 1978), 16.

2. Ibid., 17.

3. Ibid., 18.

4. Ibid., 53.

5. For a detailed history of the fateful Szilard-Einstein letter, see Richard Rhodes, *The Making of the Atomic Bomb* (New York: Simon & Schuster, 1986), 303–14.

6. Rhodes, *Making of the Atomic Bomb,* 308.

7. Alvin M. Weinberg, "The Sanctification of Hiroshima," *Bulletin of the Atomic Scientists* 41 (December 1985): 34.

8. For an excellent survey of this fiction, see Martin Ceadel, "Popular Fiction and the Next War, 1918–1939" in *Class, Culture and Social Change: A New View of the 1930s,* ed. Frank Gloversmith (Sussex: Harvester Press, 1980), 161–84.

9. The first survey of nuclear war fiction was Sam Moskowitz, "The Atom Smashers: Fiction's Prophetic Parallel to Fact," *Fantasy Fiction Field* 210 (October 6, 1945): 1–2. Subsequent scholarship is summarized in Paul Brians' indispensable bibliographic and analytic study of several hundred stories and novels, *Nuclear Holocausts: Atomic War in Fiction, 1895–1984* (Kent, Ohio: Kent State University Press, 1987), 95–100.

10. This noteworthy novel was rediscovered and its significance explored by Merritt Abrash, "Through Logic to Apocalypse: Science-Fiction Scenarios of Nuclear Deterrence Breakdown," *Science-Fiction Studies* 13 (July 1986): 129–38.

11. John J. O'Neill, "Enter Atomic Power," *Harper's* 181 (June 1940): 1–10; R. M. Langer, "Fast New World," *Collier's,* July 6, 1940, 18–19, 54–55; William L. Laurence, "The Atom Gives Up," *Saturday Evening Post,* September 7, 1940, 12–13, 60–63. For an incisive discussion of these articles and their implications, see Stephen Hilgartner, Richard C. Bell, and Rory O'Connor, *Nukespeak: The Selling of Nuclear Technology in America* (New York: Penguin, 1983), 16–21.

12. Part VIII, *Liberty,* October 19, 1940. Further references will be to parts as follows: Prologue by Edward Hope, August 24, 4–9; I, August 31, 16–22; II, September 7, 20–27; III, September 14, 32–38; IV, September 21, 34–41; V, September 28, 37–44; VI, October 5, 49–54; VII, October 12, 56–61; VIII, October 19, 52–59; IX, October 26, 53–58; X, November 2, 56–60; XI, November 9, 52–57; XII, November 16, 52–55. These parts appear as chapters, with corresponding numbers, in the 1979 reprint published by Prentice-Hall, which includes an informative introduction by Terry Miller.

13. Rhodes, *Making of the Atomic Bomb,* 510; Henry DeWolf Smyth, *Atomic Energy for Military Purposes: The Official Report on the Development of the Atomic Bomb under the Auspices of the United States Government, 1940–1945* (Princeton: Princeton University Press, 1945), 65; Barton J. Bernstein, "Oppenheimer and the Radioactive-Poison Plan," *Technology Review* 88 (May/June 1985): 14–17.

14. Groff Conklin, *The Best of Science Fiction* (New York: Crown, 1946), vii.

15. Ibid., xxiv.

16. John W. Campbell, Jr., *The Atomic Story* (New York: Henry Holt, 1947), 123; Leslie R. Groves, *Now It Can Be Told: The Story of the Manhattan Project* (New York: Harper & Brothers, 1962), 146–48, 325; Brians, *Nuclear Holocaust,* 7; *The Secret History of the Atomic Bomb,* edited by Anthony Cave Brown and Charles B. MacDonald (New York: The Dial Press, 1977), 203–9; "Drop That Post!" *Saturday Evening Post,* September 8, 1945, 4.

17. "Writer Charges U.S. with Curb on Science," *New York Times,* August 14, 1941.

18. Sam Moskowitz, *Explorers of the Infinite* (Cleveland: Meridian Books, 1963), 292–93.

19. *Newsweek,* August 20, 1945, 68.

20. "1945 Cassandra," *New Yorker,* August 25, 1945, 15; Albert I. Berger, "The *Astounding* Investigation," *Analog* (September 1984): 125–37.

21. Campbell, *Atomic Story,* 118.

9. Atomic Decision

1. James Franck et al., "A Report to the Secretary of War, June 11, 1945," in Robert Jungk, *Brighter than a Thousand Suns: A Personal History of the Atomic Scientists* (New York: Harcourt, Brace, 1958), 349, 354.

2. Szilard, *Leo Szilard,* 211.

3. United States Strategic Bombing Survey, *Japan's Struggle to End the War* (Washington, D.C.: U.S. Government Printing Office, July 1, 1946), 13; William D. Leahy, *I Was There* (New York: Whittlesey House, 1950), 441; "Ike on Ike," *Newsweek,* November 11, 1963, 108.

4. P. M. S. Blackett, *Fear, War, and the Bomb: Military and Political Consequences of Atomic Energy* (New York: Whittlesey House, 1948), 139.

5. The pioneering work in exploring how atomic bombs were used as a diplomatic weapon against the Soviet Union, thus destroying the wartime alliance with the United States, is Gar Alperovitz's 1965 *Atomic Diplomacy: Hiroshima and Potsdam.* The book was subjected to fierce attack that caricatured its thesis and harped on its errors, though these were typical of scholarship that blazes new trails in uncharted territory. As more and more of the relevant documents have come to light, recent scholarship has moved ever closer to Alperovitz's position, which is refined and strengthened in his book's updated edition (New York: Penguin, 1985). Even works that take issue with Alperovitz use much of his scholarship and sometimes end up closer to his argument than they purport to be. Well-documented accounts of how deeply the struggle against the Soviet Union influenced the decision to drop the bombs in early August include Ronald W. Clark, *The Greatest Power on Earth: The International Race for Nuclear Supremacy* (New York: Harper & Row, 1980); Gregg Herken, *The Winning Weapon: The Atomic Bomb in the Cold War 1945–1950* (New York: Vintage, 1982); Robert Messer, *The End of an Alliance: James F. Byrnes, Roosevelt, Truman and the Origins of the Cold War* (Chapel Hill, N.C.: University of North Carolina Press, 1982); Peter Wyden, *Day One: Before Hiroshima and After* (New York: Simon & Schuster, 1984); and Rhodes, *Making of the Atomic Bomb.*

6. Leo Szilard, "A Personal History of the Atomic Bomb," *University of Chicago Roundtable,* September 25, 1949, 14–15.

7. Herken, *Winning Weapon,* 17–18.

8. Several of the sources cited in note 5 present the evidence in great detail; the 1985 edition of Alperovitz, *Atomic Diplomacy* makes the most thorough case.

9. Robert H. Ferrell, ed., *Off the Record: The Private Papers of Harry S. Truman* (New York: Harper & Row, 1980), 53; Gar Alperovitz, "More on Atomic Diplomacy," *Bulletin of the Atomic Scientists* 41(December 1985): 39; Ronald E. Powaski, *March to Armageddon: The United States and the Nuclear Arms Race, 1939 to the Present* (New York: Oxford University Press, 1987), 20–21.

10. Anthony Cave Brown, *The Last Hero: Wild Bill Donovan* (New York: Times Books, 1982), 39.

11. Ferrell, *Off the Record,* 53; Robert L. Messer, "New Evidence on Truman's Decision," *Bulletin of the Atomic Scientists* 41 (August 1985): 50–56.

12. Herken, *Winning Weapon,* 44; Messer, *End of an Alliance,* 105.

13. R. J. C. Butow, *Japan's Decision to Surrender* (Stanford, Calif.: Stanford University Press, 1954), 153–54; Herken, *Winning Weapon,* 21; Len Giovanitti and Fred Freed, *The Decision to Drop the Bomb* (New York: Coward, 1965), 273, 333.

14. Martin J. Sherwin, *A World Destroyed: The Atomic Bomb and the Grand Alliance* (New York: Vintage, 1977), 202.

15. Martin J. Sherwin, "Old Issues in New Editions," *Bulletin of the Atomic Scientists* 41 (December 1985): 44.

16. Szilard, *Leo Szilard,* 209.

17. ". . . most useful": Ferrell, *Off the Record,* 56; "in history": Harry S. Truman, *Year of Decision* (Garden City, N.J.: Doubleday, 1955), 421.

10. The Rise of Nuclear Culture

1. Paul Boyer, *By the Bomb's Early Light: American Thought and Culture at the Dawn of the Atomic Age* (New York: Pantheon, 1985), xviii. My relatively brief account of nuclear culture in the first few years after Hiroshima owes a great debt to this book, whose author agonized about how he could possibly synthesize in 440 pages the colossal mass of American cultural material about nuclear weapons in the 1945–1950 period.

2. Editorial, *Newark Star-Ledger,* August 7, 1945; Editorial, *New York Herald Tribune,* August 7, 1945; Holmes quoted by Boyer, *By the Bomb's Early Light,* 3; Anne O'Hare McCormick, "The Promethean Role of the United States," *New York Times,* August 8, 1945; Norman Cousins, "Modern Man Is Obsolete," *Saturday Review of Literature,* August 18, 1945, 5 (in October 1945 an influential book-length version of this essay was published under the same title); *Time,* August 20, 1945, 19.

3. Boyer, *By the Bomb's Early Light,* xix, 3–26.

4. *Newsweek,* August 20, 1945, 37.

5. Kaltenborn quoted by Boyer, *By the Bomb's Early Light,* 5; "Outlook for 1946— Second Year of Atomic Age May Be Complacent but Bomb Race Looms," *Wall Street Journal,* December 27, 1945, 6; "The inevitability": Blackett, *Fear, War, and the Bomb,* 202.

6. Boyer, *By the Bomb's Early Light,* 5.

7. Hanson W. Baldwin, "The Atom Bomb and Future War," *Life,* August 20, 1945, 18, 20; "Arnold Reveals Secret Weapons," *New York Times,* August 18, 1945; "Push Button War," *Newsweek,* August 27, 1945, 25; *Washington Post* quoted by Boyer, *By the Bomb's Early Light,* xix; *Time,* December 3, 1945, 28.

8. "The 36-Hour War," *Life,* November 19, 1945, 27–35.

9. Boyer, *By the Bomb's Early Light,* 10–11; Lloyd J. Graybar and Ruth Flint Graybar, "America Faces the Atomic Age: 1946," *Air University Review* 35 (January/February 1984): 68–77; Charles K. Wolfe, "Nuclear Country: The Atomic Bomb in Country Music," *Journal of Country Music* 7 (January 1978): 7–21.

10. For discussions of the profound effects of *Hiroshima* on American culture, see Michael Yavenditti, "John Hersey and the American Conscience: The Reception of 'Hiroshima,'" *Pacific Historical Review* 43 (February 1974): 31–39; and Boyer, *By the Bomb's Early Light,* 203–10.

11. Bernard Brodie, "The Weapon," in *The Absolute Weapon: Atomic Power and World Order,* ed. Bernard Brodie (New Haven: Yale Institute of International Studies, 1946), 76.

12. Dexter Masters and Katharine Way, eds., *One World or None* (New York: McGraw-Hill, 1946).

13. Lewis Mumford, "Gentlemen," 5, 6.

14. Theodore Sturgeon, "Memorial," in *The Other Side of the Moon,* ed. August Derleth (New York: Pellegrini & Cudahy, 1949), 423–24, 430. Later anthologies including this story have a revised text, with a reference to the hydrogen bomb.

11. The Baruch Plan: American Science Fiction

1. Herken, *Winning Weapon,* 171.

2. "The Soviet-Anglo-American Communique, December 27, 1945," in *The International Control of Atomic Energy: Growth of a Policy* (Washington, D.C.: U.S. Goverment Printing Office, n.d.), 125.

3. See, for example, the Soviet proposal of June 11, 1947, which sets forth in detail the powers of the International Control Commission to carry out "inspection of atomic energy facilities" all the way from "mining atomic raw materials" through "production of atomic materials and atomic energy." The text is available in a U.S. Department of State publication, *The International Control of Atomic Energy: Policy at the Crossroads* (Washington, D.C.: U.S. Government Printing Office, 1948), 105–7.

4. All citations are to the text of *The International Control of Atomic Energy: Growth of a Policy,* 138–47.

5. Although Baruch's speech did not state explicitly that the "punishment" would be an atomic attack, any other interpretation is difficult to sustain. As P. M. S. Blackett cogently explained, the "punishment" of a major power for allegedly violating this atomic disarmament plan could be carried out only by war, and if the supposed culprit were the Soviet Union, only by atomic war (*Fear, War, and the Bomb,* 149, 155–56, 170). Furthermore, the men who framed the Baruch Plan intended the punishment to be atomic; the adjective "immediate," which Truman himself insisted upon including, was meant to remove any ambiguity (Herken, *Winning Weapon,* 164–69, 368). Admiral Chester Nimitz pointed out the absurd "incongruity" in the plan: "the atom bomb is necessary to enforce an agreement to outlaw its use" (Powaski, *March to Armageddon,* 43).

6. Richard G. Hewlett and Oscar E. Anderson, Jr., *The New World, 1939/1946,* vol. 1 of *A History of the United States Atomic Energy Commission* (University Park, Pa.: Pennsylvania State University Press, 1962), 580–81.

12. Nuclear Scenarios

1. *Space Weapons: A Handbook of Military Astronautics,* edited by the editors of *Air Force Magazine* (New York: Praeger, 1959), 56.

2. George F. Kennan, *The Nuclear Delusion: Soviet-American Relations in the Atomic Age* (New York: Pantheon, 1983), 177–78.

3. See my discussion of these juvenile technological fantasies as expressions of American culture in *Robert A. Heinlein,* 10–11, *et passim,* and "America as Science Fiction: 1939," *Science-Fiction Studies* 9 (March 1982): 38–50.

13. Early Warnings

1. Letter to R. Dale Mullen, January 12, 1976, quoted in Albert I. Berger, "Love, Death, and the Atomic Bomb: Sexuality and Community in Science Fiction, 1935–1955," *Science-Fiction Studies* 8 (November 1981): 289–90.

2. *Space Weapons,* 56.

3. Ellen W. Schrecker, *No Ivory Tower* (New York: Oxford University Press, 1986), 220–21.

4. For an incisive analysis of this wish fulfillment in other science fiction, see Martha A. Bartter, "Nuclear Holocaust as Urban Renewal," *Science-Fiction Studies* 13 (July 1986): 148–58.

14. Triumphs of Nuclear Culture

1. Excellent discussions of this propaganda campaign can be found in Hilgartner, Bell, and O'Connor, *Nukespeak*, 2–82, and Boyer, *By the Bomb's Early Light*, 107–40, 289–302; for "The Sunny Side of the Atom" and the Dagwood comic book, see Boyer, 291, 299; for *Our Friend the ATOM*, see Hilgartner, Bell, and O'Connor, 39.

2. Allen Dulles, *The Craft of Intelligence* (New York: Harper & Row, 1963), 149; H. Bruce Franklin, *Back Where You Came From* (New York: Harper's Magazine Press, 1975), 107–12.

3. Richard G. Hewlett and Francis Duncan, *Atomic Shield, 1947/1952*, vol. 2 of *A History of the United States Atomic Energy Commission* (University Park, Pa.: Pennsylvania State University Press, 1969); William L. Laurence, *Hell Bomb* (New York: Knopf, 1951), 113. Portions of this book promoting the hydrogen bomb were published widely in the weekly magazines, for example, as "Truth about the Hydrogen Bomb," *Saturday Evening Post*, June 24, 1950, 228–29, and "What about the H-Bomb," *Life*, January 8, 1951, 57–60.

4. On the 1951 HUAC hearings, see John Cogley, *Report on Blacklisting: I. Movies* (n.p.: The Fund for the Republic, 1956), 92–117; an appendix by Dorothy B. Jones, "Communism and the Movies: A Study of Film Content," quantifies the dramatic changes in film content, with at least thirteen militantly anti-Communist films released in 1952 (231). On movies advocating nuclear weapons, including the government-sponsored documentaries made in Hollywood, see A. Costandina Titus, "Selling the Bomb: Hollywood and the Government Join Forces at Ground Zero," *Halcyon* 7 (1985): 16–29; on the bomber movies, see pages 113–17.

5. The cultural content of 1950s science-fiction horror and disaster films has been widely discussed. See Susan Sontag, "The Imagination of Disaster," in *Against Interpretation and Other Essays* (New York: Farrar, Straus & Giroux, 1965), 200–225; Andrew Dowdy, *The Films of the Fifties: The American State of Mind* (New York: William Morrow, 1975), 159–71; Biskind, *Seeing Is Believing*, 101–160, especially his discussion of *Them!*, 123–33; and Michael Rogin, *Ronald Reagan, the Movie and Other Episodes in Political Demonology* (Berkeley: University of California Press, 1987), 263–67.

6. A penetrating exposé of the fraudulent "missile gap" can be found in Edgar Bottome's *The Missile Gap: A Study of the Formulation of Military and Industrial Policy* (Cranbury, N.J.: Fairleigh Dickinson University Press, 1970); portions of this work appear in Bottome's very useful brief history of the arms race, *The Balance of Terror* (Boston: Beacon Press, 1971; rev. and exp., 1986).

7. Andrew Sarris, review of "Dr. Strangelove," *Village Voice*, February 13, 1964; reprinted in his *Confessions of a Cultist: On the Cinema, 1955/1969* (New York: Touchstone, 1971), 121; Sarris further blunts the point by falsely reporting that the billboard reads "Peace Is Our Business."

8. John M. Carroll, *Secrets of Electronic Espionage* (New York: Dutton, 1966), 134–35; David Wise and Thomas B. Ross, *The U-2 Affair* (New York: Random House, 1963), *passim;* Gettleman et al., *Vietnam and America*, 237–53; Franklin, *Back Where You Came From*, 107–19.

9. For discussions of the significance of sex in the film, see F. Anthony Macklin, "Sex and Dr. Strangelove," *Film Comment* 3 (Summer 1965): 55–57; Norman Kagan, *The Cinema of Stanley Kubrick* (New York: Holt, Rinehart & Winston, 1972), 136–37; Thomas Allen Nelson, *Kubrick: Inside a Film Artist's Maze* (Bloomington, Ind.: Indiana University Press, 1982), 89–95.

15. Arms Control?

1. John McPhee, *The Curve of Binding Energy* (New York: Farrar, Straus & Giroux, 1974), 183–84.

2. See H. Bruce Franklin, "Future Imperfect," *American Film* 8 (March 1983): 46–49, 75–76.

3. My analysis of this ebb in nuclear consciousness partially parallels Boyer's in his epilogue to *By the Bomb's Early Light*.

4. Robert Scheer, *With Enough Shovels: Reagan, Bush and Nuclear War* (New York: Random House, 1982), 36–65, 144–46; Pipes quotation in Simon Rosenblum, "The Russians Aren't Coming," *Search for Sanity: The Politics of Nuclear Weapons and Disarmament*, ed. Paul Joseph and Simon Rosenblum (Boston: South End Press, 1984), 303; Powaski, *March to Armageddon*, 184–89; for an extensive account, see Jerry W. Sanders, *Peddlers of Crisis: The Committee on the Present Danger and the Politics of Containment* (Boston: Beacon Press, 1983).

5. *Arms Control and Disarmament Agreements, 1982 Edition* (Washington, D.C.: U.S. Arms Control and Disarmament Agency, 1982), 50.

6. Daniel O. Graham, *We Must Defend America* (Chicago: Regnery Gateway, 1983), 106; Ronald Reagan, "Peace and National Security," *U.S. Department of State Bulletin*, vol. 8, no. 2073 (Washington, D.C.: U.S. Government Printing Office, 1983).

16. War in Space?

1. David Perlman, "Losing Sight of Space," *San Francisco Chronicle,* September 6, 1981; Richard D. Lyons, "Military Planners View the Shuttle as Way to Open Space for Warfare," *New York Times,* March 29, 1981; "Air Force Forms Space Command on Military Uses," *Wall Street Journal,* June 22, 1982; "Allen to Head Jet Propulsion Lab," *New York Times,* July 23, 1982.

2. General Graham's *High Frontier* appeared first in *Survive* magazine and as a report published by Project High Frontier, Washington, D.C., 1982. For an indispensable history of Star Wars, see these writings by William J. Broad: "Star Wars: Pentagon Lunacy," *New York Times,* May 13, 1982; "X-ray Laser Weapon Gains Favor," *New York Times,* November 15, 1983; "Star Wars' Research Forges Ahead," *New York Times,* February 5, 1985; "Reagan's 'Star Wars' Bid: Many Ideas Converging," *New York Times,* March 4, 1985; "Science Showmanship: A Deep 'Star Wars' Rift," *New York Times,* December 16, 1985; " 'Star Wars' Traced to Eisenhower Era," *New York Times,* October 28, 1986; " 'Star Wars' Push Dimming Prospect for Exotic Arms," *New York Times,* March 9, 1987; "Space Weapon Idea Now Being Weighed Was Assailed in '82," *New York Times,* May 4, 1987; and *Star Warriors* (New York: Simon & Schuster, 1985).

3. Clarence A. Robinson, Jr., "Advance Made on High-Energy Laser," *Aviation Week and Space Technology,* February 23, 1981, 25; Broad, *Star Warriors,* 121.

4. Ben Bova, *The High Road* (Boston, Houghton Mifflin, 1981), ch. 25; *Assured Survival* (Boston: Houghton Mifflin, 1984); *Privateers* (New York: Tor, 1985); J. E. Pournelle, ed., *There Will be War* (New York: Tor, 1983) and *There Will be War: Volume II* (New York: Tor, 1984). For further background on the use of science fiction to promote Star Wars, see William J. Broad, "Science Fiction Authors Choose Sides in 'Star Wars,' " *New York Times,* February 26, 1985, and Thomas M. Disch, "The Road to Heaven: Science Fiction and the Militarization of Space," *Nation,* May 10, 1986,

650–56; for the influence of Pournelle's and other science fiction on the Livermore researchers, see Broad, *Star Warriors,* 119, 126, 131.

5. See McGeorge Bundy, "The President's Choice: Star Wars or Arms Control," *Foreign Affairs* 63 (Winter 1984/85); Wolfgang K. H. Panofsky, "The Strategic Defense Initiative: Perception vs. Reality," *Physics Today* (June 1985): 34–45; Sidney D. Drell, Philip J. Farley, and David Holloway, *The Reagan Strategic Defense Initiative: A Technical, Political, and Arms Control Assessment* (Stanford, Calif.: Center for International Security and Arms Control, 1984); Ashton B. Carter, *Directed Energy Missile Defense in Space* (Washington, D.C.: Office for Technology Assessment, Congress of the United States, April 1984); John Tirman, ed., *The Fallacy of Star Wars: Based on Studies Conducted by the Union of Concerned Scientists* (New York: Vintage, 1984); Paul B. Stares, *The Militarization of Space: U.S. Policy, 1945–84* (Ithaca, N.Y.: Cornell University Press, 1985); and Union of Concerned Scientists, *Empty Promise: The Growing Case against Star Wars* (Boston: Beacon Press, 1986). Throughout the 1980s, many important analyses have appeared in *Bulletin of the Atomic Scientists.* Some of my own analysis is presented in "Invasion from Space," *Beyond* 1 (January 1982): 6, 7, 16, and in "Don't Worry, It's Only Science Fiction," *Issac Asimov's Science Fiction Magazine* 8 (Mid-December 1984): 27–39.

6. For fine analyses of *Murder in the Air* and its possible effects on Reagan's outlook, see Michael Rogin, *Ronald Reagan, the Movie and Other Episodes in Political Demonology* (Berkeley: University of California Press, 1987), 1–3; and "Spies, National Security, and the 'Inertia Projector': The Secret Service Films of Ronald Reagan," *American Quarterly* 39 (Fall 1987): 355–80. Tesla quoted in " 'Death Ray' for Planes," *New York Times,* September 22, 1940: sec. 2, 7.

7. "Reagan Says Video Games Provide the Right Stuff," *Wall Street Journal,* March 9, 1983.

8. The psychosexual significance of such pure machines is probed by orthopsychiatrist Robert Planck in "The Golem and the Robot," *Literature and Psychology* 15 (Winter 1965): 12–28.

17. The Age of the Automatons?

1. Nikola Tesla, "The Problem of Increasing Human Energy," *The Century Magazine,* June 1900, 182.

2. Unlike most heads of modern "defense" corporations, Maxim did his own cultural work. His best-selling *Defenseless America* (New York: Hearst's International Library, 1915) was turned into *The Battle Cry of Peace,* a 1915 hit that showed movie-goers the grisly spectacle of America ravaged by bloodthirsty invaders, obviously Germans; he later invented the popular game "War."

3. For accounts of the decision-making apparatus, see Paul Bracken, *The Command and Control of Nuclear Forces* (New Haven: Yale University Press, 1983); Robert C. Aldridge, *First Strike! The Pentagon's Plan for Nuclear War* (Boston: South End Press, 1983); Peter Pringle and William Arkin, *SIOP: The Secret U.S. Plan for Nuclear War* (New York: Norton, 1983); and Daniel Ford, *The Button: The Pentagon's Command and Control System—Does It Work?* (New York: Simon & Schuster, 1985). On the necessity for autonomous computer decisions in Star Wars, see Broad, *Star Warriors,* 90, 197; Graham, *We Must Defend America,* 64; and *Empty Promise:* chs. 3, 4.

4. Pringle and Arkin, *SIOP,* 220–21.

5. A. O. Sulzberger, Jr., "Error Alerts U.S. Forces To a False Missile Attack," *New*

York Times, November 11, 1979; Gary Hart and Barry Goldwater, "Recent False Alerts from the Nation's Missile Attack Warning System," *Report to the Senate Committee on Armed Services* (Washington, D.C.: U.S. Government Printing Office, October 9, 1980), 5.

6. Hart and Goldwater, "Recent False Alerts," 5–7; Richard Halloran, "Computer Error Falsely Indicates a Soviet Attack," *New York Times,* June 6, 1980; Richard Burt, "False Nuclear Alarms Spur Urgent Effort to Find Flaws," *New York Times,* June 13, 1980; "Missile Alerts Traced to 46¢ Item," *New York Times,* June 18, 1980.

7. Hart and Goldwater, "Recent False Alerts," 4–5.

8. "NASA Exec Says Light Cord Tangle Triggered Rupture," Newark *Star-Ledger,* August 5, 1987; William Broad, "Orbiting Debris Threatens Space Missions," *New York Times,* August 4, 1987.

9. Hart and Goldwater, "Recent False Alerts," 4.

18. Recall?

1. Paul Brians, "Red Holocaust: The Atomic Conquest of the West," *Extrapolation* 28 (Winter 1987): 319–29.

Bibliography of
Fiction Discussed

Ahern, Jerry. *The Survivalist*. Fifteen volumes. New York: Zebra, 1981–1987.

Allhoff, Fred. *Lightning in the Night*. *Liberty*. August 24, August 31, September 7, September 14, September 21, September 28, October 5, October 12, October 19, October 26, November 2, November 9, November 16, 1940.

Anderson, Poul, and F. N. Waldrop. "Tomorrow's Children." *Astounding Science-Fiction*. March 1947.

Barnes, James. *The Unpardonable War*. New York: Macmillan, 1904.

Barney, John Stewart. *L.P.M.: The End of the Great War*. New York: Putnam, 1915.

Barton, Samuel. *The Battle of the Swash and the Capture of Canada*. New York: Charles T. Dillingham, 1888.

Benjamin, Park. "The End of New York." *Fiction Magazine*. October 31, 1881. Reprinted in *Stories by American Authors*, Vol. 5. New York: Scribner, 1884.

Bester, Alfred. "Disappearing Act." *Star Science Fiction Stories No. 2*. Ed. Frederik Pohl. New York: Ballantine, 1953.

Bova, Ben. *Privateers*. New York: Tor, 1985.

Bradbury, Ray. "The Million-Year Picnic." *Planet Stories*. Summer 1946.

_____. "There Will Come Soft Rains." *Collier's*. May 6, 1950.

Brin, David. *The Postman*. New York: Bantam, 1985.

Bryant, Peter. *See* George, Peter.

Burdick, Eugene, and Harvey Wheeler. *Fail-Safe*. New York: McGraw-Hill, 1962. [*Saturday Evening Post*. October 13, October 20, October 27, 1962.]

Bywater, Hector C. *The Great Pacific War: A History of the American-Japanese Campaign of 1931–1933*. Boston: Houghton Mifflin, 1925.

Cartmill, Cleve. "Deadline." *Astounding Science-Fiction*. March 1944.

Clarkson, Helen. *The Last Day: A Novel of the Day after Tomorrow*. New York: Torquil, 1959.

Cloete, Stuart. *The Blast*. *Collier's*. April 12, April 19, 1947.

Coppel, Alfred. *Dark December*. Greenwich: Fawcett, 1960.

Countdown to Midnight. Ed. H. Bruce Franklin. New York: DAW Books, New American Library, 1984.

"Coverdale, Sir Henry Standish." *The Fall of the Great Republic (1886–1888)*. Boston: Roberts Bros., 1885.

Cozzens, James Gould. *Guard of Honor*. New York: Harcourt, Brace, 1948.

Crane, William Ward. "The Year 1899." *Overland Monthly*. 2d ser. 21 (June 1893): 579–89.

Cromie, Robert. *The Crack of Doom*. 3d ed. London: Diby, Long, 1895.

Davenport, Benjamin Rush. *Anglo-Saxons, Onward! A Romance of the Future*. Cleveland: Hubbell, 1898.

Davis, Chandler. "Nightmare." *Astounding Science-Fiction*. May 1946.

———. "To Still the Drums." *Astounding Science-Fiction*. October 1946.

Del Rey, Lester. "Nerves." *Astounding Science-Fiction*. September 1942.

Dick, Philip K. *Dr. Bloodmoney, or How We Got Along after the Bomb*. New York: Ace Books, 1965.

———. *The Penultimate Truth* (1964). New York: Bluejay Books, 1984.

Donnelly, Ignatius. *The Golden Bottle; Or, The Story of Ephraim Benezet of Kansas*. New York: D. D. Merrill, 1892.

Dooner, Pierton W. *Last Days of the Republic*. San Francisco: Alta California Publishing House, 1880.

Douhet, General Giulio. "La guerra del 19—." *The Command of the Air*. Trans. Dino Ferrari. New York: Coward-McCann, 1942.

Drumm, D. B. *Traveler*. Eleven volumes. New York: Dell, 1984–1987.

Ellison, Harlan. "A Boy and His Dog." *New Worlds*. April, 1969.

———. "I Have No Mouth, and I Must Scream." *If*. March 1967.

Engel, Leonard, and Emanuel Piller. *World Aflame: The Russian-American War of 1950*. New York: Dial Press, 1947.

Fitzpatrick, Ernest H. *The Coming Conflict of Nations; Or the Japanese American War*. Springfield, Ill., 1909.

Frank, Pat. *Alas, Babylon*. Philadelphia: Lippincott, 1959.

Gallun, Raymond. "Atomic Fire." *Amazing Stories*. April 1931.

Gann, W. D. *The Tunnel Thru the Air; or, Looking Back from 1940*. New York: Financial Guardian, 1927.

George, Peter [as "Peter Bryant"]. *Red Alert*. New York: Ace Books, 1958.

Giesy, John Ulrich. *All for His Country. Cavalier Weekly*. February 21, February 28, March 7, March 14, 1914. New York: Macaulay, 1915.

"Giles, Gordon A." [Otto Binder]. "The Atom Smasher." *Amazing Stories*. October 1938.

Godfrey, Hollis. *The Man Who Ended War*. Boston: Little, Brown, 1908.

Grendon, Edward. "The Figure" (1947). *A Treasury of Science Fiction*. Ed. Groff Conklin. New York: Crown, 1948.

Griffith, George. *The Lord of Labour*. London: F. V. White, 1911.

Haines, Donal Hamilton. *The Last Invasion*. New York: Harper & Brothers, 1914.

———. *Clearing the Seas; Or The Last of the Warships*. New York: Harper & Brothers, 1915.

Haines, William Wister. *Command Decision*. Boston: Little, Brown, 1947.

Haldeman, Joe. *The Forever War*. New York: St. Martin's Press, 1974.

———. "To Howard Hughes: A Modest Proposal." *Magazine of Fantasy and Science Fiction*. November 1974.

Hamilton, Edmond. "Day of Judgment" (1946). *The Last Man on Earth*. Ed. Isaac Asimov, Martin Greenberg, and Charles Waugh. New York: Fawcett, 1982.

Hancock, H. Irving. *The Invasion of the United States; Or, Uncle Sam's Boys at the Capture of Boston*. Philadelphia: Henry Altemus, 1916.

———. *In the Battle for New York; Or, Uncle Sam's Boys in the Desperate Struggle for the Metropolis*. Philadelphia: Henry Altemus, 1916.

_____. *At the Defense of Pittsburgh; Or, The Struggle to Save America's "Fighting Steel" Supply.* Philadelphia: Henry Altemus, 1916.

_____. *Making the Stand for Old Glory; Or, Uncle Sam's Boys in the Last Frantic Drive.* Philadelphia: Henry Altemus, 1916.

Harding, Richard. *The Outrider.* Five volumes. New York: Pinnacle Books, 1984–1985.

Heinlein, Robert A. "Blowups Happen." *Astounding Science-Fiction.* September 1940.

_____. "Solution Unsatisfactory." *Astounding Science-Fiction.* May 1941.

_____. *Stranger in a Strange Land.* New York: Putnam, 1961.

Heller, Joseph. *Catch-22.* New York: Simon & Schuster, 1961.

_____. *We Bombed in New Haven.* New York: Knopf, 1968.

Irving, Washington. "The Men of the Moon" (1809). H. Bruce Franklin. *Future Perfect.* Rev. ed. New York and London: Oxford University Press, 1978.

Jenkins, Will F. *The Murder of the U.S.A.* New York: Crown, 1946.

Johnson, Denis. *Fiskadoro.* New York: Knopf, 1985.

Johnson, Samuel. *The History of Rasselas, Prince of Abissinia.* London, 1759.

Johnstone, William. *Ashes.* Six volumes. New York: Zebra, 1983–1987.

Jones, D. F. *Colossus.* New York: Putnam, 1966.

Lea, Homer. *The Valor of Ignorance.* New York: Harper & Brothers, 1909.

Leiber, Fritz. "A Bad Day for Sales." *Galaxy Science Fiction.* July 1953.

_____. "Coming Attraction." *Galaxy Science Fiction.* November 1950.

Leinster, Murray. *See* Jenkins, Will F.

Lewis, Sinclair. *It Can't Happen Here!* Garden City: Doubleday, 1935.

London, Jack. "A Thousand Deaths." *The Black Cat.* May 1899.

_____. "The Unparalleled Invasion." *McClure's Magazine.* July 1910.

"Lorelle." "The Battle of the Wabash." *Californian* 2 (October 1880): 364–76.

Manson, Marsden. *The Yellow Peril in Action.* San Francisco: Britton and Rey, 1907.

McDougall, Walt. "The Last Conflict, the Horror That Awoke the Nation." *American Magazine* 78 (November 1914): 33–36.

Merril, Judith. *Shadow on the Hearth.* Garden City: Doubleday, 1950.

_____. "That Only a Mother." *Astounding Science-Fiction.* June 1948.

Miller, Walter M., Jr., *A Canticle for Leibowitz.* Philadelphia: Lippincott, 1959. [Originally published as three stories in *Magazine of Fantasy and Science Fiction:* "A Canticle for Leibowitz," April 1955; "And the Light Is Risen," August 1956; "The Last Canticle," February 1957.]

"Minor, John W." *Bietigheim.* New York: Funk & Wagnalls, 1886.

Moffett, Cleveland. *The Conquest of America; A Romance of Disaster and Victory.* New York: Doran, 1916.

Moore, Ward. "Lot." *Magazine of Fantasy and Science Fiction.* May 1953.

_____. "Lot's Daughter." *Magazine of Fantasy and Science Fiction.* October 1954.

Muller, Julius W. *The Invasion of America: A Fact Story Based on the Inexorable Mathematics of War.* New York: Dutton, 1916.

Mundo, Oto. *The Recovered Continent; A Tale of the Chinese Invasion.* Columbus, Ohio: Harper-Osgood, 1898.

Nathanson, Isaac. "World Aflame." *Amazing Stories.* January 1935.

Neal, W. C. "Who Shall Dwell." *Playboy.* July 1962.

Newcomb, Simon. *His Wisdom, the Defender.* New York: Harper & Brothers, 1900.

Nicolson, Harold. *Public Faces.* Boston and New York: Houghton Mifflin, 1933. [London: Constable, 1932.]

Norton, Roy. *The Vanishing Fleets.* New York: Appleton, 1908.

Noyes, Pierrepont B. *The Pallid Giant.* New York: Fleming H. Revell, 1927. Reissued as *Gentlemen: You Are Mad!* New York: Baxter Freres, 1946.

O'Brien, Tim. *The Nuclear Age*. New York: Knopf, 1985.

Odell, S. W. *The Last War; Or, The Triumph of the English Tongue*. Chicago: Charles H. Kerr, 1898.

Palmer, J. H. *The Invasion of New York; Or, How Hawaii Was Annexed*. New York: F. Tennyson Neely, 1897.

Pournelle, J. E., ed. *There Will Be War*. New York: Tor, 1983.

———. *There Will Be War: Volume II*. New York: Tor, 1984.

Purple Heart, The. In *Best Film Plays of 1943–44*. New York: Garland, 1977.

Reed, Samuel Rockwell. *The War of 1886, Between the United States and Great Britain*. Cincinnati: Robert Clarke & Co., 1882.

Ridenour, Louis N. "Pilot Lights of the Apocalypse." *Fortune*. January 1946.

Robbins, David. *Endworld*. Six volumes. New York: Leisure Library, 1986–1987.

Robinson, Frederick. *The War of the Worlds; A Tale of the Year 2,000 A.D.* N.p: Frederick Robinson, 1914.

Roshwald, Mordecai. *Level 7*. New York: McGraw-Hill, 1959.

———. *A Small Armageddon*. London: Heinemann, 1962.

Rousseau, Victor. "The Atom Smasher." *Astounding Science-Fiction*. May 1930.

Sedberry, J. Hamilton. *Under the Flag of the Cross*. Boston: C. M. Clark, 1908.

See, Carolyn. *Golden Days*. New York: McGraw-Hill, 1986.

Serviss, Garrett P. *Edison's Conquest of Mars. New York Evening Journal,* January 12 through February 10, 1898; reprinted, with introduction by A. Langley Searles. Los Angeles: Carcosa House, 1947.

Shelley, Mary Wollstonecraft. *Frankenstein; Or, The Modern Prometheus*. London, 1818.

Shiel, M. P. *The Yellow Danger*. New York: R. F. Fenno, 1899.

Shute, Nevil. *On the Beach*. New York: Morrow, 1957.

Simak, Clifford. "Lobby." *Astounding Science-Fiction*. April 1944.

Smith, E. E. "Doc." *The Skylark of Space*. Serialized, with sequels, beginning in *Amazing Stories*, August 1928.

Spinrad, Norman. "The Big Flash." *Orbit 5*. New York: Putnam, 1969.

Spohr, Carl W. *The Final War. Wonder Stories* 3 (March 1932): 1111–29, 1187–89; (April 1932): 1267–86.

Stacy, Ryder. *Doomsday Warrior*. Five volumes. New York: Zebra, 1984–1985.

"Stochastic." *The Stricken Nation*. New York, 1890.

Stockton, Frank R. *The Great War Syndicate*. New York: Dodd, Mead, 1889.

Sturgeon, Theodore. "Memorial." *Astounding Science-Fiction*. April 1946.

———. "Thunder and Roses." *Astounding Science-Fiction*. November 1947.

Szilard, Leo. "The Voice of the Dolphins." *The Voice of the Dolphins and Other Stories*. New York: Simon & Schuster, 1961.

Tenn, William. "Brooklyn Project." *Planet Stories*. Fall 1948.

"The 36-Hour War." *Life*. November 19, 1945.

Train, Arthur Cheney, and Robert Williams Wood. *The Man Who Rocked the Earth. Saturday Evening Post*. November 14, 1914 through January 9, 1915. Garden City: Doubleday, Page, 1915.

Twain, Mark. *A Connecticut Yankee in King Arthur's Court*. New York: Charles L. Webster, 1889.

Vincent, Harl. "Power Plant." *Astounding Science-Fiction*. November 1939.

Vonnegut, Kurt, Jr. *Cat's Cradle*. New York: Holt, Rinehart & Winston, 1963.

———. *Galápagos*. New York: Delacorte/Seymour Lawrence, 1985.

———. *God Bless You, Mr. Rosewater*. New York: Holt, Rinehart & Winston, 1965.

_____. "Report on the Barnhouse Effect." *Collier's*. February 11, 1950.

_____. *Slaughterhouse-Five; or, The Children's Crusade: A Duty-Dance with Death.* New York: Dell, 1971. [Originally published by Delacorte/Seymour Lawrence, 1969.]

Walker, J. Bernard. *America Fallen! The Sequel to the European War.* New York: Dodd, Mead, 1915.

Wallace, King. *The Next War: A Prediction.* Washington, D.C.: Martyn Publishing House, 1892.

Waterloo, Stanley. *Armageddon: A Tale of Love, War, and Invention.* Chicago: Rand McNally, 1898.

Wells, H. G. *The War in the Air.* London, 1908.

_____. *The War of the Worlds* (London, 1898). *The Time Machine and The War of the Worlds.* Ed. Frank D. McConnell. New York: Oxford University Press, 1977.

_____. *The World Set Free: A Story of Mankind.* New York: Dutton, 1914.

Woltor, Robert. *A Short and Truthful History of the Taking of California and Oregon by the Chinese in the Year A.D. 1899.* San Francisco, 1882.

Wylie, Philip. "Blunder: A Story of the End of the World." *Collier's*. January 12, 1946.

_____. *The Paradise Crater. Blue Book Magazine.* October, 1945: 2–27.

Zelazny, Roger. "Damnation Alley." *Galaxy.* October 1967.

Films Discussed

Above and Beyond (1952)
Air Force (1943)
Amerika (1987)
Atomic Cafe (1982)
The Atomic City (1952)
The Battle Cry of Peace (1915)
The Beast from 20,000 Fathoms (1953)
Beneath the Planet of the Apes (1970)
Bombardier (1943)
Bombers B-52 (1957)
A Boy and His Dog (1976)
Buck Rogers in the 25th Century (1979)
Command Decision (1948)
The Court Martial of Billy Mitchell
   (1955)
Damnation Alley (1977)
The Day After (1983)
The Day the Earth Stood Still (1951)
Desert Bloom (1986)
Destination Moon (1950)
Dr. Strangelove, or, How I Learned to
   Stop Worrying and Love the Bomb
   (1964)
Fail-Safe (1964)
Fighting Devil Dogs (1938)
Five (1951)
Flight (1929)
The Flying Torpedo (1916)
The Forbin Project (1970)

Glen and Randa (1971)
Hell and High Water (1954)
I Wanted Wings (1940)
The Incredible Shrinking Man (1957)
International Squadron (1941)
Invasion USA (1952)
Invasion: USA (1985)
Logan's Run (1976)
Murder in the Air (1940)
On the Beach (1959)
Our Friend the ATOM (1956)
Planet of the Apes (1968)
The Purple Heart (1944)
Red Dawn (1984)
Star Wars (1977)
Strategic Air Command (1955)
Superman IV: The Quest for Peace
   (1987)
Test Pilot (1938)
Testament (1983)
Them! (1954)
Things To Come (1936)
Thirty Seconds Over Tokyo (1944)
Top Gun (1986)
Twelve O'Clock High (1949)
Victory through Air Power (1943)
WarGames (1983)
The World, the Flesh, and the Devil
   (1959)

Index

Abbot, Henry Larcom, 213n.2
Above and Beyond (film), 182
Abrahamson, James, 199
Abrash, Merritt, 222n.10
Absolute Weapon, The (Brodie ed.), 158, 170
Adelman, Kenneth, 195
"Aerial Navigation (A Priori)" (Stedman),
    82–83
Aerial torpedoes. *See* Robot planes
"Aerial Warfare" (Hearne), 84
Aerospace industry, 103, 113
    and freeze movement, 197
*Against the Annexation of the Hawaiian
    Islands* (Mitchell), 92
Agee, James, 108, 220n.22
Ahern, Jerry, 211
Aircraft carriers, Mitchell on, 97
Air Force (film), 114–15
Airplanes
    as failed ultimate weapon ("Memorial"),
        158–59
    first use of in bombing, 85–86
    in future-war fiction, 83, 84–85
    Maxim on, 84
    in WWI, 86
Air power
    in colonial conflict, 85–86, 88–89
    doubts on, 81, 84
    early enthusiasm for, 81–84
    Edison for, 76
    first use of balloons, 82
    first use of bomber airplanes, 85–86
    and Franklin on balloons, 81
    in future-war fiction, 33, 38, 41, 45, 47,
        49, 83, 84–85, 87

in Korean War, 117
rationale for, 122, 126
over sea power (future-war fiction), 33
in Vietnam, 117–19
in WWI, 49, 50, 86–87, 88
Air power, strategic. *See also* Nuclear
    weapons
    and bombing of cities, 89, 96, 101–2,
        103–11. *See also* Civilian populations,
        attacks on
    Douhet on, 89–90
    films on, 101, 108, 113–17, 182, 186
    Heller and Vonnegut on, 119–27
    and Billy Mitchell, 91, 95–96, 97, 103
    and 1930s economy, 103
    in Spanish Civil War, 89
    and superweapon world view, 153–54
    and surface forces, 88
    three leading proponents of, 87–88
    U.S. attitudes toward, 101, 102, 135
    U.S. development of, 90
    in WWII, 89, 103–111, 117, 118
Airships, in future-war fiction, 33, 38, 41, 45,
    47, 49, 83. *See also* Balloons; Zeppelins,
    London bombed by
Alas, Babylon (Frank), 127, 183
Aldridge, Robert C., 228n.3
Alien invaders, in Irving's satire, 21
Allen, Lew, 199
Allen, Richard, 195
All for His Country (Giesy), 41, 44–45, 52, 85
Allhoff, Fred, 138–41
Alperovitz, Gar, 223nn.5,8,9
*America Fallen! The Sequel to the European
    War* (Walker), 46

239

American dictatorship
and future-war fiction, 52
in "Solution Unsatisfactory," 144, 145–46
American hegemony (Pax Americana). *See
also* Imperialism
and atomic bombing of Japan, 156
with atom bomb for perpetual peace, 154
in Baruch Plan, 140
and Billy Mitchell, 99–100
as science-fiction scenario, 25–26, 28,
29–32, 33, 42–44, 52, 168
in "Solution Unsatisfactory," 140, 142–43
American Legion Magazine, 113
American Magazine, 53, 47, 83, 108
American Mercury, 108
American Revolution, 9
American Turtle, 9–10
Amerika (television miniseries), 138
Analog, 200
Anderson, Oscar E., Jr., 225n.6
Anderson, Poul, 159
*Anglo-Saxons, Onward! A Romance of the
Future* (Davenport), 31
Annihilation
as taken for granted, 20
threat of, 3, 4, 148, 167
Antarctica, treaty to demilitarize, 192
Anti-Ballistic Missile (ABM) treaty (1972),
193, 196
Anti-feminism. *See* Feminism
Anti-imperialism
campaign against (1914–17), 45
in *Connecticut Yankee,* 62
defeat of, 22
Antipopulist novels, 27
Antiutopian literature, *Level 7* as, 184
Arkin, William, 228nn.3,4
*Armageddon: A Tale of Love, War, and
Invention* (Waterloo), 32, 52, 186
Arms control, WWI and goal of, 75
Arms race
and *Edison's Conquest,* 67–68
pre-WWI, 75
Arms race, nuclear, 166, 191–92. *See also*
Disarmament, nuclear
acceleration of (late 1970s and 1980s),
195–96
and arms control, 191
Catholic bishops' letter on, 211
co-opting of resistance to, 197–98
and Cuban missile crisis, 192
elimination of challenges to, 180
films and novels on, 182, 183–87, 191,
193–95, 211–12
hydrogen bomb in, 167

as insanity, 4
popular movement against, 195, 196–97,
211–12
reversing of, 192–93
science fiction in, 169
scientists' report on, 149
and Star Wars, 201, 202. *See also* Star
Wars
and third generation nuclear weapons, 199
treaties on, 192–93. *See also* Treaties
U.S. escalation of, 166–68
Army Air Forces in World War II, The, 102
Arnold, "Hap," 114, 156, 157
Ashes (Johnstone), 211
Associated Sunday Magazines, 41, 52, 153
Assured Survival (Bova), 200
Astounding Science-Fiction, 142, 147, 158,
170
*At the Defense of Pittsburgh; Or, the Struggle
to Save America's "Fighting Steel"
Supply* (Hancock), 46
Atlantic Monthly, 108
"Atom Gives Up, The" (Laurence), 137, 181
Atomic Attack (television program), 178
Atomic bomb. *See also* Nuclear weapons
America's first use of, 5
discoveries leading to, 133–34, 147
effort to develop, 138
exaggeration of power of, 165
as first generation, 199
and hydrogen bomb, 167
and imagination vs. reality, 16
main technical problems of, 147
Soviet test of, 167, 180
Wells' projecting of, 68
Atomic bombing of Japan, 149–54
and doctrine of mass guilt, 160
and Dresden bombing, 120
and *Man Who Rocked the Earth,* 51
moral considerations in, 143, 150, 152
opposition to, 149–50, 151
rationale for, 16, 150
and threat of obliteration, 148
U.S. reaction to, 4, 155–56, 166, 202–3.
See also Nuclear culture
Atomic bomb test, Bikini (1946), 61
Atomic Café (film), 181
Atomic City, The (film), 182
Atomic (nuclear) energy
and Laurence, 181
in *Lightning in the Night,* 139–40
media discussion of, 137–38, 139–41
selling of, 180
Wells' prediction of, 132, 133
"Atomic Fire" (Gallun), 135

Atomic Shield, 1947/1952, 181

"Atom Smasher" (comic strip), 147

"Atom Smasher, The" ("Gordon A. Giles"),
135

"Atom Smasher, The" (Rousseau), 135

Atoms for Peace, 140, 180

Automatons, 204–5
and human identity, 209–10
Tesla's conception of, 205–6
for thermonuclear war, 5, 207–10

Aviation Week and Space Technology, 185,
200

B-1 bomber, 141, 196

B-17 Flying Fortresses, 75, 98–99, 103, 105,
108, 113–14, 114–15, 116

B-29 Superfortresses, 44, 75, 105, 108, 109,
110, 112, 153–54, 165, 166

B-36 bombers, 166

B-47 bombers, 166, 185, 186

B-50 bombers, 112

B-52 bombers, 112, 119, 166, 167, 185

Bacon, Francis, 11

Bacteriological (germ) warfare
and Baruch Plan, 164
in future-war fiction, 37, 38
U.S. ratifies Geneva Accord on, 193

"Bad Day for Sales, A" (Leiber), 174

Bailey, W. E., 60

Baker, Newton D., 87

Baldwin, Hanson, 156, 224n.7

Ballistic missiles. *See* Intercontinental ballistic
missiles

Balloons
in Civil War, 5
first ascent in, 81
first military use of, 82

Bangkok, trial bombing of, 109

Barcelona, bombing of, 89, 101

Barnes, James, 27–28, 215n.2

Barnes, John S., 214n.19

Barney, John Stewart, 49–50, 145

Barras, Paul, 13

Barton, Samuel, 24–26

Bartter, Martha A., 225n.4(ch.13)

Baruch, Bernard, 163, 225n.5

Baruch Plan (June 1946), 140, 157, 162–64,
225n.5
and Bikini tests, 164–65

Battle of Britain, 104

Battle Cry of Peace, The (film), 228n.2

Battleships (dreadnoughts)
Billy Mitchell's bombing of, 94–95
vs. submarines, 68–69

Battle of the Swash and the Capture of Canada
(Barton), 24–26

"Battle of the Wabash, The" ("Lorelle"), 34

Beam weapons, 202–3
and Edison, 71
in *The Final War*, 137
and future-war fiction, 50, 51, 65, 66–67,
83, 202, 205
in Irving's satire, 21
1945 discussion of, 157, 203
for space (1980s), 200
in Star Wars, 207

Beast from 20,000 Fathoms, The (film), 182

Bell, Richard C., 222n.11, 226n.1

"Bell Tower, The" (Melville), 42

Beneath the Planet of the Apes (film), 194

Benjamin Park, 22, 23, 39

Berger, Albert I., 222n.20, 225n.1(ch.13)

Bernstein, Barton J., 222n.13

Bester, Alfred, 174

Bietigheim ("Minor"), 45

"Big Flash, The" (Spinrad), 197–98

Bikini atomic bomb test (1946), 61, 157, 165

Binder, Otto (pseud. Gordon A. Giles), 135

Biological warfare, and Baruch Plan, 164. *See
also* Bacteriological (germ) warfare

Birdsall, Steve, 220n.24

"Birthmark, The" (Hawthorne), 42

Biskind, Peter, 221n.9, 226n.5

Blackett, P. M. S., 151, 223n.4, 224n.5,
225n.5

Black Peril literature, 35–36

Blast, The (Cloete), 174–75

"Blitz, the," 104

"Blowups Happen" (Heinlein), 135

Blue Book Magazine, 147

"Blunder" (Wylie), 159

Boeing Company, 44, 98, 103, 112, 114

Bombardier (film), 115

"Bomber gap," 181, 183, 195

Bombers, 113. *See also Air power*
films about, 108, 113–17, 182
intercontinental, 3
revulsion toward, 119, 120
superstealth in Gann novel, 102

Bombers B-52 (film), 116, 182

Bombing of cities. *See* Civilian populations,
attacks on

Bombing of Germany, The (Rumpf), 120

Book-of-the-Month Club, 108

Boothe, Clare, 40

Bottome, Edgar, 226n.6

Bova, Ben, 200, 227n.4(ch.16)

Boyer, Paul, 224nn.1,2,3,5,6,7,9,10,
227n.3(ch.15)

"Boy and His Dog, A" (Ellison), 194
Boy and His Dog, A (film), 194
Bracken, Paul, 228n.3
Bradbury, Ray, 159, 174
Brians, Paul, 222nn.9,16, 229n.1
Brin, David, 211
"Brinkmanship," 192
Britain. *See* Great Britain
Broad, William J., 227n.2,4(ch.16),
    228n.4(ch.16), n.3(ch.17), 229n.8
Brodie, Bernard, 158, 170, 224n.11
"Brooklyn Project" (Tenn), 174
Brown, Anthony Cave, 222n.16, 223n.10
Browning, Robert Pack, 216n.12
Brzezinski, Zbigniew, 168
Buck Rogers in the 25th Century (film), 194
Bullard, Robert Lee, 138
Bulletin of the Atomic Scientists, 158, 185
Bulosan, Carlos, 218n.2
Bundy, McGeorge, 228n.5
Burdick, Eugene, 183
Burgess, Robert F., 213n.2
Burhans, Clinton, Jr., 221n.16
Burow, R. J. C., 223n.13
Burt, Richard, 229n.6
Bush, George, 195
Bushnell, David, 9–10
Byrnes, James F. (Jimmy), 151, 152
Bywater, Hector, 102, 219n.5

Caidin, Martin, 220n.30
Campbell, John W., Jr., 142, 147,
    222nn.16,21
Canticle for Leibowitz, A, 127, 183, 184
Capitalism
    in future-war fiction, 26, 45, 47
    and progress, 18
Capitalism, industrial
    and air-power ideology, 82
    and *Connecticut Yankee,* 62
    and Fulton, 10, 11, 18
    and ideological contradictions, 18
    and mass reading audience, 19
    War of the Worlds and culture of, 64–65
    Wells on, 132
Capra, Frank, 97
Carnegie Institution, 74
Carroll, John M., 226n.8
Carter, Ashton B., 228n.5
Cartmill, Cleve, 135, 147–48, 150
Cartwright, Edmund, 13
Casey, William, 195
Catch-22 (Heller), 119, 123–27, 183

Cate, James Lea, 219nn.4,11,
    220nn.12,15,23,25,28,29,30,31
Catholic Bishops, Pastoral Letter of
    ("Challenge of Peace"), 189, 211
Cat's Cradle (Vonnegut), 119
Ceadel, Martin, 222n.8
Censorship. *See* Secrecy ("security")
Central Intelligence Agency, 185, 195
"Challenge of Peace, The" (Pastoral Letter of
    U.S. Catholic Bishops, 1983), 189, 211
Chambliss, W. C., 213n.3
Chapelle, Howard Irving, 214n.18
Chennault, Claire, 98, 219n.22
China
    imagined U.S. war with, 34–37, 38–39, 85,
        161
    Japanese bombing of, 90, 105
    U.S. bombing of, 109
Chinese Exclusion Act (1882), 35
Christmas bombing of Vietnam (1972), 118
Chu, Limin, 214n.6
Citizens Advisory Council on National Space
    Policy, 200
Civil defense
    in 1949–57 propaganda films, 181
    in *Shadow on the Hearth,* 178
Civilian populations, attacks on
    in *Battle of the Swash,* 25
    in colonial conflict, 86–87, 88–89
    in Douhet's theory, 89–90
    efforts to prevent, 87, 88, 101, 103
    films' rejection of, 101–2
    in future-war fiction, 20, 25, 44, 87
    Billy Mitchell's rationale for, 96, 97
    in Spanish Civil War, 89
    "strategic" air power as, 87
    U.S. public's attitude toward, 101, 102, 105
    in WWI, 86, 89
    and WWII bombing, 103–10
Civil War, U.S.
    and Britain as enemy, 24
    and efforts to formalize combat, 88
    and Fulton's weapons, 17
    modern warfare in, 5, 19
    observation balloons of, 5
Clareson, Thomas D., 214n.1
Clark, Ronald W., 223n.5
Clarke, I. F., 214n.1, 216n.15
Clarkson, Helen, 179, 183
Clearing the Seas (Haines), 46
Cloete, Stuart, 174–75
Coexistence, 192
Cogley, John, 226n.4
Cold War. *See also* Arms race, nuclear; Soviet
    Union

and anti-nuclear dissent, 173. *See also*
  Secrecy ("security")
  and atomic bombing of Japan, 151
  propaganda films for, 182
  and suppression of dissent, 180
Collier's, 96, 97, 137, 140, 159, 174
Colonial conflict
  air bombing in, 85–86, 88
  and Billy Mitchell, 100
  U.S. army in, 92
Colossus (Jones), 209
"Columbiads" (Fulton underwater cannon),
  16
"Coming Attraction" (Leiber), 174
*Coming Conflict of Nations; Or the Japanese
  American War, The* (Fitzpatrick),
  39–40
*Command of the Air, The (Il dominio
  dell'aria)* (Douhet), 89
Command Decision (Haines), 116
Command Decision (film), 116
Committee on the Present Danger, 195
Communism. *See also* Soviet Union
  in future-war fiction, 27, 32, 42
  as U.S. enemy, 180
Compton, Arthur, 142
Conant, James, 153
Conklin, Groff, 145, 222n.14
Connecticut Yankee in King Arthur's Court, A
  (Twain), 62–64
Conquest of America, The (Moffett), 53, 85
Coolidge, Calvin, and Mitchell court-martial,
  97
Cooper, Gary, 97, 116
Coppel, Alfred, 183
Corn, Joseph J., 83, 217nn.4,6
Coronet, 108
Cosmopolitan, 65, 66
Counterinsurgency
  airplanes in, 86
  by U.S. in Philippines, 92
Court Martial of Billy Mitchell, The (film), 97,
  116
Cousins, Norman, 155, 224n.2
"Coverdale, Sir Henry Standish," 27
Cozzens, James Gould, 116
Crack of Doom, The (Cromie), 50, 131
Crane, William Ward, 35–36, 37
Craven, Wesley Frank, 219nn.4,11,
  220nn.12,15,23,25,28,29,30,31
Cromie, Robert, 50, 131, 215n.10
Cruise missiles, 168, 201
Cuba, 21, 22, 31, 40, 67
  imagined U.S. war with, 21, 36
  U.S. attacks on, 92, 100, 192

Cuban missile crisis (1962), 192
Cult of the superweapon
  attacks on, 71
  and fallacy of last move, 26
  in future-war fiction, 5, 31, 37
  in *The Last Day*, 179
  and nuclear arms race, 168
  official secrecy required by, 42
Cultural forces, and material forces, 5, 6

Dagwood Splits the Atom (comic book), 180
"Damnation Alley" (Zelazny), 194
Damnation Alley (film), 194
Daniels, Josephus, 71, 74, 93
  quoted, 54
Dark December (Coppel), 183
Davenport, Benjamin Rush, 7, 31
Davis, Burke, 217n.14, 218nn.1,7
Davis, Chandler, 172–73
Day After, The (television film), 211
Day the Earth Stood Still, The (film), 182
"Day of Judgment" (Hamilton), 159
"Deadline" (Cartmill), 147–48, 150
Decision-making process for nuclear war,
  automation of, 207–10
Defenseless America (Maxim), 205
Del Rey, Lester, 135
Democracy
  future-fiction safeguarding of, 20
  in *Level 7*, 185
  vs. nuclear deterrence, 160
  and nuclear-war decisions, 206
  and radioactive weapons (Heinlein), 142
  and "To Still the Drums," 172
Demologos (Fulton steam warship), 14, 16–17
Desert Bloom (film), 211
de Seversky, Alexander, 108, 220n.21
Design for Survival (Power), 117–18
Desmond, Shaw, 76, 217n.27
Destination Moon (film), 113
Destruction of Dresden, The (Irving), 122
Detente, 192, 195, 196
Deterrence
  of Japanese ambitions by B-17s, 99
  "persuasive," 118
  WWI arms race for, 75
Deterrence, nuclear, 158
  fallacy of, 135, 136
  and morality of mass slaughter, 160
  and *Murder of the U.S.A.*, 160, 170–71
  and protracted nuclear war, 195–96
  as revenge, 160, 170–71
  and Star Wars, 201
  and "Thunder and Roses," 170–72

De Voto, Bernard, 216n.13
"Diamond Lens, The" (O'Brien), 42
Dick, Philip K., 183, 186–87, 194
Dirigibles, in future-war fiction, 30, 33, 84, 85
"Disappearing Act" (Bester), 174
Disarmament, as goal after WWI, 75
Disarmament, nuclear. *See also* Arms race, nuclear
    and Baruch Plan, 161–64
    and Soviet atomic bomb test, 167
    Soviet efforts toward, 163, 166–67
Disarmament Conference of 1932, 87
Disch, Thomas M., 227n.4(ch.16)
Disney, Walt, 108, 113, 180
Doenitz, Karl, 106
Donnelly, Ignatius, 29–30
Donovan, William, 151
Doolittle, Jimmy, 105
Doomsday Warrior (Stacy), 211
Dooner, Pierton, 33–34
Douglas, Gordon, 116
Douhet, Guilio, 87, 89–90, 91, 104, 126, 134, 218n.20(ch.4)
Dowdy, Andrew, 226n.5
Dower, John W., 215n.7, 219n.10, 220n.20
Dreadnoughts. *See* Battleships
Drell, Sidney D., 228n.5
Dresden
    bombing of, 107, 119, 122
    and *Slaughterhouse-Five*, 120–23
Dr. Strangelove, or How I Learned to Stop Worrying and Love the Bomb (film), 117, 127, 183, 184, 185–86, 187, 193, 206
Drumm, D. B., 211
Drysdale, William, 23
Dulles, Allen, 226n.2
Duncan, Francis, 226n.3

Eaker, Ira C., 122
Eastman Kodak, 94
Edison, Thomas, 54–55, 91
    and American genius, 183
    anti-AC campaign of, 58–59
    and antinuclear novels, 187
    electric weapons envisioned by, 59–61
    in future-war fiction, 47, 50, 85, 215n.2
    and industrialized war, 69–72, 133, 153
    as lone wizard, 169
    and media, 69, 93
    and military-industrial complex, 69, 74, 113
    quoted, 54
    and Sims-Edison, 55–58

on superweapon as ultimate peacekeeper, 75–77, 158
    weapons contributions of, 58, 72–74
Edison's Conquest of Mars (Serviss), 50, 66–68, 69, 71, 76, 205
Education, and nuclear arms race, 211
Einstein, Albert, 133
Eisenhower, Dwight D.
    on atomic bombing of Japan, 150
    on military-industrial complex, 113, 123–24
Ellison, Harlan, 193, 194, 209–10
"End of New York, The" (Benjamin), 22, 39, 157
Endworld (Robbins), 211
Engel, Leonard, 159
"Enter Atomic Power" (O'Neill), 137
Epcot Center, Reagan speech at, 203
Epstein, Edward Z., 221n.5
ERCS (Emergency Rocket Communications System), 207
Eroticism toward weapons, 186
    in *Air Force*, 114
    in air-training films, 114
    in *Armageddon*, 32–33
    in *Dr. Strangelove*, 186
    in SAC-bomber films, 117, 186
    and space war, 203
Escalation
    in *The Final War*, 136
    and ICBMs, 167
    of WWI air attacks, 86
Europe
    anti-nuclear war movement in, 196
    in Puritan vision, 31

Fail-Safe (Burdick and Wheeler), 183, 192, 206
Fail-Safe (film), 183, 192, 206
Fallacies and myths
    in assumption of peace through new weapons, 71
    of deterrence theory, 135, 136
    in Fulton's thinking, 17
    of the last move, 26, 169, 205
    of lone genius, 52, 54, 132, 169
    of the omnipotent individual, 52
Fall of the Great Republic, The ("Coverdale"), 21, 27
False alarms about nuclear attack, 207–8
"Fanciful Predictions of War" (Crane), 35
Fantasies
    postnuclear, 174–77
    of power, 202
    of security, 202

Farley, Philip J., 228n.5
Fascism
  in Barney novel, 49, 50
  in *Catch-22*, 125
  Douhet's theory as, 89
"Fast New World" (Langer), 137
Fate of the Earth, The (Schell), 171, 184
Feminism
  campaign against (1914–17), 45
  and future-war fiction, 27, 47, 49, 50, 85
  Lea against, 40
Fermi, Enrico, 142
Ferrell, Robert H., 215n.12, 223nn.9,11
Fiction. *See* Films and newsreels; Future-war
  fiction; Science fiction
Fighting Devil Dogs (film), 102
"Figure, The" (Grendon), 159
Films and newsreels. *See also individual films*
  antinuclear, 183, 185–86, 194, 211–12
  on bombing and bombers, 94, 108, 113–17,
    182
  science-fiction-horror, 182
  on Yellow Peril, 101–2
Final War, The (Spohr), 135–37, 168
Finnegan, John Patrick, 215n.8
First-strike
  MX intended for, 168
  and preemptive strikes (Noyes), 135
First World War. *See* World War I
Fiskadoro (Johnson), 211
Fitzpatrick, Ernest H., 39–40
Five (film), 182
Flight (film), 97
Flying and Popular Aviation, 108
Flying Fortresses (B-17s), 98–99, 103, 105,
    108, 113–14, 114–15, 116
Flying Torpedo, The (film), 101–2
Forbin Project, The (film), 209
"Force, the" (in Barnes novel), 28
Ford, Daniel, 228n.3
Forever War, The (Haldeman), 194
Fortune, 159
France
  bombing of cities in, 105–6
  and bombing of Germany, 104, 105
  colonial air-power use by, 85–86, 88
  as fictional attacker, 21
  and Fulton, 12–15, 16
Franck, James, 149, 223n.1
Franco-Prussian War, 19, 88
Frank, Michael B., 216n.12
Frank, Pat, 183
Frankenstein comparison, 43, 150, 156, 158, 168
Frankenstein; or, The Modern Prometheus
  (Shelley), 42, 168

Franklin, Benjamin, and air power, 81, 83,
    217n.1
Franklin, Jane, 221n.13
Frazar, Everett, 57
Freed, Fred, 223n.13
Freedom
  automatons' assuming of, 209
  in *Level 7*, 185
  nuclear culture as loss of, 169
  and radioactive weapons (Heinlein), 142
  weapons as false means to, 212
Freeze movement, 196–97
Frontier culture
  in "Solution Unsatisfactory," 144
  and Truman, 142
Fulton, Robert, 10–18, 54, 55, 213nn.5,7,
    214nn.10,13,15,16
  and antinuclear novels, 187
  and Baruch Plan, 162
  and *Connecticut Yankee*, 62
  and defensive role as offensive, 93
  epigraph by, 9
  ideology of, 10, 11–13, 17–18, 81, 82, 91,
    153
  as lone inventor, 74
  sales documents of, 56
  and *Star Wars* rationale, 15–16
  and submarine, 10
  as utopian, 83
Fulton the First (steam warship), 16
Future (magazine), 178
Future-war fiction, 19–22. *See also* Science
  fiction
  aerial weapons in, 30, 33, 38, 41, 45, 47,
    49, 83, 84–85, 87
  and *Air Force*, 114
  Anglo-American alliance in, 30–33, 40, 44,
    132
  and anti-arms control propaganda, 195
  and antinuclear novels, 187
  atomic weapons in, 20, 26, 50, 84, 131–33,
    134–37, 141–45
  and automatons, 205
  beginning of, 19–20
  and bombing of cities, 101–2
  Britain as enemy in, 22–23, 24–29
  Germany as enemy in, 45–47, 49
  and Heinlein, 142
  and Irving's satire, 21
  Lightning in the Night as, 138–41
  and middle America, 52–53
  nuclear-disarmament terror in, 194
  post-WWI impact of, 75
  role reversal in, 156
  Russia as enemy in, 29–30, 31–33

Future-war fiction (*continued*)
superweapon-geniuses in, 48–52
and ''36-Hour War,'' 157
ultimate weapon in, 25–26, 28, 33, 153
and U.S. imperialism, 20, 21–22, 30–31, 153
and Vietnam victory through airpower, 118
and Wells, 65
Yellow Peril in, 20, 33–42, 85

Gable, Clark, 113, 116
Galaxy Science Fiction, 183
Galapagos (Vonnegut), 119
Galileo, 146
Gallun, Raymond, 135
Gann, W. D., 102
General Electric, 74
General Mills, 157
Geneva Accord (1925), 193
Genocide
in future-war fiction, 20, 85
in nuclear deterrence (*Murder of the U.S.A.*), 160
by U.S., 92, 100
warplane as instrument of, 117, 119
War of the Worlds as preview of, 65
in Yellow Peril literature, 37
''Gentlemen, You Are Mad!'' (Mumford), 4, 158
Gentlemen: You Are Mad! (re-issue of *Pallid Giant*), 158
George, Peter, 183
Germany, imagined U.S. war with, 32, 43, 45–47, 49, 85
Germany, Nazi
bombing by, 89, 104
bombing of, 89, 104–7
in *Lightning in the Night,* 138–41
Germ warfare. *See* Bacteriological (germ) warfare
Gettleman, Marvin E., 221n.13, 226n.8
Gibson, James William, 221n.13
Giesy, John Ulrich, 41, 44–45, 52
quoted, 19
''Giles, Gordon A.'' (Otto Binder), 135
Gimpel, H. J., 213n.3
Giovanitti, Len, 223n.13
Glen and Randa (film), 194
Gloversmith, Frank, 222n.8
God Bless You, Mr. Rosewater (Vonnegut), 120
Goddard, George, 94
Godfrey, Hollis, 50–51, 52
Godzilla, 182
Gold, Horace, 182–83

Golden Bottle: Or, The Story of Ephraim Benezet of Kansas, The (Donnelly), 29–30
Golden Days (See), 212
Goldwater, Barry, and nuclear false alarms, 208, 229nn.5,6,7,9
''Good News of High Frontier, The'' (Heinlein), 200
Graham, Daniel O., 198, 200, 227n.6, 227n.2(ch.16)
Gray, Colin, 195
Graybar, Lloyd J., 224n.9
Graybar, Ruth Flint, 224n.9
Great Britain
air power of, 86, 88–89, 90, 103, 104
atomic weapons in fiction of, 131, 134
colonialist air-power use by, 88–89
Fulton in, 11, 15
imagined U.S. war with, 21, 22–23, 24–29, 36, 43–44, 85
test-ban treaty by, 192
as U.S. ally, 31
as U.S. enemy, 5, 23–24
and Wells' *War of the Worlds,* 64–65
Yellow-Peril literature in, 37
Great Pacific War: A History of the American-Japanese Campaign of 1931–1933, The (Bywater), 102
Great War Syndicate, The (Stockton), 26, 47, 52, 163
Grendon, Edward, 159
Grey, C. G., 217n.16
Griffith, D. W., 101–2
Griffith, George, 131–32
Griggs, John, 221n.5
Groves, Leslie, 180, 222n.16
Guard of Honor (Cozzens), 116
Guernica, 89
''guerra del' 19—, La'' (Douhet), 89
Gulf of Tonkin incidents, 185

Hacker, Louis M., 214n.3
Hague Conference (1907), 86
Haines, Donal Hamilton, 46
Haines, William Wister, 116
Haldeman, Joe, 194
Halloran, Richard, 229n.6
Hamilton, Edmond, 159
Hancock, H. Irving, 46
Hansell, Haywood S., Jr., 110
Harding, Richard, 211
Harper's, 137, 140
Hart, Gary, and nuclear false alarms, 208, 229nn.5,6,7,9

Hawaii, U.S. annexation of, 39, 92
Hawks, Howard, 114
Hayworth, Rita, 165
Hearne, R. P., 84
Hegemony, American. *See* American
    hegemony (Pax Americana)
Heinlein, Robert A., 52, 135, 141–46, 150,
    200
Heller, Joseph, 112, 119, 123–27
Hell and High Water (film), 182
Henry Altemus Company, 46
Heritage Foundation, 200
Herken, Gregg, 223nn.5,7,12,13,
    225nn.1,5(ch.11)
Hersey, John, 157
Hertz Foundation, 200
Hewlett, Richard G., 225n.6, 226n.3
Hicks, W. Joynson, 217nn.7,8
High Frontier, 198, 207
High Frontier (Graham), 200
High Road, The (Bova), 200
Hilgartner, Stephen, 222n.11, 226n.1
Hindle, Brooke, 213n.1(ch.1)
Hiroshima. *See also* Atomic bombing of Japan
    atomic bombing of, 16, 51
    and Dresden bombing, 120
    as reserved from incendiary raids, 111
Hiroshima (Hersey), 157
History, re-writing of, in *The Penultimate
    Truth*, 186–87
*History of New York by Diedrich
    Knickerbocker, A* (Irving), 21
History of Rasselas, Prince of Abissinia, The
    (Johnson), 81
His Wisdom, the Defender (Newcomb), 48–49,
    52, 83, 83–84, 145
Hitler, Adolf, in *Lightning in the Night*, 139,
    140
Holland, John P., 10
Holloway, David, 228n.5
Holmes, John Hayne, 155, 224n.2
Hoover, Herbert, 87
Hoover Institution on War, Revolution and
    Peace, 200
Hope, Edward, 222n.12
Hopkins, George E., 219nn.2,3,6,10, 220n.18
House Un-American Activities Committee
    and Chandler Davis, 173
    film-industry hearings of, 182
Howe, Lord Richard, 9
Howe, Sir William, 9
Hubler, Richard G., 220n.27
Hull, Cordell, 101
Human creativity, vs. automatons, 209–10
Hurley, 217n.14, 218nn.3,4,19(ch.5)

Hutcheon, Wallace S., Jr., 213n.3,
    214nn.12,14
Hydrogen bomb. *See* Thermonuclear
    (hydrogen) bomb

ICBMs. *See* Intercontinental ballistic missiles
"If Japan Awakens China" (London), 37
"I Have No Mouth, and I Must Scream"
    (Ellison), 209–10
Immigrants, and future-war fiction, 27–28, 40
Imperialism. *See also* Spanish-American War
    and Anglo-American alliance, 31
    and future-war fiction, 20, 21–22, 30–31,
        153
    Japanese, 99
    in Maxim's expectations, 84
    John Mitchell against, 92
    in "Solution Unsatisfactory," 145
    War of the Worlds as critique of, 65
*In the Battle for New York; Or, Uncle Sam's
    Boys in the Desperate Struggle for the
    Metropolis* (Hancock), 46
Incredible Shrinking Man, The (film), 182
Industrial capitalism. *See* Capitalism, industrial
Industrialization, 11
*Influence of Sea Power upon History, 1660–
    1783* (Mahan), 22
Ingenhousz, Jan, 217n.1
Ingersoll-Rand Corporation, 74
Inspection, on-site, in Soviet proposals, 163
Intercontinental ballistic missiles (ICBMs),
    3–4, 167–68, 191
    and ABM treaty, 193
    Nazi scientists work on, 166
    and nuclear-war decision making, 206
    U.S.-Soviet race for, 183
Intercontinental bomber, America's first use
    of, 5
Intermediate-Range Nuclear Forces Treaty,
    212
International Squadron (film), 113
*Invasion of New York; Or, How Hawaii Was
    Annexed, The* (Palmer), 39
*Invasion of the United States; Or, Uncle Sam's
    Boys at the Capture of Boston, The*
    (Hancock), 46
Invasion USA (1952 film), 182
Invasion: USA (1985 film), 21
Irish republicanism, and future-war fiction, 27
Irving, David, 122, 220n.17
Irving, Washington, 21, 64
Italy
    bombing attacks by, 85, 86, 90, 125, 126
    and Douhet, 89–90

It Can't Happen Here! (Lewis), 124
I Wanted Wings (film), 114
I Was a Communist for the FBI (film), 116

Japan
  air power of, 90
  imagined U.S. war with, 36, 38–45, 85,
    102, 153
  and Billy Mitchell, 93, 97–99
  U.S. bombing of, 105, 107–8, 108–11, 150
  U.S. bombing threat to, 98–99
  WWII surrender of, 151–53
Japan, atomic bombing of. *See* Atomic
  bombing of Japan
Japanese-American internment in WWII, 37,
  40
Jarrell, Randall, quoted, 79
Jastrow, Robert, 168
Jefferson, Thomas, 15
Jenkins, Will F. (pseud. Murray Leinster),
  159–61, 170, 183
Jet Propulsion Laboratory, 199
Johnson, Denis, 211
Johnson, Samuel, 81, 83, 84, 217n.2
Johnson, Van, 115
Johnstone, William, 211
Joliot-Curie, Irene and Frederic, 133
Jones, D. F., 209
Jones, Dorothy B., 226n.4
Jones, Lloyd S., 219n.21
Joseph, Paul, 227n.4(ch.15)
Jungk, Robert, 223n.1

Kagan, Norman, 226n.9
Kaltenborn, H. V., 156, 224n.5
Kampelman, Max, 168, 195
Kennan, George, 167, 225n.2(ch.12)
Kennett, Lee, 217nn.11,12,13,14,15,
  218n.19(ch.4), 219n.7,
  220nn.14,16,30,32
Kimmel, Husband, 98
Kirkpatrick, Jeane, 195
Korean "police action," U.S. airpower in,
  117
Kramer, Stanley, 183
Kubrick, Stanley, 183, 185–87

Langer, R. M., 137, 222n.11
Laser-guided bomb, America's first use of, 5
"Last Conflict, the Horror That Awoke the
  Nation, The" (McDougall), 53
Last Day, The (Clarkson), 179, 183, 184

Last Days of the Republic (Dooner), 21, 33–34
Last Invasion, The (Haines), 46
*Last War; Or, The Triumph of the English
  Tongue, The* (Odell), 30, 31
Laurence, William L., 137, 181, 222n.11,
  226n.3
Lawrence Livermore Laboratory, 200
Lawson, Ted, 115
Lea, Homer, 40–41
Leahy, William D., 96, 150, 223n.3
Lee, Ezra, 9
Lee, William, 214n.17
Leiber, Fritz, 174
Leinster, Murray (Will F. Jenkins), 159
LeMay, Curtis, 109, 110, 121
Leonard, Thomas C., 216n.17
Level 7 (Roshwald), 127, 183, 183–84,
  184–85, 186, 187
Levine, Isaac Don, 218n.3
Lewis, Sinclair, 124
Liberty (magazine), 96, 97, 102–3, 137, 138,
  139
Libraries, readers' names requested of, 146
Libya, air attacks against, 85, 89, 126
Life, 156, 157, 183
Lightning in the Night (Allhoff), 138–41, 153,
  162, 163
Literature. *See* Films and newsreels; Future-
  war fiction; Science fiction
Littauer, Ralph, 221n.13
"Lobby" (Simak), 135
"Locksley Hall" (Tennyson), 82, 83–84, 153
Logan's Run (film), 194–95
London, Jack, 37–39, 50, 53, 85, 102, 161
Look, 108
"Lorelle" (author), 34
"Lot" (Moore), 175–76
"Lot's Daughter" (Moore), 175, 176–77
L.P.M.; The End of the Great War (Barney),
  49–50, 52, 85, 145
Luce, Clare Boothe, 40
Lumet, Sidney, 183
Lyons, Richard D., 227n.1(ch.16)

McClure's, 53
McCormick, Anne O'Hare, 224n.2
MacDonald, Charles B., 222n.16
McDougall, Walt, 83
McFadden, Bernarr, 137
McKee, Alexander, 220n.17
McNutt, Paul V., 108
McPhee, John, 227n.1(ch.15)
Magazine of Fantasy and Science Fiction, 184

Mahan, Alfred Thayer, 22, 32, 33, 214n.3
  and submarine vs. dreadnought, 68–69
Making the Stand for Old Glory; Or, Uncle
  Sam's Boys in the Last Frantic Drive
  (Hancock), 46
Maneuverable Reentry Vehicle (MARV), 5,
  167
Manhattan Project, 42, 134, 141, 142, 147,
  152, 180, 181
Manifest Destiny, 20, 92
Manson, Marsden, 39
Manvell, Roger, 221n.5
Man Who Ended War, The (Godfrey), 50–51,
  52, 131
Man Who Rocked the Earth, The (Train),
  51–52, 53, 83, 131
Marshall, Edward, 217nn.25,28
Marshall, George C., 99
Martian Chronicles (Bradbury), 159
MARV (Maneuverable Reentry Vehicle), 5,
  167
Masters, Dexter, 224n.12
Material forces and conditions
  and cultural forces, 5
  and superweapons, 5
  and total war, 87
Matsuo, Kinoaki, 99, 219n.25
Maxim, Sir Hiram, 84, 217n.9
Maxim, Hudson, 41–42, 74, 205, 228n.2
Maxim machine gun, 84
Media
  atomic energy/weapons discussed in,
    137–38, 139–41
  and Edison, 69, 93
  for middle America (early 1900s), 52–53
  and Billy Mitchell, 93, 94–95, 95–96,
    102–3, 113, 115
Meier, Hugo A., 213n.4
"Memorial" (Sturgeon), 158–59
"Men from the Moon" (Irving), 21, 64
Merril, Judith, 177–79
Messer, Robert L., 223nn.5,11
Mexico, imagined U.S. war with, 21
MGM, 113–14
Militarism, 4, 40, 45
Military-industrial complex, 113
  and Edison, 69, 74, 113
  Eisenhower on, 123–24
  in future-war fiction, 26
  Naval Consulting Board as initiation of, 74
Miller, Terry, 222n.12
Miller, Walter M., Jr., 183
"Million-Year Picnic, The" (Bradbury), 159
Millis, Walter, 215n.9
Mines ("torpedoes")

  in *Connecticut Yankee*, 64
  of Fulton, 10, 14, 15
"Minor, John W.", 45
MIRV (Multiple Independently targeted
  Reentry Vehicle), 3, 5, 167, 191
"Missile Command" (video game), 203
"Missile gap," 183, 195
Missiles
  Arnold's predictions on (1945), 156–57
  in future-war fiction, 44, 85
  Mitchell's foreshadowing of, 96
Missiles, intercontinental. *See* Intercontinental
  ballistic missiles
Mitchell, John (Senator; father of Billy),
  91–92
Mitchell, William (Billy), 77, 87, 91, 92–97,
  99–100, 104, 218nn.11,12,14,15,16,17,
  18,20(ch.5), 219nn.26,27
  and bombing of cities, 117
  and Bywater's critique, 102
  court martial of, 97, 103
  and Douhet, 126
  and future-war fiction, 134
  and independent air force, 112, 138
  and Japan, 97–99
  and *Liberty* magazine, 137, 138
  and media, 93, 94–95, 95–96, 102–3, 113,
    115
  quoted, 79
  and ultimate peacekeeping weapon, 153
  in *Victory through Air Power*, 108
"Modern Man is Obsolete" (Cousins), 155
Moffett, Cleveland, 46–47, 215n.2
Moore, Ward, 175–77
Morality in warfare
  American consciousness of, transformed,
    102
  and atomic bombing of Japan, 143, 150,
    152
  and bombing in Asia vs. Europe, 108–9
  and bombing of civilians, 101, 103, 105,
    106. *See also* Civilian populations,
    attacks on
  and nuclear deterrence, 160
  and strategic airpower leaders, 87–88
  in *Thirty Seconds Over Tokyo*, 115–16
Morella, Joe, 221n.5
Morison, Samuel Eliot, 221n.7
Morris, Richard R., 213n.3
Morrow, William, 226n.5
Moskowitz, Sam, 222nn.9,18
Movies. *See* Films and newsreels
Mullen, R. Dale, 225n.1(ch.13)
Multiple Independently targeted Reentry
  Vehicle (MIRV), 3, 5, 167, 191

Mumford, Lewis, 4, 158, 213n.1(Intro.), 224n.13
Mundo, Oto, 36–37
Munsey, Frank A., 44
Murder in the Air (film), 202
Murder of the U.S.A., The (Jenkins), 159–61, 162, 164, 170–71, 183
Mussolini, Benito, and MacFadden, 137
Mussolini, Vittorio, 125, 221n.17
Mutually Assured Destruction (MAD), 4, 5, 77, 136, 170, 186
MX missiles (Peacekeeper), 168, 196
My Lai, 118–19
Myths. *See* Fallacies and myths

Nagasaki
  atomic bombing of, 16
  as reserved from incendiary raids, 111
Napalm (jellied gasoline), 118
  and *The Last War*, 30
  in *Slaughterhouse-Five*, 120
Napoleon, 14–15, 16
NASA, 199
Nathanson, Isaac, 135
National Security Decision Document *13*, 195
Native Americans
  as fictional attackers, 36
  warfare against, 21, 92
Nautilus (Fulton submarine), 14, 17
Nautilus, USS, 17
Naval Consulting Board, 71, 72, 74
Neal, W. C., 184
Neider, Charles, 216n.13
Nelson, Thomas Allen, 226n.9
"Nerves" (Del Rey), 135
Neutron bombs, 167–68
Newcomb, Simon, 48–49, 83, 145
Newman, John, 214n.1
Newsweek, 137
New York Times Magazine, 70, 168
New York World's Fair (1939–40), 137–38, 168
Next War: A Prediction, The (Wallace), 35
Nicaragua, 21, 32
  U.S. bombing of (1927), 97
Nicolson, Harold, 134
Nimitz, Chester, 225n.5
Nitze, Paul, 195
Nonproliferation treaty, 192, 193
Norton, Roy, 7, 41–44, 50, 52, 153
Noyes, Pierrepont B., 135, 158
NSAM 273 (plan), 185
Nuclear Age, The (O'Brien), 211–12

Nuclear arms race and arms control. *See* Arms race, nuclear
Nuclear culture, 155, 157–61
  and automatons, 204
  as all-consuming, 187
  and American people's will, 196–97
  antinuclear dissent suppressed, 173, 180, 182
  antinuclear films (late 1950s), 182
  antinuclear films and novels (post-Sputnik), 127, 183–87
  antinuclear films and novels (1965–79), 193–95
  antinuclear films and novels (recent), 211–12
  civil-defense propaganda, 181
  Cold War propaganda films, 182
  and deterrence, 158, 160
  and Heinlein, 141
  and science fiction, 168–69
  science-fiction catastrophe and horror movies, 158–61, 182–83
  and science fiction as dissent, 173–79
  and "security" (censorship), 146–48, 173. *See also* Secrecy ("security")
  and Soviet Union, 159, 173
  survivalist fiction, 174–75, 211
Nuclear disarmament. *See* Disarmament, nuclear
Nuclear energy. *See* Atomic (nuclear) energy
Nuclear-free zones, 192
Nuclear war, protracted, 195–96
Nuclear weapons. *See also* Atomic bomb
  in *Lightning in the Night*, 140
  question of defense against, 156–57
  rationale for, 141, 142–43
  and Saundby on Dresden bombing, 122–23
  in science fiction, 20, 26, 50, 84, 131–33, 134–37, 141–45, 147
  thermonuclear (hydrogen) bomb, 167
  third generation of, 199–200

O'Brien, Tim, 211–12
O'Connor, Rory, 222n.11, 226n.1
Odell, S. W., 30
Official History of the Army Air Forces, The, 110
Omni (magazine), 200
On the Beach (Shute), 127, 183, 184
On the Beach (film), 127, 183, 184
O'Neill, John J., 137, 146, 222n.11
One World or None, 158
Oppenheimer, J. Robert, 142
Orion (space battleship), 191

Orriss, Bruce W., 220n.3
Our Friend the ATOM (Disney book and film), 180
Outer Space Treaty, 192, 196, 199
Outrider, The (Harding), 211

Pallid Giant, The (Noyes), 135, 158, 168
Palmer, J. H., 39
Panofsky, Wolfgang K. H., 228n.5
Paradise Crater, The (Wylie), 147
Paramount Studios, 114
Parker, Eleanor, 182
Pastoral Letter of U.S. Catholic Bishops ("Challenge of Peace"), 189, 211
Pauling, Linus, 184
Pax Americana. *See* American hegemony
"Peace is our profession," 81, 100, 185
Peck, Gregory, 116
Penultimate Truth, The (Dick), 183, 186–87
Perlman, David, 227n.1(ch.16)
Pershing II missile, 168, 196
Philadelphia, Pennsylvania, aerial attack in (1985), 95
Philip, Cynthia Owen, 213nn.4,6,8, 214n.11
Philippines, U.S. war against, 92, 100
Philmus, Robert M., 215n.10
Piller, Emanuel, 159
"Pilot Lights of the Apocalypse" (Ridenour), 159
Pipes, Richard, 195, 227n.4(ch.15)
Planck, Robert, 228n.8
Planet of the Apes (film), 194
"Plausibility of denial," 185
Poison gas
    and Baruch Plan, 164
    in Douhet's theory, 90
    Edison for, 76
    in future-war fiction, 28
    U.S. ratifies Geneva Accord on, 193
    and WWI bombers, 87
Popular Mechanics, 137
Populism
    campaign against (1914–17), 45
    and future-war fiction, 27, 29–30
Possony, Stefan, 200
Postman, The (Brin), 211
Postnuclear fantasies, 174–77
Potsdam conference, 151
Pournelle, Jerry, 200–201, 227n.4(ch.16)
Powaski, Ronald E., 223n.9, 227n.4(ch.15)
Power, fantasy of, 202
Power, Thomas S., 117–18, 191, 221n.12
"Power Plant" (Vincent), 135
Powers, Richard Gid, 32, 214n.5

Preemptive strikes, Noyes on, 135
Presidential Directive 59, 195
Presidential rule, in future-war fiction, 42. *See also* American dictatorship
Pringle, Peter, 228nn.3,4
Privateers (Bova), 200
"Problem of Increasing Human Energy, The" (Tesla), 205
Progress, faith in, 18
Project MX-774, 166
Public Faces (Nicolson), 134
Purple Heart, The (film), 115

Quester, George H., 217n.15, 218n.19(ch.4), 219nn.9,22,25

Radioactive dust, 142
    in "Solution Unsatisfactory," 142, 143–44
Radioactive weapons. *See* Nuclear weapons
"Rappaccini's Daughter" (Hawthorne), 42
Reader's Digest, 108
Reagan, Ronald
    and Committee on the Present Danger, 195
    in *International Squadron,* 113
    in *Murder in the Air,* 202
    and Outer Space Treaty, 196
    quoted, 189
    and Star Wars, 198, 200, 203, 227n.6
    in training movie, 109–10, 220n.27
Reagan administration, vs. freeze movement, 196–97
Recovered Continent; A Tale of the Chinese Invasion (Mundo), 36–37
Red Alert (George), 127, 183
Red Dawn (film), 21
Red Menace, 42
Reed, Samuel Rockwell, 27
"Report on the Barnhouse Effect" (Vonnegut), 119
Republic Aviation Company, 108
Reynolds, William, 14
Rhodes, Richard, 222nn.5,6,13, 223n.5
Ridenour, Louis N., 159
RKO, 115
Robbins, David, 211
Robert A. Heinlein: America as Science Fiction (Franklin), 141
Robinson, Frederick, 41
Robinson, Clarence, Jr., 227n.3(ch.16)
Robot planes. *See also* V-1 missiles
    Mitchell's plan for, 96
    U.S. use of, 106
Rocket bomb, Nazi
    and atom bomb, 149

Rocket bomb, Nazi (*continued*)
    V-1, 104
    V-2, 104, 119
Rogin, Michael, 226n.5, 228n.6
Roosevelt, Franklin D.
    and air war, 87, 89, 101, 103, 219n.1
    Einstein-Szilard letter to, 133–34
    and fire-bombing of Japan, 98, 219n.22
    military-buildup campaign by, 138
Rosenberg, Julius and Ethel, 148, 173
Rosenblum, Simon, 227n.4(ch.15)
Roshwald, Mordecai, 183, 184–85
Ross, Thomas B., 226n.8
Rotterdam, bombing of, 103–4
Rousseau, Victor, 135
Rumpf, Hans, 120
Rush, C. W., 213n.3
Russell, Bertrand, 184
Russia. *See also* Soviet Union
    imagined U.S. war with, 21, 29–33, 41,
        138–40, 144–45
    Japanese defeat of (1905), 36
Rutherford, Lord, 133

Salamo, Lin, 216n.12
Salmond, Sir John, 89
SALT I treaty (1972), 193
SALT II treaty (1979), 193, 195, 196
Sanders, Jerry W., 227n.4(ch.15)
San Juan, E., Jr., 218n.2
Sarris, Andrew, 226n.7
Saturday Evening Post, 51, 53, 96, 113, 137,
    140, 146, 181
Saturday Review, 155, 158
Saundby, Sir Robert, 122
Schaffer, Ronald, 218n.13, 219n.22,
    220nn.28,30
Scheer, Robert, 227n.4(ch.15)
Schell, Jonathan, 171, 184
Schickel, Richard, 220n.22
Schrecker, Ellen W., 225n.3(ch.13)
Schwellenbach, Lewis, 103
Science fiction. *See also* Future-war fiction
    air power in, 87
    antinuclear, 173–79, 183–87, 193–95
    and atomic "security" measures, 146–47
    atomic weapons in, 20, 26, 50, 84, 131–33,
        134–37, 141–45, 147–48, 158–61
    and bombing of cities, 101–2
    Baruch Plan as, 162
    and Heinlein, 145
    and nuclear culture, 157, 168–69
    and nuclear-effects films, 182–83
    space-oriented, 113

Star Wars boosted by, 200
SDI. *See* Star Wars
Searles, A. Langley, 216n.16
Second World War. *See* World War II
Secrecy ("security")
    and cult of the superweapon, 42
    in Heinlein's portrayal, 141
    media discussions prior to, 137
    nuclear annihilation as result of
        ("Blunder"), 159
    and suppression of atomic-energy
        knowledge, 146–48
    and suppression of dissent, 148, 173
Security
    fantasy of, 202
    and modern weapons, 3
    and Soviet disarmament proposals, 166–67
    weapons as false means to, 212
Sedberry, J. Hamilton, 41
See, Carolyn, 212
Serviss, Garrett P., 50, 66–69, 216n.16
Shadow on the Hearth (Merril), 177, 178–79
Shelley, Mary, 168
Sherman, William Tecumseh, 92
Sherry, Michael, 219nn.10,11,22,24,
    220nn.1,14,22,28,30
Sherwin, Martin J., 223n.14, 224n.15
Shiel, M. P., 37, 38
Shield
    nuclear weapons as, 181, 202
    in *Star Trek*, 203
    Star Wars as, 201, 202, 204
Short and Truthful History of the Taking of
    California and Oregon in the Year A.D.
    1889, A (Woltor), 34–35
Shultz, George, 195
Shute, Nevil, 183
Simak, Clifford, 135
Sims, W. Scott, 56
Sims-Edison Electric Torpedo Company,
    55–58, 74
Skylark of Space, The (Smith), 135
Slaughterhouse-Five; or, The Children's
    Crusade: A Duty-Dance with Death
    (Vonnegut), 119, 120–23
SLBMs (submarine-launched ballistic
    missiles), 3, 17, 191, 196, 201, 206
Small Armageddon, A (Roshwald), 183
Smith, A. Merriman, 217n.5
Smith, E. E. "Doc," 135
Smith, Jacob H., 92
Smyth, Henry D., 142, 222n.13
Snell, Bradford C., 221n.15
Social control, in films about bombers, 116
Socialism

campaign against (1914–17), 45
and future-war fiction, 27–28, 29, 47, 49,
  50, 85
Social sacrifice, in *Lightning in the Night,*
  138–39
Soddy, Frederick, 132
Sokolosky, George, 138
"Solution Unsatisfactory" (Heinlein), 52,
  141–45, 153
Sontag, Susan, 226n.5
Soviet Union. *See also* Arms race, nuclear;
  Cold War; Detente
  atomic bomb tested by, 167, 180
  and Baruch Plan, 162–64
  final European WWII campaign by, 106–7
  imagined U.S. war with, 21, 138–40,
    144–45, 159, 175–79, 181–83, 184–87,
    207–8, 211
  intervention in (1917–21), 23
  and Japanese surrender, 151–52
  and nuclear arms race, 166–67
  nuclear capability of, 180–81
  as nuclear enemy, 159, 180
  SAC missions over, 181, 185
  Sputnik launched by, 183
  and U.S. suppression of dissent, 173
Space battleship (Orion), 191
Space Command, 199
Space program
  as ICBM cover, 172
  militarization of, 199
Spaceship, in *Edison's Conquest,* 66–67
Spain, imagined U.S. war with, 21, 22, 23, 39
Spanish-American War, 22, 31, 67
  in *Armageddon,* 32
  and Britain, 31
  and *Edison's Conquest,* 67, 68
  and Billy Mitchell, 92
Spanish Civil War, bombing of cities in, 89,
  90, 103, 104
Sperry, Elmer, 74
Sperry Electric Company, 74
Spinrad, Norman, 194, 197–98
Spohr, Carl W., 135–37
Sprague, Frank J., 74
Sprague Company, 74
Sputnik, 183
Stacy, Ryder, 211
Stallone, Sylvester, 121
Stares, Paul B., 228n.5
Stark, Harold, 98, 219n.22
Star Trek, 203
Star Wars (Strategic Defense Initiative), 5,
  201–3
  announcement of, 198, 200

arguments against, 201
and automated decision making, 207
and *The Final War,* 137
and Fulton, 15
Heinlein for, 141, 200
hidden agenda behind, 202
High Frontier version of, 198
in 1945 vision, 157
science-fiction promoting of, 200–201
and space-war development, 199–200
and survivalist literature, 211
and video-arcade games, 203
Star Wars (film), 204
Stealth bombers in fiction, 102
Steam warship, of Fulton, 10, 14, 16–17, 82
Stedman, Edmund C., 82–83, 217n.3
Stein, Bernard L., 216n.14
Sterling, Yates, 138
Stewart, Jimmy, 117, 121, 186
Stimson, Henry, 149, 151, 153
"Stochastic" (author), 24
Stockton, Frank, 26, 47, 52
  quoted, 19
Stone, I. F., 221n.11
Stranger in a Strange Land (Heinlein), 141
Strategic Air Command (SAC), 112, 166
  and "Peace Is Our Profession," 185
Strategic Air Command (film), 116–17, 182,
  186
Strategic Arms Limitations Treaties, 193
Strategic Defense Initiative. *See* Star Wars
Strategy of Technology, The (Pournelle and
  Possony), 200
Stricken Nation, The ("Stochastic"), 24
Sturgeon, Theodore, 158–59, 170, 224n.14
Submarine-launched ballistic missile (SLBM),
  3, 17, 191, 196, 201, 206
Submarines, 5
  as failed ultimate weapon ("Memorial"),
    158–59
  of Fulton, 10, 13–14, 17, 82
  and Fulton/Edison, 55
  and future-war fiction, 26, 28
  in Revolutionary War, 9–10
  storage batteries for, 61
  in WWI, 68–69, 71
Suid, Lawrence H., 221n.6
Sulzberger, A. O., Jr., 229n.5
Superfortresses (B-29s), 44, 105, 108, 109,
  110, 112, 153–54, 165, 166
Superman IV: The Quest for Peace (film), 212
Superweapons
  American monopoly on (fiction), 44
  and American people, 5, 6, 135, 196–97
  atomic bomb as, 163

Superweapons (*continued*)
  atomic energy as (magazine articles), 138
  automatons as, 203–4
  better world promised by, 168
  domination by, 4, 169
  Edison for, 75–77
  fallacies in vision of (Fulton), 17, 18
  in history of imagination, 4–6
  and imagination vs. reality, 16, 201
  jet bombers as, 116–17, 119
  and Billy Mitchell, 91
  opposition to, 6
  rationalization of, 10
  and security, 3, 146, 148
  Star Wars as, 198
  as ultimate (fiction), 25–26
  as ultimate peacekeeper, 153–54
  and vision of superweapon use, 17
  war made impossible through (fiction), 28,
    33, 43
  war made impossible through (Edison),
    76–77, 158
  and WWII outcome, 112
  "working" of, 201
Superweapons, cult of. *See* Cult of the
  superweapon
Survivalism, 174–75, 211
Survivalist (Ahern), 211
Sweetser, Arthur, 216n.17
Swift, Jonathan, 11
Szilard, Gertrud Weiss, 221n.1
Szilard, Leo, 133–34, 149, 151, 152, 153,
  221n.1, 223nn.2,6, 224n.16
  The Voice of the Dolphins by, 183

Taylor, Robert, 182
"Team B," 195
Teitler, Stuart, 214n.1
Teller, Edward, 133, 153, 195, 200, 202
Tenn, William, 174
Tennyson, Alfred, 82, 83–84, 84–85
Tesla, Nikola, 58–59, 62, 202, 205–6,
  228nn.1,6
Testament (film), 178, 211
Test ban, comprehensive, 199
Test-ban treaty, atmospheric (1963), 192, 195
Test-ban treaty, threshold (1974), 193, 196
Test Pilot (film), 113–14
"That Only a Mother" (Merril), 177–78
Them! (film), 182
"There Will Come Soft Rains" (Bradbury),
  174
Thermonuclear (hydrogen) bomb, 167
  Laurence on, 181

  as second generation, 199
Thermonuclear missiles, in U.S. submarines,
  3, 17
Thermonuclear weapons
  America's first use of, 5
  and future-war fiction, 26
Things To Come (Wells), 65
Things To Come (film), 84–85
Thirty Seconds Over Tokyo, 115–16
"36-Hour War, The," 157, 183
Thompson, Richard Austin, 215n.6
"Thousand Deaths, A" (London), 50
"Throw-weight gap," 195
"Thunder and Roses" (Sturgeon), 170–72
Tibbets, Paul, 182
Time, 137, 155
Tirman, John, 228n.5
Titus, A. Constandina, 226n.4
"To the Friends of Mankind" (Fulton), 12–13
"To Howard Hughes: A Modest Proposal"
  (Haldeman), 194
Tokyo
  fire bombing of, 110
  1942 raid on, 105, 115
  1942 raid on (film), 115–16
"Tomorrow's Children" (Anderson), 159
Top Gun (film), 114
Torpedo boats
  in *Battle of the Swash*, 25
  and Fulton/Edison, 55
Torpedoes
  as failed ultimate weapon ("Memorial"),
    158–59
  of Fulton, 10, 14, 15, 16, 17, 82
  and Fulton/Edison, 55
  in future-war fiction, 39, 47
  in Revolutionary War, 9–10
  of Sims-Edison, 55–58
Torpedo War, and Submarine Explosions
  (Fulton), 15
"To Still the Drums" (Davis), 172–73
Town and Country, 108
Tracy, Spencer, 113, 115
Train, Arthur Cheney, 51–52
Traveler (Drumm), 211
Treaties
  ABM, 193, 196
  on demilitarizing Antarctica, 192
  Geneva Accord (1925), 193
  INF, 212
  nonproliferation, 192, 193
  on nuclear-free zones, 192
  nullification of, 196
  Outer Space, 192, 196, 199
  SALT I, 193

SALT II, 193, 195, 196
  test-ban, 192, 193, 195, 196
*Treatise on the Improvement of Canal
  Navigation, A* (Fulton), 11, 12
Trenchard, Sir Hugh, 87, 88–89, 91, 104, 134
Truman, Harry S.
  and American mythology, 142
  and atomic bombing of Japan, 122, 151,
    152, 153, 156
  and atomic weapons, 135
  and Baruch Plan, 225n.5
  and future-war fiction, 53, 153
  and Tennyson lines, 84, 153, 224n.17
Truman administration, and Soviets as enemy,
  159
Trumbo, Dalton, 115
Tullock, T. G., 217n.7
Tulsa, Oklahoma, air attacks on ghetto in, 95
*Tunnel Thru the Air; or, Looking Back from
  1940* (Gann), 102
Turtle (American Turtle), 9–10
Twain, Mark, 62–64, 216n.14
Twelve O'Clock High (film), 116

U-2 overflight (1960), 185
UN Commission for the Control of Atomic
  Energy, 162–63
Under the Flag of the Cross (Sedberry), 41,
  85
United Nations, 162
  nuclear-weapons resolutions in, 197
"Unparalleled Invasion, The" (London),
  37–39, 53, 85, 102, 161
Unpardonable War, The (Barnes), 27–28
Uphoff, Norman, 221n.13
Urey, Harold, 145
Utopianism
  of Fulton, 83
  in future-war fiction, 38–39

V-1 missiles, 104
V-2 rockets, 104, 119
Valor of Ignorance, The (Lea), 40–41
Vanishing Fleets, The (Norton), 41–44, 50,
  52, 131, 153
Vertical-takeoff airships, in future-war fiction,
  45
Victory through air power, gospel of, 85
Victory through Air Power (de Seversky),
  108
Victory through Air Power (film), 108, 113
Video games, 79, 203
Vietnam, U.S. bombing of in 1945, 109

Vietnam War
  airpower in, 117–19
  and *The Forever War,* 194
  and nuclear-weapons opposition, 195
  and Second World War, 121
Vincent, Harl, 135
Voice of the Dolphins (Szilard), 183
Vonnegut, Kurt, 119–23, 221n.14

Wachhorst, Wyn, 215n.1
Waldrop, F. N., 159
Walker, J. Bernard, 46
Wallace, King, 35
Walsh, W. T., 216n.22
War, glorification of, 4
War in the Air, The (Wells), 65, 84
War of 1812, 16, 23, 24
*War of 1886, Between the United States and
  Great Britain, The* (Reed), 27
War Games (film), 209, 211
Warner Brothers, 113, 114
War Weary Bomber project, 96, 106
War of the Worlds, The (Wells), 64–66, 68, 71
*War of the Worlds; A Tale of the Year 2,000
  A.D., The* (Robinson), 41, 85
Waterloo, Stanley, 32, 52
Way, Katharine, 224n.12
Wayne, John, 121
Weapons. *See* Beam weapons; Nuclear
  weapons; Submarines; Superweapons;
  Torpedoes; *other specific weapons*
Weapons technology, and human progress, 13
Weart, Spencer R., 221n.1
We Bombed in New Haven (Heller play), 127
Weeks, John, 96
Weinberg, Alvin M., 222n.7
Wells, H. G., 64–66, 68, 71, 84–85, 132–33
Westinghouse, George, 58–59
Westinghouse Company, 74
Wheeler, Harvey, 183
"When the Air Raiders Come" (Mitchell), 97
White, Walter F., 218n.10
Whitnall, Harold O., 220n.20
"Who Shall Dwell" (Neal), 184
Wigner, Eugene, 133, 134, 142
Wilhelm, Kate, 193–94
Wilson, Charles E., 124
Wilson, Woodrow, 87
"Window of vulnerability," 195
Winged Defense (Mitchell), 96
Winged Gospel, The (Corn), 83
Wise, David, 226n.8
"Wizard of Staten Island", 28
Wolfe, Charles K., 224n.9

Wolfe, Homer C., 220n.19
Woltor, Robert, 34–35
Women's rights. *See* Feminism
Wonder Stories, 135
Wood, Robert Williams, 51–52
World, the Flesh, and the Devil, The (film), 184
"World Aflame" (Nathanson), 135
World Aflame: The Russian-American War of 1950 (Engel and Piller), 159
World Set Free: A Story of Mankind, The (Wells), 65, 84, 132–33, 134, 168
World's Fair (1939), 137–38, 168
World War I, 68–69
  and air power, 49, 50, 86–87, 88
  arms race before, 75
  and German use of Fulton weapons, 17
  Billy Mitchell in, 92
  WWII foreshadowed in, 72
World War II
  Catch-22 interpretation of, 123
  and Cold War, 121
  in Hollywood films, 108, 113–16
  home front in, 74–75

strategic bombing in, 89, 103–11, 117, 118
superweapon role in, 112
Wu, William F., 215n.6
Wyatt, Harold F., 217n.7
Wyden, Peter, 223n.5
Wylie, Philip, 147, 159

"Xevius" (video game), 203

Yavenditti, Michael, 224n.10
"Year 1899, The" (Crane), 35
Yellow Danger, The (Shiel), 37, 38
"Yellow Peril, The" (London), 37–38
Yellow Peril in Action, The (Mansen), 39
Yellow Peril literature, 20, 33–42, 85
  in film, 101–2, 114
  and U.S. bombing of Japan, 107
Young, Marilyn, 221n.13

Zelazny, Roger, 194
Zeppelins, London bombed by, 72, 86